D1034042

ELEMENTARY MATHEMATICAL and COMPUTATIONAL TOOLS for ELECTRICAL and COMPUTER ENGINEERS USING MATLAB®

Second Edition

ELEMENTARY MATHEMATICAL and COMPUTATIONAL TOOLS for ELECTRICAL and COMPUTER ENGINEERS USING MATLAB®

Second Edition

Jamal T. Manassah

CRC is an imprint of the Taylor & Francis Group, an informa business

CRC Press
Taylor & Francis Group
6000 Broken Sound Parkway NW, Suite 300
Boca Raton, FL 33487-2742

CRC Press is an imprint of Taylor & Francis Group, an Informa business

Printed in the United States of America on acid-free paper
10 9 8 7 6 5 4 3 2 1

International Standard Book Number-10: 0-8493-7425-1 (Hardcover)
International Standard Book Number-13: 978-0-8493-7425-8 (Hardcover)

Library of Congress Cataloging-in-Publication Data

Manassah, Jamal T.
Elementary mathematical and computational tools for electrical and computer engineers using MATLAB / Jamal T. Manassah. -- 2nd ed.
p. cm.
Includes bibliographical references and index.
ISBN-13: 978-0-8493-7425-8 (alk. paper)
ISBN-10: 0-8493-7425-1 (alk. paper)
1. Electric engineering--Mathematics. 2. Computer science--Mathematics. 3. MATLAB. I. Title.

TK153.M362 2007
510.2'46213--dc22 2 006018608

Visit the Taylor & Francis Web site at
http://www.taylorandfrancis.com

and the CRC Press Web site at
http://www.crcpress.com

Author

Jamal T. Manassah has been a professor of electrical engineering at the City College of New York since 1981. He earned his B.Sc. degree in physics from the American University of Beirut, and his M.A. and Ph.D. in theoretical physics from Columbia University. Dr. Manassah was a member of the Institute for Advanced Study. His current research interests are in theoretical and computational quantum and nonlinear optics, and in photonics.

Preface to the First Edition

This book is mostly based on a series of notes for a primer course in electrical and computer engineering that I taught at the City College of New York School of Engineering. Each week, the class met for an hour of lecture and a three-hour computer laboratory session where students were divided into small groups of 12 to 15 students each. The students met in an informal learning community setting, a computer laboratory, where each student had the exclusive use of a PC. The small size of the groups permitted a great deal of individualized instruction, which was a key ingredient to cater successfully to the needs of students with heterogeneous high school backgrounds.

A student usually takes this course in the second semester of his or her freshman year. Typically, the student would have completed one semester of college calculus, and would be enrolled in the second course of the college calculus sequence and in the first course of the physics sequence for students in the physical sciences and engineering.

My purpose in developing this book is to help bring the beginner engineering student's analytical and computational skills to a level of competency that would permit him or her to participate, enjoy, and succeed in subsequent electrical and computer engineering courses. My experience indicates that the lack of mastery of fundamental quantitative tools is the main impediment to a student's progress in engineering studies.

The specific goals of this book are:

1. To make you more comfortable applying the mathematics and physics that you learned in high school or in college courses, through interactive activities.
2. To introduce you, through examples, to many new practical tools of mathematics, including discrete variables material that is essential to your success in future electrical engineering courses.
3. To instruct you in the use of a powerful computer program, MATLAB®, which was designed to be simultaneously user-friendly and powerful in tackling efficiently the most demanding problems of engineering and sciences.
4. To give you, through the applications and examples covered, glimpses of some of the fascinating problems that an electrical or computer engineer solves in the course of completing many of his or her design projects.

My experience indicates that you can achieve the above goals through the following work habits that I usually recommend to my own students:

- Read carefully the material from this book that is assigned to you by your instructor for the upcoming week, and make sure to solve the suggested preparatory exercises in advance of the weekly lecture.

- Attend the lecture and follow closely the material presented, in particular the solutions to the more difficult preparatory exercises and the demonstrations.
- Following the lecture, make a list of questions on the preparatory material to which you still seek answers, and ask your instructor for help and clarification on these questions, preferably in the first 30 minutes of your computer lab session.
- Complete the in-class exercises during the computer lab session. If you have not finished solving all in-class exercises, make sure you complete them on your own, when the lab is open, or at home if you own a computer, and certainly before the next class session, along with the problems designated in the book as homework problems and assigned to you by your instructor.

In managing this course, I found it helpful for both students and instructors to require each student to solve all problems in a bound notebook. The advantage to the student is to have easy access to his or her previous work, personal notes, and reminders that he or she made as the course progressed. The advantage to the instructor is to enhance his or her ability to assess, more easily and readily, an individual student's progress as the semester progresses.

This book may be used for self-study by readers with perhaps a little more mathematical maturity acquired through a second semester of college calculus. The advanced reader of this book who is familiar with numerical methods will note that, in some instances, I did not follow the canonical order for the sequence of presentation of certain algorithms, thus sacrificing some optimality in the structure of some of the elementary programs included. This was necessitated by the goal I set for this book, which is to introduce both analytical and computational tools simultaneously.

The sections of this book that are marked with asterisks include material that I assigned as projects to students with either strong theoretical interest or more mathematical maturity than a typical second semester freshman student. Although incorporated in the text, they can be skipped in a first reading. I hope that, by their inclusion, I will facilitate to the interested reader a smooth transition to some new mathematical concepts and computational tools that are of particular interest to electrical engineers.

This text greatly benefited from course material previously prepared by my colleagues in the departments of electrical engineering and computer science at City College of the City University of New York, in particular, P. Combettes, I. Gladkova, B. Gross, and F. Thau. They provided either the starting point for my subsequent efforts in this course, or the peer critique for the early versions of this manuscript. I owe them many thanks and, of course, do not hold them responsible for any of the remaining imperfections in the text.

The preparation of this book also owes a lot to my students. Their questions and interest in the material contributed to many modifications in the order and in the presentation of the different chapters. Their desire for working out more applications led me to expand the scope of the examples and exercises included in the text. I am grateful to all of them. I am also grateful to Erwin Cohen, who introduced me to the fine team at CRC Press, and to Jerry Papke whose stewardship of the project from start to end at CRC Press was most supportive and pleasant. The editorial and production teams at CRC in particular, Samar Haddad, the project editor, deserve credit for the quality of the final product rendering. Naomi Fernandes and her colleagues at The MathWorks Inc. kindly provided me with a copy of the new release of MATLAB for which I am grateful.

I dedicate this book to Azza, Tala, and Nigh whose support and love always made difficult tasks a lot easier.

Jamal T. Manassah
New York, January 2001

Preface to the Second Edition

Comments received from readers and my own students invariably suggested the desirability for expanding Chapter 1 in a manner that will reduce the need to refer as often to the MATLAB® online help or to a MATLAB user manual. It was also suggested that the reference value of this book might be enhanced if a chapter on symbolic manipulation with MATLAB was added. This second edition is responsive to both of these requests.

The MATLAB syntax was also updated to conform to the current version of MATLAB. I have also added a number of special sections designated to the more advanced students or for those students seeking additional challenges. Typically, these sections were assigned to outstanding students who supplemented the regular course homework with special assignments. The continuous need to challenge these gifted students is specially rewarding when teaching in an urban university where many promising future scientific talents are discovered only after these students are enrolled in college-level courses.

Now that my modest experiment with introducing an integrated, empirical, and interdisciplinary style to teaching beginner engineering students applied mathematics is approximately 10 years in progress, it may be useful to reflect on the successes and failures of this particular effort:

- Almost everybody in my academic circle now recognizes the usefulness of supplementing and enhancing the regular rigorous standard mathematical curriculum for engineering students with a hands-on elementary applied course.
- Those of my colleagues who still prefer to have the different material taught in a more compartmentalized manner and at a later stage of a student career argue that students may not be willing or able to devote the blocks of time required to integrate the skills that both the analytical and numerical techniques demand within the same course. While I accept the premise that my adopted approach certainly requires a higher level of motivation and time commitment by students, it is also true that students who were willing to invest the required time in a course based on this text ended up ready to plunge in more advanced engineering courses much earlier than others. I am particularly gratified by the performance of those of my students who, following the successful completion of this text went on to advanced engineering courses and were able to concentrate and excel in the new engineering material covered

there while many of their classmates were bogged down acquiring the computational tools which those selected had already mastered.

- There is still no consensus in my own academic circle as to whether it is advisable to let students explore through computer experiments new mathematical results prior (or even much prior) to the formal introduction of the underlying theoretical principles in a more structured environment. My own observations are that a student's intuitive skills are strongly enhanced by accumulated mathematical empirical discovery and experience; however, it will be probably up to psychologists working in learning theories to come up with a more definitive answer to the value of an empirical approach in teaching applied mathematics.
- Teachers who adopted this text successfully informed me that they incorporated a certain amount of flexibility in different students' assignments to reflect the different levels of skills present in a normal student population distribution. My own experience was not different.

My advice to beginner students of engineering using this text is to take their time and finish each chapter with its solved examples and suggested problems before moving on to the next chapter. I do not recommend to the beginner to merely skim through the core material. The reader will quickly discover that many computational skills are acquired and internalized best when developed slowly and cumulatively.

The use of this text as reference material by a number of practicing scientists and engineers is often brought to my attention by some pleasant comments in a faculty meeting, professional conference, or social function. This encouragement by peers is most appreciated.

This new edition benefited from comments, emails, and letters that I received from many students, readers, and colleagues. I am grateful for all those who took the time to make suggestions, point out misprints, and/or argued the pedagogical approach that I adopted in presenting certain material. I owe special thanks to Herman Cummins and Robert Brenart for their valuable comments on this second edition's book proposal, and to Allison Taub, my editor at Taylor & Francis who took the initiative in actively and enthusiastically championing the publication of a second edition of this text, and in supporting its preparation at this time. Courtney Esposito and her colleagues at The MathWorks Inc. continue to provide me with copies of the most current releases of MATLAB and their supporting documentation for which I am most grateful. Julie Spadaro, project editor at Taylor & Francis, assured a quality final product, successfully coordinating production teams on different continents.

My wife Azza and daughters Tala and Nigh urged me to prepare a second edition of this book because partially, I suspect, they liked the fact that the mail folder that I bring home with me every day now had many communications (even fan letters!) from other than the dozen or so colleagues, friends, and editors who regularly read, comment, and respond to my research papers. But above all, I know that deep in their hearts they supported me unconditionally in what they knew was to me a labor of love. To them I rededicate this book.

Jamal T. Manassah
New York, September 2006

Contents

1

Starting with MATLAB and Exploring Its Graphics Capabilities

MATLAB can be thought of as a library of programs that will prove very useful in solving many electrical engineering computational problems. MATLAB is an ideal tool for numerically assisting you in obtaining answers, which is a major goal of engineering analysis and design. This program is very useful in research on circuit analysis, device design, signal processing, filter design, control system analysis, antenna design, microwave engineering, photonics engineering, computer engineering, and every other subfields of electrical engineering. It is also a powerful graphic and visualization tool of beneficial use for both quantitative courses and in your future career. In this chapter, we start our exploration of MATLAB; in subsequent chapters, we will add to our knowledge about this powerful tool of the modern engineer.

1.1 First Steps

The first step in using MATLAB is to know how to call it. It is important to remember that although the front-end and the interfacing for machines with different operating systems are sometimes different, once you are inside MATLAB, all programs and routines are written in the same manner. Only those few commands that are for file management and for interfacing with external devices such as printers may be different for different operating systems. After entering MATLAB, usually by double clicking on the MATLAB program icon, you should see the prompt `>>`, which means the program interpreter is waiting for you to enter instructions. The interpreter goes to work when you press the Return key. In case you wish to go to a new line for the purpose of continuing your instructions, you should enter ellipses (...) before you hit the Return key.

In case the program is already opened, and to make sure that nothing is saved from a previous session, type and enter:

```
>>clear all
```

This command removes all variables from the workspace. Also type and enter (we shall henceforth say just enter):

```
>>clf
```

This command creates a graph window (if one does not already exist) or clears an existing graph window. The command **`figure`** creates a new figure window, leaving the existing ones unaltered.

The command **`quit`** stops MATLAB.

Because it is not the purpose of this text to explain the function of every MATLAB command, how would you get information on a particular syntax? The MATLAB program has extensive help documentation available with simple commands. For example, if you wanted help on a function called **`roots`** (we will use this function often), you would type **`help roots`**.

Note that the help facility cross-references other functions that may have related uses. This requires that you know the function name. If you want an idea of the available help files in MATLAB, type **`help`**. This gives you a list of topics included in MATLAB.

To get help on a particular topic such as elementary matrices and matrix manipulation, type **`matlab\elmat`**. This gives you a list of all relevant functions pertaining to that area. Now you may type **`help`** for any function listed there. For example, try **`help size.`**

1.2 Basic Algebraic Operations and Functions

The MATLAB environment can be used, on the most elementary level, as a tool to perform simple algebraic manipulations and function evaluations.

Example 1.1

Exploring the calculator functions of MATLAB: The purpose of this example is to show how to manually enter data and how to use the basic algebraic operations of MATLAB. Note that the statements will be executed immediately after they are typed and entered (no equal sign is required).

Enter the text that follows the >> prompt to find out the MATLAB responses to the following:

```
>> format short
>> 2.54376+2.32e1    %2.32e1=2.32*(10^1)
ans =
     25.7438
>> 5.45^2
ans =
```

```
      29.7025
>> ceil(7.58)
ans =
      8
>> 2*sin(pi/4)
ans =
      1.4142
```

The % symbol is used so that one can type comments in a program. (Comments following the % symbol are ignored by the MATLAB interpreter.) Then enter:

```
>> format short e
>> 2.54376+2.32e1
ans =
   2.5744e+001
>> 5.45^2
ans =
   2.9703e+001
>> ceil(7.58)
ans =
      8
>> 2*sin(pi/4)
ans =
   1.4142e+000
```
□

The last command gave twice the sine of $\pi/4$. Note that the argument of the function was enclosed in parentheses directly following the name of the function. Therefore, if you wanted to find $\sin^3(x)$, the proper MATLAB syntax would be

```
>>sin(pi/4)^3
```

The numeric functions of MATLAB are:

`ceil(x)`	rounds the number to nearest integer toward ∞
`fix(x)`	rounds the number to nearest integer toward 0
`floor(x)`	rounds the number to nearest integer toward $-\infty$
`round(x)`	rounds the number to nearest integer

The numeric formats of MATLAB are:

`format short`	(default value) gives the number with four decimal digits
`format long`	gives the number with 16 digits
`format short e`	gives the number with four decimals plus an exponent
`format long e`	gives the number with 15 digits plus an exponent
`format bank`	gives the number with 2 digits after the decimal (useful for financial calculations)

To facilitate its widespread use, MATLAB has all the standard elementary mathematical functions as built-in functions. Type **help elfun** to obtain a listing of these functions. Remember that this is just a subset of the available functions in the MATLAB library.

```
>>help elfun
```

The response to the last command will give you a list of these elementary functions, some of which may be new to you, but all of which will be used in your future engineering studies. Many of these functions will also be used in later chapters of this book.

In the following table, the MATLAB expressions for the most common mathematical functions are given:

MATLAB form	Name	Mathematical form
`sqrt(x)`	squareroot	$\sqrt{x}$
`exp(x)`	exponential	$\exp(x) = e^x$
`log(x)`	natural logarithm	$\ln(x)$
`log10(x)`	base10 logarithm	$\log(x) = (\ln(x)/\ln(10))$
`sin(x)`	sine	$\sin(x)$
`cos(x)`	cosine	$\cos(x)$
`sec(x)`	secant	$\sec(x)$
`tan(x)`	tangent	$\tan(x) = (\sin(x)/\cos(x))$
`cot(x)`	cotangent	$\cot(x) = (1/\tan(x))$
`asin(x)`	inverse sine	$\sin^{-1}(x) = \arcsin(x)$
`acos(x)`	inverse cosine	$\cos^{-1}(x)$
`asec(x)`	inverse secant	$\sec^{-1}(x)$
`atan(x)`	inverse tangent	$\tan^{-1}(x)$
`acot(x)`	inverse cotangent	$\cot^{-1}(x)$
`sinh(x)`	hyperbolic sine	$\sinh(x) = (e^x - e^{-x})/2$
`cosh(x)`	hyperbolic cosine	$\cosh(x) = (e^x + e^{-x})/2$
`sech(x)`	hyperbolic secant	$\text{sech}(x) = 1/\cosh(x)$
`tanh(x)`	hyperbolic tangent	$\tanh(x) = \sinh(x)/\cosh(x)$
`coth(x)`	hyperbolic cotangent	$\coth(x) = 1/\tanh(x)$
`asinh(x)`	inverse hyperbolic sine	$\sinh^{-1}(x) = \ln\left(x + \sqrt{x^2+1}\right)$ where $-\infty < x < \infty$
`acosh(x)`	inverse hyperbolic cosine	$\cosh^{-1}(x) = \ln\left(x + \sqrt{x^2-1}\right)$ where $x \geq 1$
`asech(x)`	inverse hyperbolic secant	$\text{sech}^{-1}(x) = \ln\left(\frac{1}{x} + \sqrt{\frac{1}{x^2} - 1}\right)$ where $0 < x \leq 1$
`atanh(x)`	inverse hyperbolic tangent	$\tanh^{-1}(x) = \frac{1}{2}\ln\left(\frac{1+x}{1-x}\right)$ where $-1 < x < 1$
`acoth(x)`	inverse hyperbolic cotangent	$\coth^{-1}(x) = \frac{1}{2}\ln\left(\frac{x+1}{x-1}\right)$ where $x > 1$ or $x < -1$

EXAMPLE 1.2

Assigning and calling values of parameters: In addition to inputting data directly to the screen, you can assign a symbolic constant or constants to represent data and perform manipulations on them.

For example, enter and note the answer to each of the following:

```
>> format short
>> a=2.65;
>> b=3.765;
>> c=a*(a+b)
c =
   16.9997
>> d=a*b
d =
    9.9772
>> e=a/b
e =
    0.7039
>> f=a^3/b^2
f =
    1.3128
>> g=a+3*b^2
g =
   45.1757
```

(The spacing shown above is obtained by selecting Compact for Numerical Display in the Command Window **Preferences** in the **File** pull-down menu.)

□

If we desire at any point to know the list of the current variables in a MATLAB session, we enter:

```
>>who
```

Note on variable names: Variable names must begin with a letter, contains letters, digits, and underscore characters, and must contain less than 32 characters. MATLAB is case-sensitive.

Question: From the above, can you deduce the order in which MATLAB performs the basic algebraic operations?

Answer: The order of precedence is as follows:

1. Parentheses, starting with the innermost
2. Exponentiation from left to right

3. Multiplication and division with equal precedence, also evaluated from left to right
4. Addition and subtraction, with equal precedence, also evaluated from left to right

IN-CLASS EXERCISE

Pb. 1.1 Using the values, $a = 0.875$ and $b = 1.5786$, find the values of the following expressions (give your answers in the engineering short format):

$$h = \sin(a)\sin^{3/2}(b)$$
$$j = a^{1/3}\, b^{-3/7}$$
$$k = \left(\sin^{-1}(a/b)\right)^{2/3}$$
$$l = \sinh(a)\, a^{b}$$

1.3 Plotting Points

In this section, you will learn how to use some simple MATLAB graphics commands to plot points. We use these graphics commands later in the text for plotting arrays and for visualizing their properties. To view all the functions connected with two-dimensional (2-D) graphics, type

```
>>help plot
```

All graphics functions connected with three-dimensional curves graphics can be looked up by typing

```
>>help plot3
```

A point P in the x–y plane is specified by two coordinates: the x-coordinate measures the horizontal distance of the point from the y-axis, while the y-coordinate measures the vertical distance above or below the x-axis. These coordinates are called Cartesian coordinates, and any point in the plane can be described in this manner. We write for the point, $P(x, y)$.

Other representations can also be used to locate a point with respect to a particular set of axes. For example, in the polar representation, the point is specified by an r-coordinate that measures the distance of the point from the origin, while the θ-coordinate measures the angle which the line passing through the origin and this point makes with the positive x-axis. The angle is measured anticlockwise from the positive x-axis.

The purpose of the following two examples is to learn how to represent points in a Cartesian plane and to plot them using MATLAB.

EXAMPLE 1.3

Plot the point *P*(3, 4). Mark this point with a red asterisk.

Solution: Enter the following:

```
>> x1=3;
y1=4;
plot(x1,y1,'*r')
```
□

Note that the semicolon is used in the above commands to suppress the echoing of the values of the inputs. The `'*r'` is used to mark the location and the color of the point that we are plotting.

The symbols for the most common markers and colors for display in MATLAB plot-functions are:

Markers	Symbol	Color	Symbol
Circle	o	Blue	b
Cross	x	Cyan	c
Diamond	d	Green	g
Dot	.	Magenta	m
Pentagram	p	Yellow	y
Plus sign	+	Red	r
Square	s	White	w
Star	*	Black	k

EXAMPLE 1.4

Plot the second point, *R*(2.5, 4), on the graph while keeping point *P* of the previous example on the same graph. Mark the new point with a small green circle.

Solution: If we went ahead, defined the coordinates of *R*, and attempted to plot the point *R* through the following commands:

```
>>x2=2.5;
y2=4;
plot(x2,y2,'og')
```

We would find that the last plot command erases the previous plot output. Thus, what should we do if we want both points plotted on the same graph? The answer is to use the **hold on** command after the first plot.

The following illustrates the instructions that you should have entered instead of entering the above:

```
>>hold on
x2=2.5;
y2=4;
plot(x2,y2,'og')
hold off
```

The `hold off` turns off the `hold on` feature. □

NOTES

1. There is no limit to the number of plot commands you can type before the hold is turned off.
2. An alternative method for viewing multiple points on the same graph is available: we may instead, following the entering of the values of **x1, y1, x2, y2,** enter:

```
>>plot(x1,y1,'*r',x2,y2,'og')
```

This has the advantage, in MATLAB, if no color is specified, that the program assigns automatically a different color to each point.

1.3.1 Axes Commands

You may have noticed that MATLAB automatically adjusts the scale on a graph to accommodate the coordinates of the points being plotted. The axis scaling can be manually enforced by using the command

```
>>axis([xmin xmax ymin ymax])
```

Make sure that the minimum axis value is less than the maximum axis value or an error will result.

In addition to being able to adjust the scale of a graph, you can also change the aspect ratio of the graphics window. This is useful when you wish to see the correct x to y scaling. For example, without this command, a circle will look more like an ellipse, even if we have chosen ymax − ymin = xmax − xmin.

EXAMPLE 1.5

Plot the vertices of the square, formed by the points $(-1, -1)$, $(-1, 1)$, $(1, -1)$, $(1, 1)$, keeping the same aspect ratio as in a graph paper, i.e., 1.

Solution: Enter the following:

```
>>x1=-1;y1=-1;
>>x2=1;y2=-1;
>>x3=-1;y3=1;
>>x4=1;y4=1;
>>plot(x1,y1,'o',x2,y2,'o',x3,y3,'o',x4,y4,'o')
>>axis([-2 2 -2 2])
>>axis square                   %makes the aspect ratio 1
```
□

Note that prior to the **axis square** command, the square looked like a rectangle. If you want to go back to the default aspect ratio, type **axis normal**.

In specific instances, you may need to choose other than the default tick marks chosen by MATLAB, you can achieve this by adding the following command:

```
set(gca,'Xtick',[xmin:dx:xmax],'Ytick',[ymin:dy:ymax])
```

Here **gca** stands for get current axes, and *dx* and *dy* are respectively the spacings between tick marks that you desire in the *x*- and *y*-directions.

You can also make the graph easier to read, by using the command:

```
>>grid
```

This command displays gridlines at the tick marks.

If we repeat Example 1.5 with these new features added, the set of instructions will now read:

```
>>x1=-1;y1=-1;x2=1;y2=-1;x3=-1;y3=1;x4=1;y4=1;
>>plot(x1,y1,'o',x2,y2,'o',x3,y3,'o',x4,y4,'o')
>>axis([-2 2 -2 2])
>>axis square
>>set(gca,'Xtick',[-2:0.5:2],'Ytick',[-2:0.5:2])
>>grid
```

The tick marks were chosen in this instance to be separated by $\frac{1}{2}$ units in both the *x*- and *y*-directions.

1.3.2 Labeling a Graph

To add labels to your graph, the functions **xlabel**, **ylabel**, and **title** can be used as follows:

```
>>xlabel('x-axis')
>>ylabel('y-axis')
>>title('points in a plane')
```

If you desire to also add a caption anywhere in the graph, you can use the MATLAB command

```
>>gtext('caption')
```

and place it at the location on the graph of your choice, by clicking the mouse on the desired location when the crosshair is properly centered there.

Different fonts and symbols can be used in labeling MATLAB graphs. LaTeX symbols are used by MATLAB for this purpose (see Appendix E).

1.3.3 Plotting a Point in 3-D

In addition to being able to plot points on a plane (2-D space), MATLAB is also able to plot points in a three-dimensional space (3-D space). For this, we use the **plot3** function.

EXAMPLE 1.6

Plot the points $P(1, 1, 1)$, $Q(4, 5, 6)$ and $R(2, 5, 3)$. Show a grid in your graph.

Solution: Enter the following commands:

```
>>x1=1;x2=4;x3=2;
>>y1=1;y2=5;y3=5;
>>z1=1;z2=6;z3=3;
>>plot3(x1,y1,z1,'o',x2,y2,z2,'*',x3,y3,z3,'d')
>>axis([0 6 0 6 0 6])
>>grid
```
□

NOTE You can also plot multiple points in a 3-D space in exactly the same way as you did on a plane. Axis adjustment can still be used, but the vector input into the **axis** command must now have six entries, as follows:

```
axis([xmin xmax ymin ymax zmin zmax])
```

You can similarly label your 3-D figure using `xlabel`, `ylabel`, `zlabel`, and `title`.

The `grid` command is also valid in conjunction with the `plot3` command.

1.4 M-files

In the last section, we found that to complete a figure with a caption, we had to enter several commands one by one in the command window. Typing errors will be time-consuming to fix because if you are working in the command window, and if you make an error, you will need to retype all or part of the program. Even if you do not make any mistakes (!), all of your work may be lost if you inadvertently quit MATLAB and have not taken the necessary steps to save the contents of the important program that you just finished developing. This will be time-consuming. Especially, if you are simulating a process, and all that you want to change in successive runs are the parameters of the problem. To preserve large sets of commands, you can store them in a special type of file called an *M-file*.

MATLAB supports two types of *M-files*: *script* and *function M-files*. To hold a large collection of commands, we use a *script M-file*. The *function M-file* is discussed in Chapter 3. To make a *script M-file*, you need to open a file using the built-in MATLAB Menu. First select **New** from the **File** menu. Then select the *M-file* entry from the pull-down menu. After typing the *M-file* contents, you need to save the file. For this purpose, use the **save as** command from the **File** window. A field will pop up in which you can type in the name you have chosen for this file. In the pop-up lower window, indicate that it is an .m file.

NOTES

1. Avoid naming a file by a mathematical abbreviation, the name of a mathematical function, a MATLAB command, or a number.
2. To run your *script M-file*, just enter in the command window the filename omitting the **.m** extension in the file name at its end at the MATLAB prompt. To be able to access this file from the command window, make sure that the folder in which you saved this file is in the MATLAB **Path**. If this is not the case use the **Set Path** in the **File** pull-down menu and follow the screen prompts.

EXAMPLE 1.7

For practice, go to the **File** pull-down menu, select **New** and **M-file** to create the following file that you will name **myfile_17**.

Enter in the editor window, the following set of instructions:

```
x1=1;y1=.5;x2=2;y2=1.5;x3=3;y3=2;
plot(x1,y1,'o',x2,y2,'+',x3,y3,'*')
axis([0 4 0 4])
xlabel('xaxis')
ylabel('yaxis')
title('3points in a plane')
```

After creating and saving this file as **myfile_17** in your work folder, go to the MATLAB command window and enter **myfile_17**. MATLAB will execute the instructions in the same sequence as the statements stored in that file. □

1.5 MATLAB Simple Programming

In this section, we introduce the programming flow control commands `for`, `if`, and `while`.

1.5.1 The `for` Iterative Loop

The power of computers lies in their ability to perform a large number of repetitive calculations. To do this without entering the value of a parameter or variable each time that these are changed, all computer languages have control structures that allow commands to be performed and controlled by counter variables, and MATLAB is no different. For example, the MATLAB `for` loop allows a statement or a group of statements to be repeated a prescribed number (i.e., positive integer) of times.

EXAMPLE 1.8

Generate the square of the first ten integers.

Solution: Edit and execute the following *script M-file:*

```
for m=1:10
  x(m)=m^2;
end
```
□

In this case, the number of repetitions is controlled by the index variable **m**, which takes on the values **m=1** to **10** in intervals of 1. Therefore, 10 assignments are made. What the above loop is doing is sequentially assigning the different values of **m^2** (i.e., m^2) for each element of the "**x**-array." An array is just a data structure that can hold multiple entries. An array can be 1-D such as in a vector or 2-D such as in a matrix. More will be said about vectors and matrices in subsequent chapters. At this time, assume 1-D and 2-D arrays as pigeonholes with numbers or ordered pair of numbers respectively assigned to them.

To find the value of a particular slot of the array, such as slot 3, enter:

```
>>x(3)
```

To read all the values stored in the array, enter:

```
>>x
```

Question: What do you get if you enter **m**, following the execution of the above program?
Answer: The value 10, which is the last value of m read by the counter.

Homework Problem

Pb. 1.2 A couple establishes a college savings account for their newly born baby girl, deposit $2000 in this account, and commit themselves to deposit an equal amount at each future birthday of the new born. They chose to invest this money with an investment firm that guarantees a minimum annual return of 5%. What will be the minimum balance of the account when the young lady turns 18? (Include in your calculation the amount deposited on her 18th birthday.)

1.5.2 `if-else-end` Structures

If a sequence of commands must be conditionally evaluated based on a relational test, the programming of this logical relationship is executed with some variation of an **if-else-end** structure.

A. The simplest form of this structure is:

```
if expression
        Commands evaluated if expression is True
else
        Commands evaluated if expression is False
end
```

NOTES

1. The commands between the **if** and **else** statements are evaluated for all elements in the expression which are true.
2. The conditional expression uses the Boolean logical symbols & (AND), | (OR), and ~ (NOT) to connect different propositions.

Example 1.9

Find for integer a, $0 < a \leq 10$, the values of C, defined as follows:

$$C = \begin{cases} ab & \text{for } a > 5 \\ \frac{3}{2} ab & \text{for } a \leq 5 \end{cases}$$

and $b = 15$.

Solution: Edit and execute the following *script M-file:*

```
>>b=15;
     for a=1:10
          if a>5
               C(a)=a*b;
          else
               C(a)=(a*b)*(3/2);
          end
     end
```

Check that the values of C that you will obtain by typing **C** are:

```
22.5 45 67.5 90 112.50 90 105 120 135 150
```
□

B. When there are three or more alternatives for the conditional, the **if-else-end** structure takes the form:

```
if expression 1
        Commands 1 evaluated if expression 1 is True
elseif expression 2
        Commands 2 evaluated if expression 2 is True
```

elseif *expression 3*
Commands 3 evaluated if expression 3 is True
...

....
else
Commands evaluated if no other expression is True
end

In this form, only the commands associated with the first True expression encountered are evaluated, ensuing relational expressions are not tested.

Example 1.10

Find for integers a, $1 \le a \le 15$, the value of D defined by

$$D = \begin{cases} a^3 & \text{for } a \le 5 \\ 2a & \text{for } 5 < a < 10 \\ a+7 & \text{for } 10 \le a \le 15 \end{cases}$$

Solution: Edit and execute the following *script M-file:*

```
for a=1:15
      if a<=5
            D(a)=a^3;
      elseif a>5 & a<10
            D(a)=2*a;
      else
            D(a)=7+a;
      end
end
D
```
□

Homework Problem

Pb. 1.3 For the values of integer a going from 1 to 10, find the values of C such that

$$C = \begin{cases} a^2 & \text{if } a \text{ is even} \\ a^3 & \text{if } a \text{ is odd} \end{cases}$$

Use the **stem** command to graphically show C. (Hint: Look up in the help file the function **mod**.)

1.5.3 The `while` Loop

The **while** loop is used when the iteration process has to terminate when *a priori* specified condition is satisfied. The difference with the **for** command is that there the number of iterations is specified in advance.

The syntax for the **while** loop is as follows:

```
Initial assignment
while relational condition
      expression
end
```

NOTE Special care should be exercised when using the **while** loop, as in some cases the looping may never stop.

EXAMPLE 1.11

The unknown quantity x has the form 2^n, where n is an integer. We desire to find the largest such number subject to the condition that $x < 75{,}345$.

Solution: Using the above syntax of the **while** loop, we enter:

```
>>x=2;
      while 2*x<75345
            x=2*x;
      end
  x
```

The unknown variable is initially assigned the value 2, it has this value until it encounters the statement **x=2*x,** which should be read that $x(n + 1) = 2*x(n)$, the value of x is then changed to **xnew**. Before each pass through the loop, the value of x is checked to see whether $x < 75{,}345$. If this condition is satisfied, the next iteration is carried through; otherwise the looping is stopped. The final result to our problem is the value of x before any more loop is executed.

If the statement of the problem is changed, and we had asked instead the question to list all x's that satisfied the condition $x < 75{,}345$. We would have entered instead of the above the following instructions:

```
>>x=2;
      while x<75345
      disp(x)
            x=2*x;
      end
```

The **disp** (display) command displays the results of all iterations satisfying the desired condition. □

Homework Problem

Pb. 1.4 Using **while**, find the largest number x which can be written in the form

$$x = 3^n + 2n + 3$$

and such that n is an integer and $x < 35{,}724$.

1.6 Arrays

As mentioned earlier, think of an array as a string of ordered numbers, i.e., to determine an array, we need to specify the value and position of each element in this string.

As pointed out earlier, an element in the array is addressed by writing the name of the array followed by the position of the element within parentheses. For example, **x(3)** will be the value of the third element in the array **x**.

The basic array functions are:

length(x)	finds the length of the array **x** (i.e., number of elements)
find(x)	computes a new array where the successive elements indicate the positions where the components of the array **x** are nonzero
max(x)	gives the value of the largest element in the array **x**
min(x)	gives the value of the smallest element in the array **x**
sum(x)	gives the sum of all elements in the array **x**

EXAMPLE 1.12

Enter the array $x = [0\ 3\ 6\ -2\ 11\ 0\ 7\ 9]$, and find its third element, length, nonzero elements, value of its largest and smallest elements, and sum of all its elements.

Solution:

```
>> x=[0 3 6 -2 11 0 7 9];
>> x(3)
ans =
     6
>> length(x)
ans =
     8
>> find(x)
```

```
ans =
     2     3     4     5     7     8
>> max(x)
ans =
     11
>> min(x)
ans =
     -2
>> sum(x)
ans =
     34
```
□

1.6.1 Array Relational Operations

MATLAB has six relational operations that compare the elements of two arrays of the same length.

The relational operators are:

= =	equal to
~ =	not equal to
<	smaller than
<=	smaller or equal to
>	larger than
>=	larger than or equal

If we combine within parentheses the symbols for two arrays with a relational operator, the result will be a new array consisting of 0's and 1's, where the 1's are in the positions where the relation between the two elements in the same position from the two arrays is satisfied, and 0's otherwise.

EXAMPLE 1.13

Study the following printout:

```
x = [1    4    7    5    11    2];
y = [3   -2    7    2    12   -4];
>>   z1=(x==y)
z1   =
       0    0    1    0    0    0
>>   z2=(x~=y)
z2   =
       1    1    0    1    1    1
>>   z3=(x<y)
z3   =
       1    0    0    0    1    0
>>   z4=(x<=y)
```

```
z4 =
     1    0    1    0    1    0
>> z5=(x>y)
z5 =
     0    1    0    1    0    1
>> z6=(x>=y)
z6 =
     0    1    1    1    0    1
>> z7=(z1==z2)
z7 =
     0    0    0    0    0    0
>> z8=(z3==z6)
z8 =
     0    0    0    0    0    0
>> z9=(z4==z5)
z9 =
     0    0    0    0    0    0
```

Comments: The values of z7, z8, and z9 are identically zeros. This means that no corresponding elements of z1 and z2, z3 and z6, z4 and z5, respectively are ever equal, i.e., each pair represents mutually exclusive conditions. □

1.6.2 Array Algebraic Operations

In Section 1.5, we used **for** loops repeatedly. However, this kind of loop-programming is very inefficient and must be avoided as much as possible in MATLAB. In fact, ideally, a good MATLAB program will always minimize the use of loops because MATLAB is an interpreted language — not a compiled one. As a result, any looping process is very inefficient. Nevertheless, at times we use the **for** loops, when necessitated by pedagogical reasons.

To understand array operations more clearly, consider the following:

```
a=1:3 % a starts at 1, goes to 3 in increments of 1.
```

If the increment is not 1, you must specify the increment; for example,

```
b=2:0.2:6 % b starts at 2 ends at 6 in steps of 0.2
```

To distinguish arrays operations from either operations on scalars or on matrices, the symbol for multiplication becomes **.***, that of division **./**, and that of exponentiation .^. Thus, for example,

```
c=a.*b % takes every element of a and multiplies
% it by the element of b in the same array location
```

Similarly, for exponentiation and division:

```
d=a.^b
e=a./b
```

If you try to use the regular scalar operations symbols, you will get an error message.

Note that array operations such as the above require the two arrays to have the same length (i.e., the same number of elements). To verify that two arrays have the same number of elements (dimension), use the **`length`** command. The exception to the equal length arrays rule is when one of the arrays is a scalar constant.

NOTE The expression **`x=linspace(0,10,200)`** is also a generator for an x-array. Its first element equal to 0, its last element equal to 10, and it has 200 equally spaced points between 0 and 10. Here, the number of points rather than the increment is specified; that is, **`length(x)=200`**.

1.6.3 Combining Arrays Relational and Algebraic Operations: Alternative Syntax to the `if` Statement

As an alternative to the **`if`** syntax, we can use a combination of arrays relational and algebraic operations to generate a complicated array from a simple one.

EXAMPLE 1.14

Using array operations, find for integers a, $1 \le a \le 15$, the value of D defined by

$$D = \begin{cases} a^4 & \text{for } a \le 5 \\ 2a & \text{for } 5 < a < 10 \\ a+7 & \text{for } 10 \le a \le 15 \end{cases}$$

Solution:

```
>>a=1:15;
D=(a.^4).*(a<=5)+2*a.*(5<a).*(a<10)+(a+7).*(10<=a)...
  .*(a<=15)
```
□

NOTE We have used in the above the property that the logical AND linking two separate relational operations can be represented by the array product of these relations.

Homework Problem

Pb. 1.5 The array n is all integers from 1 to 110. The array y is defined by

$$y = \begin{cases} n^{1/2} & \text{if } n \text{ divisible by 3, 5, or 7} \\ 0 & \text{otherwise} \end{cases}$$

Find the value of the sum of the elements of y.

1.6.4 Plotting Arrays

If we are given two arrays x and y, we can plot the y-array as function of the x-array. What MATLAB is actually doing is displaying all the points $P(x_i, y_i)$ and connecting each two consecutive points by a straight line.

The same options for markers and color that were available for plotting points and described earlier are also valid in this case. Additionally, we can specify here the style of the line that connects the different points.

Style	MATLAB Symbol
Solid line	-
Dashed line	--
Dash-dot line	-.
Dotted line	:

If the points are close together, the points will look connected by a continuous line making a smooth curve; we say that the program graphically interpolated the discrete points into a continuous curve.

The commands for labeling, axis and tick marks used in plotting points are also valid here. We can zoom to a particular region of the graph by using the command **zoom**, And we can read off the coordinates of particular point(s) in the graph, by using

```
[x,y]=ginput(n)
```

This command serves to read the values of the coordinates of points off a displayed graph (**n** is the number of points). The command pops out the crosshair. The operator manually zeros on the points of interest and clicks. As a result, MATLAB prints the coordinates of the points in the same order as those of the clicks of the mouse.

Example 1.15

The three arrays x, y, and z are defined as follows:

$$x = 0{:}0.1{:}10; \qquad y = x^2 - 6x - 20; \qquad z = 3x - 5$$

Plot y as function of x, and z as function of x. Show the traces as red dotted line and blue solid line, respectively, label the x- and y-axis, and label graph as Figure of Example 1.15.

Solution:

```
>>x=0:0.1:10;
y=x.^2-6*x-20;
z=2*x-5;
plot(x,y,'r:',x,z,'b-')
xlabel('x axis')
```

```
ylabel('y-z axis')
title('Figure of Example 1.15')
grid
```

In the above solution, we plotted both curves on the same graph. Often, it may be required to plot the two curves in separate graphs but in the same figure window. This can be achieved using the **subplot** command. The arguments of the subplot command **subplot(m,n,p)** specify m the number of rows partitioning the graph, n the number of columns, and p the number of the particular subgraph chosen (enumerated through the left to right, top to bottom convention). The program to plot the above arrays in two graphs in a column would read

```
>>x=0:0.1:10;
y=x.^2-6*x-20;
z=2*x-5;
subplot(2,1,1)
plot(x,y,'r:')
xlabel('x axis')
ylabel('y axis')
title('Figure of Example 1.14-a')
subplot(2,1,2)
plot(x,z,'b-')
xlabel('x axis')
ylabel('z axis')
title('Figure of Example 1.14-b')
```
□

All the plotting that we did so far has been using the linear scale in both the horizontal and vertical axes. Sometimes when the elements of an array can change values by several decades over the range of interest, it is well advised to use the different kinds of the log-plots. As the need arises we can, instead of the **plot** command, use one of the following alternatives:

semilogy creates a semilog plot with logarithmic scale on the y-axis
semilogx creates a semilog plot with logarithmic scale on the x-axis
loglog creates a log–log graph

Homework Problem

Pb. 1.6 The arrays y and z are generated from the array of positive integers $x = 1{:}20$, such that:

$$y = 2^x \quad \text{and} \quad z = 5^{1/x}$$

Plot the array z as function of the array y. (You should be able to read off the value of the coordinates of any point on the plot over the entire range of the graph.)

1.7 Data Analysis

1.7.1 Manipulation of Data

The most convenient representation for data collected from experiments is in the form of histograms. Typically, you collect data and want to sort it out in different bins. But prior to getting to this point, let us review and introduce some array-related commands that are useful in data manipulation.

Let $\{y_n\}$ be a data set. It can be represented in MATLAB by an array **y**.

`length(y)`	gives the length of the array (i.e., the number of data points)
`y(i)`	gives the value of the element of y in the ith position
`y(i:j)`	gives a new array, consisting of all elements of y between and including the locations i and j
`find(y)`	computes a new array where the successive elements indicate the positions where the components of the array y are nonzero
`max(y)`	gives the value of the largest element in the array
`min(y)`	gives the value of the smallest element in the array
`sort(y)`	sorts the elements of y in ascending order and gives a new array of the same size as y
`sum(y)`	sums the elements of y. The answer is a scalar
`mean(y)`	gives the mean value of the elements of y
`std(y)`	computes the standard deviation of the elements of y

The definitions of the mean and of the standard deviation are, respectively, given by

$$\bar{y} = \frac{1}{N}\sum_{i=1}^{N} y(i)$$

$$\sigma = \sqrt{\sum_{i=1}^{N} \frac{(y(i) - \bar{y})^2}{(N-1)}}$$

where N is the length of the array.

EXAMPLE 1.16

An old recorder at Bill's backyard in Hamilton Heights in New York downloaded the following raw data for the temperature readings on July 25 between 12:01 p.m. and 1:00 p.m.:

```
T=[95 95.5 96 95.8 97 96.5 96 95 55 96 95 97 96 ...
   96.5 96.2 125 96 95 95.5 96]
```

Bill sits to analyze the data. As a budding experimentalist, he knows that his first duty is to assess the validity of his raw data.

1. He checks that there was a reading by the recorder at every 3 min interval, he enters in his MATLAB program, to which the data were downloaded:

```
>> length(T)
ans =
            20
```

   "OK," he says, "the recorder was recording data at the prescribed time intervals."

2. Next, he wants to check for the validity of the data; he enters:

```
>> max(T)
ans =
            125
>> min(T)
ans =
            55
```

   "Hm ... even in New York the weather is not that crazy. Let me check for the validity of the rest of the data," he says.

```
>> format bank
>> z=sort(T)
z =
  Columns 1 through 8
  55.00 95.00 95.00 95.00 95.00 95.50 95.50 95.80
  Columns 9 through 16
  96.00 96.00 96.00 96.00 96.00 96.00 96.20 96.50
  Columns 17 through 20
  96.50 97.00 97.00 125.00
```

   "Well, it seems that except for the smallest and largest values of the array, the data temperature is within reasonable limits for a humid, sunny day in July at noon. Obviously, the link to the recorder is picking noise and it has to be checked by Jim, my buddy the fixer, but...," says Bill, "...can I still use any part of the data to record, in my project book, an average temperature for that one hour period?"

3. Bill decides to reject as spurious the lowest and highest values of his raw data; he enters:

```
>> TT=z(2:19);
```

4. Now he computes the mean and the standard deviation of his massaged data:

```
>> meanTT=mean(TT)
meanTT =
            95.89
```

```
>> stdTT=std(TT)
stdTT =
                    0.64
```

These are the values that Bill enters in his project's logbook. □

Homework Problem

Pb. 1.7 Your assignment is to help Bill automate his task. Use the new criterion that all raw data outside the interval

$$(\bar{T}-\sigma_T) \leq T \leq (\bar{T}+\sigma_T)$$

is rejected, and not included in the reported analysis.

1.7.2 Displaying Data

The data (i.e., the array) can be organized into a number of bins (n_b) and exhibited through the command:

```
[n,y]=hist(y,nb)
```

The array n in the output will be the number of elements in each of the bins.

Example 1.17

The students' numerical scores on a certain exam were as follows:

```
g=[65 73 87 45 91 55 85 74 55 94 62 85 72 77 81 63 ...
   35 62 85 90 57 73 68 57 51 75 74 70 77 60 77 84 ...
   73 52 73 63];
```

The instructor desires to compute the average grade and the standard deviation of the class grades distribution. Furthermore, she desires to distribute the grades into five bins and determine the number of students in each of the respective bins. Help her write a program to achieve this task.

Solution: First enter the data for the grade as an array, then enter

```
>>averageg=mean(g)
stdg=std(g)
[n,g]=hist(g,5)
```

MATLAB will return:

```
averageg =
                    70.00
```

```
stdg =
            13.75
n =
            2.00 6.00 8.00 12.00 8.00
g =
            40.90 52.70 64.50 76.30 88.10
```

The third line gives you the number of grades in each of the bins, and the last line gives you the array of the values of the average grade in each of the bins. Associated with **n**, we can associate a probability; thus

The probability of being in the third bin = **n(3)/length(g)**,

where **length(g)** represents the total number of students who sat for the exam.

If you were also requested to draw a bar chart or pie chart to display the different number of grades in each of the different bins, you would enter

bar(n) or **pie(n)** □

1.7.3 Normal Distribution

In the previous example, we have seen an example of a distribution (i.e., the distribution of grades in the different bins). When the sample is very large and the number of bins is also very large, the distribution can be approximated by a continuous function. In many instances when the sample is random, this distribution takes a special shape, called normal distribution.

In Chapter 10, we study the conditions that lead to this distribution. In this section, we limit ourselves to introducing the basic MATLAB commands, that, knowing that the distribution is normal, allow us to compute the probability for having the random variable value in the interval $x_1 < x < x_2$.

The normal probability density function is given by the following expression:

$$f_X(x) = \frac{1}{\sigma_X\sqrt{2\pi}} \exp\left(-\frac{(x-m_X)^2}{2\sigma_X^2}\right)$$

and the probability that the random variable to be in the interval $x_1 < x < x_2$ is

$$F_X(x_1 \leq x \leq x_2) = \int_{x_1}^{x_2} f_X(x)\,dx = \frac{1}{2}\left[\operatorname{erf}\left(\frac{x_2-m_X}{\sqrt{2}\sigma_X}\right) - \operatorname{erf}\left(\frac{x_1-m_X}{\sqrt{2}\sigma_X}\right)\right]$$

where erf is the error function, a standard MATLAB function, and m_X and σ_X are, respectively, the mean and standard deviation of x.

EXAMPLE 1.18

A manufacturer produces resistors. It claims in its product specifications sheet that a sampling of a very large number of its products gave a mean value 99.5 Ω for their 100 Ω nominal resistors, with a standard deviation of 1.5 Ω. A design engineer determined that the circuit that he is designing required a $100 \pm 2\,\Omega$ resistor. The question is: What is the probability that the engineer will get a resistor from this manufacturer's batch with a true value within the error tolerance that he requires?

Solution: Enter and execute the following program:

```
>> meanr=99.5;
sigmar=1.5;
r=100;
rtol=2;
r2=r+rtol;
r1=r-rtol;
F=0.5*(erf((r2-meanr)/(sqrt(2)*sigmar))...
     -erf((r1-meanr)/(sqrt(2)*sigmar)))
```

MATLAB gives:

```
F =
     0.7936
```

i.e., the chances are 79.36% that the product will meet the engineer's design requirements. □

In many engineering simulations (i.e., mathematical model for a physical system), one may desire to simulate the effects of noise. This is usually achieved by using the built-in random number generators available in MATLAB. The purpose of the next example is to acquaint you with the MATLAB in-built normally distributed random array.

EXAMPLE 1.19

Find the mean and the standard deviation; obtain the histogram, with 20 bins, for an array whose 10,000 elements are chosen from the MATLAB built-in normal distribution with zero mean and standard deviation 1.

Solution: Edit and execute the following *script M-file:*

```
y=randn(1,10000);
meany=mean(y)
stdy=std(y)
nb=20;
hist(y,nb)
```

You will notice that the results obtained for the mean and the standard deviation vary slightly from the theoretical results. This is due to the finite number of elements chosen for the array and the intrinsic limit in the built-in algorithm used for generating random numbers.

NOTE The MATLAB command for generating an N-element array of random numbers distributed uniformly in the interval [0, 1] is **rand(1,N)**. □

Homework Problem

Pb. 1.8 Return to Example 1.18, and determine what should be the sigma of the manufacturer specifications so that the probability that the engineer obtains the resistor he buys within the tolerance limits that he requires 99% of the times.

1.8 Parametric Equations

It will be helpful to the reader to review the sections of Appendix A pertaining to lines, quadratic functions, and trigonometric functions before proceeding further in this section.

1.8.1 Definition

If instead of assuming a function as a relation between an independent variable x and a dependent variable y, we assume both x and y as being dependent functions of a third independent parameter t, then we say that we have a parametric representation of the curve. This method of curve representation is described by $(x(t), y(t))$, where the parameter t varies over some finite domain (t_{min}, t_{max}).

NOTE The process of parameterization is not unique, i.e., there may be more than one way to generate the same curve from different functions $x(t)$ and $y(t)$.

EXAMPLE 1.20

The cosine and sine functions are defined respectively as the x-component and the y-component of a point P on the trigonometric circle (the circle centered at the origin of the coordinates O and having radius 1). The angle is that formed between the positive x-axis and the ray OP, measured in the anticlockwise direction. Using this parameterization of the trigonometric circle, write a program to plot it.

Solution:

```
th=linspace(0,2*pi,101)
x=cos(th);
y=sin(th);
plot(x,y)
axis([-1.5 1.5 -1.5 1.5])
axis square
```
□

1.8.2 More Examples

In the following examples, we want to identify the curves $f(x, y) = 0$ corresponding to each of the given parameterizations.

EXAMPLE 1.21

Specify the curve C defined by

$$x = 2t - 1, \quad y = t + 1 \qquad \text{where } 0 \leq t \leq 2.$$

The initial point is at $x = -1$, $y = 1$, and the final point is at $x = 3$, $y = 3$.

Solution: The curve $f(x, y) = 0$ form can be obtained by noting that from $x = 2t - 1 \Rightarrow t = (x + 1)/2$. Substituting this value of the parameter into the expression for y gives

$$y = \frac{x}{2} + \frac{3}{2}$$

This equation describes a line with slope 1/2 crossing the x-axis at $x = -3$. □

Question: Where does this line cross the y-axis?
Answer: $y = 3/2$.

EXAMPLE 1.22

The curve C is described by the following parametric equations:

$$x = 3 + 3\cos(t), \quad y = 2 + 2\sin(t) \qquad \text{where } 0 \leq t \leq 2\pi.$$

The initial point is at $x = 6$, $y = 2$, and the final point is at $x = 6$, $y = 2$. Find the $f(x, y) = 0$ equation for this curve.

Solution: The $f(x, y) = 0$ for this curve can be obtained by noting that from the parametric equations, we can write

$$\cos(t) = \frac{(x - 3)}{3} \quad \text{and} \quad \sin(t) = \frac{(y - 2)}{2}$$

Substituting these expressions into the trigonometric identity $\cos^2(t) + \sin^2(t) = 1$, we deduce the following equation for the curve C:

$$\frac{(x-3)^2}{3^2} + \frac{(y-2)^2}{2^2} = 1$$

This is the equation of an ellipse centered at $x = 3$, $y = 2$ and having major and minor radii equal to 3 and 2, respectively.

(The foci of this ellipse are at $(3 - \sqrt{5}, 2)$ and $(3 + \sqrt{5}, 2)$.) □

In-Class Exercises

Pb. 1.9

a. Show analytically that the following parametric equations:

$$x = h + a\sec(t) \quad \text{and} \quad y = k + b\tan(t) \quad \text{where} \quad -\pi/2 < t < \pi/2$$

represent a branch of the hyperbola also represented by the equation:

$$\frac{(x-h)^2}{a^2} - \frac{(y-k)^2}{b^2} = 1$$

b. Plot the hyperbola branch represented by the parametric equations of **(a)**, with $h = 2$, $k = 2$, $a = 1$, $b = 2$.

c. Find the coordinates of the vertices and the foci of this hyperbola.

d. For what values of t, will the other branch of the hyperbola be obtained?

Pb. 1.10 The parametric equations of the cycloid are given by

$$x = R\omega t + R\sin(\omega t) \quad \text{and} \quad y = R + R\cos(\omega t) \qquad \text{where } 0 < t$$

a. Show how this parametric equation can be obtained analytically by following the kinematics of a point attached to the outer rim of a wheel that is uniformly rolling, without slippage, on a flat surface.

b. Relate the above parameters to the linear speed and the radius of the wheel.

Pb. 1.11 The parametric equations of the epitrochoids and hypotrochoids families are respectively given by

(A) *Epitrochoids*:

$$\begin{cases} x = s\cos(t) - h\cos\left(\dfrac{s}{b}t\right) \\[2ex] y = s\sin(t) - h\sin\left(\dfrac{s}{b}t\right) \end{cases} \quad \text{where} \quad \begin{cases} s = a + b \\ -\pi \le t \le \pi \end{cases}$$

(B) *Hypotrochoids*:

$$\begin{cases} x = s\cos(t) + h\cos\left(\dfrac{s}{b}t\right) \\ y = s\sin(t) - h\sin\left(\dfrac{s}{b}t\right) \end{cases} \quad \text{where} \quad \begin{cases} s = a + b \\ -\pi \le t \le \pi \end{cases}$$

a. Show that the curves are completely contained in a circle of radius $r \le |s + h|$. (Assume that a, b, and h are positive.)
b. Empirically, find the number of loops in a curve if s/b is an integer.
c. Examine in details for the epitrochoids the cases: (i) $a = 0$, (ii) $a = b$, and (iii) $h = b$, and for the hypotrochoids the cases: (i) $a = 2b$ and (ii) $h = b$.

Pb. 1.12 Sketch the curve C defined by the following parametric equations:

$$x(t) = \begin{cases} t + 2 & \text{for } -3 \le t \le -1 \\ 1 - \dfrac{1}{3}\tan\left(\dfrac{\pi}{3}(1 - t^2)\right) & \text{for } -1 < t < 0 \\ -1 + \dfrac{1}{3}\tan\left(\dfrac{\pi}{3}(1 - t^2)\right) & \text{for } 0 \le t \le 1 \end{cases}$$

$$y(t) = \begin{cases} 0 & \text{for } -3 \le t \le -1 \\ -\dfrac{1}{\sqrt{3}}\tan\left(\dfrac{\pi}{3}(1 - t^2)\right) & \text{for } -1 < t < 0 \\ \dfrac{1}{\sqrt{3}}\tan\left(\dfrac{\pi}{3}(1 - t^2)\right) & \text{for } 0 \le t \le 1 \end{cases}$$

1.8.3 Oscilloscope Graphics

To most students, the familiar mode for oscilloscope operation is that in which the horizontal axis is swept linearly in time and the trace observed on the screen is the plot of the input voltage as function of time. In many important circuits characteristics measurements, it is often required that in addition to wanting to measure the ratio of the amplitudes of two sinusoidal voltages, one is required as well to determine the phase shift between these potentials. This is a practical application of parametric equations. The following set of problems provides the mathematical basis for understanding the graphical display on the screen of an oscilloscope, when the oscilloscope is in the x-y mode.

Homework Problems

Pb. 1.13 To transform the quadratic expression

$$Ax^2 + Bxy + Cy^2 + Dx + Ey + F = 0$$

into standard form (i.e., to eliminate the x-y mixed term), perform the following transformation:

$$x = x'\cos(\theta) - y'\sin(\theta)$$
$$y = x'\sin(\theta) + y'\cos(\theta)$$

Show that the mixed term in the new variables is eliminated if $\cot(2\theta) = (A - C)/B$.

Pb. 1.14 Consider the curve C represented by the parametric equations:

$$x = a\cos(t), \qquad y = b\sin(t + \varphi) \qquad \text{where } 0 \leq t < 2\pi$$

the initial point is at $(x = a, y = b\sin(\varphi))$, and the final point is at $(x = a, y = b\sin(\varphi))$.

a. Obtain the equation of the curve in the form $f(x, y) = 0$.
b. Using the results of Pb. 1.13, prove that the resulting ellipse inclination angle (i.e., the angle that its long axis makes with the x-axis) is given by:

$$\cot(2\theta) = \frac{(a^2 - b^2)}{2ab\sin(\varphi)}$$

Pb. 1.15 The simplest Lissajous curve is represented by the parametric equations

$$x = \cos(t), \qquad y = \sin(2t)$$

where $0 \leq t < 2\pi$, and the initial and final points are at $(x = 1, y = 0)$.

This curve has the shape of a figure 8 with two nodes in the x-direction and only one node in the y-direction.

What do you think the parametric equations should be if we wanted the curve to have m nodes on the x-axis and n nodes on the y-axis? Test your hypothesis by plotting the results of your guesses.

1.9 Polar Plots

MATLAB can also display polar plots. This plot is of particular interest when, *inter alia*, studying the radiation pattern from antennas. In the first example, we draw an ellipse. Its equation is

$$r = \frac{r_0}{1 - \varepsilon \cos(\theta)}$$

in polar coordinates.

Example 1.23

Plot the ellipse with an ellipticity $\varepsilon = 0.2$ and $r_0 = 1$, in a polar plot.

Solution: The following sequence of commands plot the polar plot of an ellipse:

```
>>th=0:2*pi/100:2*pi;
rho=1./(1-0.2*cos(th));
polar(th,rho)
```
□

The shape you obtain may be unfamiliar to you; but to verify that this is indeed an ellipse, view the curve in a Cartesian graph.

For that, you can use the MATLAB polar to Cartesian converter **pol2cart**, as follows:

```
>>[x,y]=pol2cart(th,rho);
plot(x,y)
axis([-1.5 1.5 -1.5 1.5])
```

Example 1.24

Graph the polar plot of a simple spiral.

Solution: The equation of a simple spiral in polar coordinates is given by

$$r = a\theta$$

Its polar plot can be viewed by executing the following *script M-file* ($a = 3$):

```
th=0:2*pi/100:2*pi;
rho=3*th;
polar(th,rho)
```
□

In-Class Exercises

Pb. 1.16 Analytically prove that the polar equation

$$r = \frac{r_0}{1 - \varepsilon \cos(\theta)}$$

where the value of ε is always between 0 and 1, results in an ellipse. Determine the center of this ellipse and its major (long) and minor (short) radii.

Pb. 1.17 Plot the three curves described by the following polar equations:

$$r = 2 - 2\sin(\theta); \qquad r = 1 - \sqrt{2}\sin(\theta); \qquad r = \sqrt{2\sin(2\theta)}$$

Specify the domain in the interval $0 \le \theta \le 2\pi$, where each equation makes sense.

Pb. 1.18 Plot the curve described in polar coordinates by the equation

$$r = |\sin(2\theta)\cos(2\theta)|$$

The above gives a flower-type curve with eight petals. How would you make a flower with 16 petals?

Pb. 1.19 Plot the curve described by $r = \sin^2(\theta)$.
This two-lobed structure shows the power distribution of a simple dipole antenna. Note the directed nature of the radiation. Can you increase the directivity of this antenna further?

Pb. 1.20 Acquaint yourself with the polar plots of the following curves (choose first $a = 1$, then experiment with other values).

a. *Straight lines*:

$$r = \frac{1}{\cos(\theta) + a\sin(\theta)} \qquad \text{for } 0 \le \theta \le \frac{\pi}{2}$$

b. *Cissoid of Diocles*:

$$r = a\frac{\sin^2(\theta)}{\cos(\theta)} \qquad \text{for } -\frac{\pi}{3} \le \theta \le \frac{\pi}{3}$$

c. *Strophoid:*

$$r = \frac{a\cos(2\theta)}{\cos(\theta)} \quad \text{for } -\frac{\pi}{3} \leq \theta \leq \frac{\pi}{3}$$

d. *Folium of Descartes:*

$$r = \frac{3a\sin(\theta)\cos(\theta)}{\sin^3(\theta) + \cos^3(\theta)} \quad \text{for } -\frac{\pi}{6} \leq \theta \leq \frac{\pi}{2}$$

1.10 3-D Plotting

1.10.1 Straight-Edge Geometric Figures

Plotting straight-edge geometric figures is the most common graphic requirement when modeling manufactured objects, drawing physical models for crystals, visualizing a 3-D geometric problem, etc. MATLAB is specially suited to draw lines and planes in 3-D. The graphic window has many tools to manipulate these objects once drawn. The object can be rotated, the camera position can be changed, the lighting can be modified, etc. It will prove practical that you acquaint yourself with these features of the graphic window.

As the number of vertices for the object increases, it becomes important to find a systematic way to keep track of the vertices, the edges, and the faces. The table below was developed for the case of a tetrahedron. In problems of similar nature, your first task is to keep a proper record and track the quantities of interest. Essentially, you would need to enter the X-, the Y-, and the Z-array of each of the geometric objects of interest. The length of the arrays is 1 for a point, at least 2 for a line and at least 3 for a plane.

The principal MATLAB commands for this kind of plotting are the **`line`** and **`fill3`** command, and their syntax are as follows: **`line(X-array, Y-array, Z-array, 'color')`** draws a colored line between collinear points, the length of the X-, Y-, and Z-array is at least 2 each, and the elements of the arrays are respectively the x-, y-, and z-components of the collinear points extending the line. **`fill3(X-array, Y-array, Z-array, 'color')`** colors the face defined by at least 3 points , the length of the X-, Y-, and Z-array is a minimum of 3 each, and the elements of the arrays are respectively the x-, y-, and z-components of the coplanar points defining the face.

Bookkeeping Table for a Tetrahedron with Vertices at *A*, *B*, *C*, *S*

Geometric object	X-array	Y-array	Z-array
Points (vertices)			
A	`Ax`	`Ay`	`Az`
B	`Bx`	`By`	`Bz`
C	`Cx`	`Cy`	`Cz`
S	`Sx`	`Sy`	`Sz`
Lines (edges)			
AB	`ABx = [Ax Bx]`	`ABy = [Ay By]`	`ABz = [Az Bz]`
AC	`ACx = [Ax Cx]`	`ACy = [Ay Cy]`	`ACz = [Az Cz]`
AS	`ASx = [Ax Sx]`	`ASy = [Ay Sy]`	`ASz = [Az Sz]`
BC	`BCx = [Bx Cx]`	`BCy = [By Cy]`	`BCz = [Bz Cz]`
BS	`BSx = [Bx Sx]`	`BSy = [By Sy]`	`BSz = [Bz Sz]`
CS	`CSx = [Cx Sx]`	`CSy = [Cy Sy]`	`CSz = [Cz Sz]`
Planes (faces)			
ABC	`ABCx = [Ax Bx Cx]`	`ABCy = [Ay By Cy]`	`ABCz = [Az Bz Cz]`
ABS	`ABSx = [Ax Bx Sx]`	`ABSy = [Ay By Sy]`	`ABSz = [Az Bz Sz]`
ACS	`ACSx = [Ax Cx Sx]`	`ACSy = [Ay Cy Sy]`	`ACSz = [Az Cz Sz]`
BCS	`BCSx = [Bx Cx Sx]`	`BCSy = [By Cy Sy]`	`BCSz = [Bz Cz Sz]`

Example 1.25

Plot the skeleton (i.e., edges only) of the tetrahedron *ABCS*, where

$$A(1,1,0), B(3,1,0), C(2,1+\sqrt{3},0), S\left(2,1+\frac{\sqrt{3}}{3},\sqrt{\frac{8}{3}}\right).$$

Solution:

```
% define the vertices
Ax=1;Ay=1;Az=0;
Bx=3;By=1;Bz=0;
Cx=2;Cy=sqrt(3)+1;Cz=0;
Sx=2;Sy=1+sqrt(3)/3;Sz=sqrt(8/3);

%plot the vertices
plot3(Ax,Ay,Az,'*',Bx,By,Bz,'o',Cx,Cy,Cz,'x',...
Sx,Sy,Sz,'d')
axis([0 3 0 3 0 2])
xlabel('x-axis')
ylabel('y-axis')
zlabel('z-axis')
title('tetrahedron')
hold on

%define the edges arrays
ABx=[Ax Bx];ABy=[Ay By];ABz=[Az Bz];
```

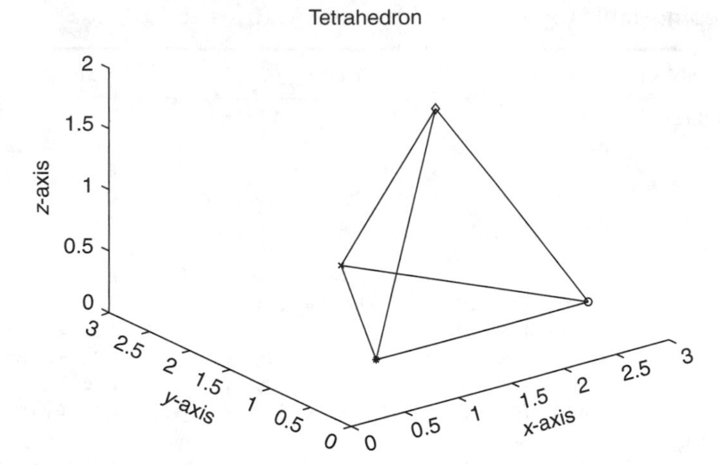

FIGURE 1.1
Graphics of Example 1.25.

```
ACx=[Ax Cx];ACy=[Ay Cy];ACz=[Az Cz];
ASx=[Ax Sx];ASy=[Ay Sy];ASz=[Az Sz];
BCx=[Bx Cx];BCy=[By Cy];BCz=[Bz Cz];
BSx=[Bx Sx];BSy=[By Sy];BSz=[Bz Sz];
CSx=[Cx Sx];CSy=[Cy Sy];CSz=[Cz Sz];
%plot the edges
line(ABx,ABy,ABz)
line(ACx,ACy,ACz)
line(ASx,ASy,ASz)
line(BCx,BCy,BCz)
line(BSx,BSy,BSz)
line(CSx,CSy,CSz)
hold off
```

The resulting graph is shown in Figure 1.1. □

EXAMPLE 1.26

Color the faces of the tetrahedron *ABCS*, where

$$A(1,1,0), B(3,1,0), C(2,1+\sqrt{3},0), S\left(2,1+\frac{\sqrt{3}}{3},\sqrt{\frac{8}{3}}\right).$$

Solution:

```
% define the vertices
Ax=1;Ay=1;Az=0;
Bx=3;By=1;Bz=0;
Cx=2;Cy=sqrt(3)+1;Cz=0;
Sx=2;Sy=1+sqrt(3)/3;Sz=sqrt(8/3);
```

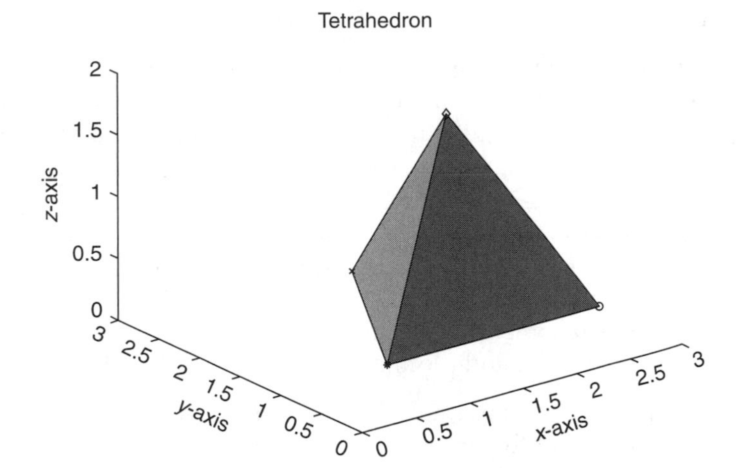

FIGURE 1.2
Graphics of Example 1.26.

```
%plot the vertices
plot3(Ax,Ay,Az,'*',Bx,By,Bz,'o',Cx,Cy,Cz,'x',...
Sx,Sy,Sz,'d')
axis([0 3 0 3 0 2])
xlabel('x-axis')
ylabel('y-axis')
zlabel('z-axis')
title('tetrahedron')
hold on

%define the faces
ABCx=[Ax Bx Cx];ABCy=[Ay By Cy];ABCz=[Az Bz Cz];
ABSx=[Ax Bx Sx];ABSy=[Ay By Sy];ABSz=[Az Bz Sz];
ACSx=[Ax Cx Sx];ACSy=[Ay Cy Sy];ACSz=[Az Cz Sz];
BCSx=[Bx Cx Sx];BCSy=[By Cy Sy];BCSz=[Bz Cz Sz];

%fill and color the faces
fill3(ABCx,ABCy,ABCz,'b')
fill3(ABSx,ABSy,ABSz,'r')
fill3(ACSx,ACSy,ACSz,'g')
fill3(BCSx,BCSy,BCSz,'y')
hold off
```

The resulting graph is shown in Figure 1.2. □

Homework Problem

Pb. 1.21 Plot the skeleton and the colored faces graphics of the cube located in the positive *x–y–z* region and having one vertex at the origin of the axes. Each edge of this cube has length of 3 units.

1.10.2 Parametric Equations for a 3-D Curve

Some examples of parametric equations representing 3-D curves are covered in this section. The parametric equations are specially suited for describing the dynamics (time dependence) of a particle's position in classical mechanics.

EXAMPLE 1.27

A helical curve can be traced by imagining a point that is revolving at a uniform speed around the perimeter of a circle, while it is also moving away from the x–y plane at some constant linear speed. Find the parametric equations describing this helix and plot its graphics.

Solution: The parametric representation of this motion can be deduced from the above description of the motion to be

$$x = a\cos(\omega t), \qquad y = a\sin(\omega t), \qquad z = bt$$

The MATLAB arrays representing this motion and that plots the helix for the values of the constants $a = 2$, $\omega = 1$, $b = 0.1$ are as follows:

```
th=linspace(0,16*pi,1000);
x=2*cos(th);
y=2*sin(th);
z=0.1*th;
plot3(x,y,z)
```
□

IN-CLASS EXERCISES

Pb. 1.22 In the helix of Example 1.27, what is the vertical distance (the pitch) between two consecutive helical turns? How can you control this distance? Find two methods of implementation.

Pb. 1.23 If instead of a circle, as in the helix, the particle were to trace in 2-D a Lissajous pattern having one node in the y-direction and three nodes in the x-direction, show the resulting 3-D trajectory, assuming that the z-parametric equation remains linear in the parameter.

Pb. 1.24 Consider the case that the motion is circular in the x–y plane with angular frequency $\omega = 1$, but now $z(t)$ is periodic in t?
Consider the cases:

(i) $z(t) = \cos(t)$,
(ii) $z(t) = \cos(2t)$.

Show the 3-D trajectory in both cases.

Example 1.28

The tip of a top having one extremity fixed at the origin of axes spins around its own axis; furthermore, this tip also moves on the surface of a sphere centered around the origin. In the so-called ideal nutation mode, the parametric equations describing the motion of the tip of the top are given by

$$x = \sin(\theta)\cos(\varphi), \quad y = \sin(\theta)\sin(\varphi), \quad z = \cos(\theta)$$

where the time dependence of the angles (θ, φ) are, under certain approximations, given by

$$\varphi = t \quad \text{and} \quad \theta = \frac{\pi}{3} + 0.15\sin(20t)$$

Plot the 3-D trajectory of the tip of the top, while showing the position of the top (represented by a line between its ends) at $t = 0$.

Solution:

```
t=0:pi/200:2*pi;
fi=t;
th=pi/3+0.15*sin(20*t);
x=sin(th).*cos(fi);
y=sin(th).*sin(fi);
z=cos(th);
plot3(0,0,0,'o');
hold on
plot3(sqrt(3)/2,0,1/2,'*')
X=[0 sqrt(3)/2];
Y=[0 0];
Z=[0 1/2];
line(X,Y,Z)
plot3(x,y,z)
axis([-1 1 -1 1 -1 1])
hold off
xlabel('x-axis')
ylabel('y-axis')
zlabel('z-axis')
title('Trajectory of the tip of a nutating top')
```
□

Question: Geometrically, what do the angles (θ and φ) in the above problem represent?
Answer: The polar and azimuthal angles of the tip of the top in spherical coordinates.

Homework Problem

Pb. 1.25 The trajectory of a certain particle is described by the following parametric equations:

$$x = (3 + \cos(u))\cos(v)$$
$$y = (3 + \cos(u))\sin(v)$$
$$z = \sin(u)$$

where $u = t^2$, $v = t$, and $0 < t < 10\pi$.

a. Plot the particle trajectory.
b. Specify the geometric surface that bounds this particle's trajectory.

1.10.3 Plotting a 3-D Surface

We now explore the three different techniques for rendering, in MATLAB, 3-D surface graphics.

- A function of two variables $z = f(x, y)$ represents a surface in 3-D geometry; for example,

$$z = ax + by + c$$

  represents a plane that crosses the vertical axis (z-axis) at c.
- There are essentially three main techniques in MATLAB for viewing surfaces: the **`surf`** function, the **`mesh`** function, and the **`contour`** function.
- In all three techniques, we must first create a 2-D array structure (like a checkerboard) with the appropriate x- and y-values. To implement this, we use the MATLAB **`meshgrid`** function.
- The z-component is then expressed in the variables assigned to implement the **`meshgrid`** command.
- We then plot the function with any of the **`surf`** command, the **`mesh`** command, or the **`contour`** command. The **`surf`** and **`mesh`** commands give a 3-D rendering of the surface, while the **`contour`** command gives contour lines, wherein each contour line represents the locus of points on the surface having the same height above or below the x–y plane. This last rendering technique is that used by map makers to represent the topography of a terrain.

Example 1.29

Plot the mesh describing the sine function whose equation is given by

$$z = \frac{\sin\left(\sqrt{x^2 + y^2}\right)}{\sqrt{x^2 + y^2}}$$

over the domain $-8 < x < 8$ and $-8 < y < 8$.

Solution: The implementation of the mesh rendering follows

```
x=[-8:.1:8];
y=[-8:.1:8];
[X,Y]=meshgrid(x,y);
R=sqrt(X.^2+Y.^2)+eps;
Z=sin(R)./R;
mesh(X,Y,Z)
```

The variable **eps** is a tolerance number $= 2^{-52}$, often used for computing expressions near apparent singularities. It was added here to avoid numerical division by zero.

To generate a contour plot, we replace the last command in the above by

```
contour(X,Y,Z,n)     % The fourth argument specifies
                     % the number of contour lines to be
                       shown
```

The resulting graphs are shown in Figure 1.3 and Figure 1.4. □

NOTES

1. If we are interested only in a particular contour level, for example, the one with elevation Z_0, we use the contour function with an option, as follows:

   ```
   contour(X,Y,Z,[Z0 Z0])
   ```

2. Occasionally, we might be interested in displaying simultaneously the mesh and contour rendering of a surface or the surface and contour of a surface. This is possible through the use of the command **meshc** or **surfc**, respectively. These commands produce either the mesh or surface graphics with the contour plot drawn below them.

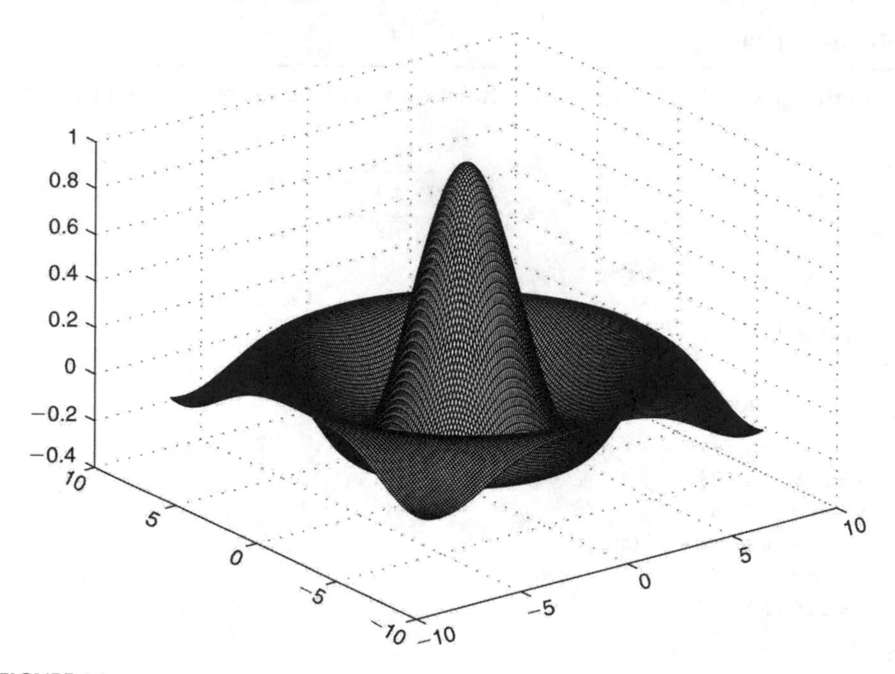

FIGURE 1.3
The **mesh** graphics of Example 1.29.

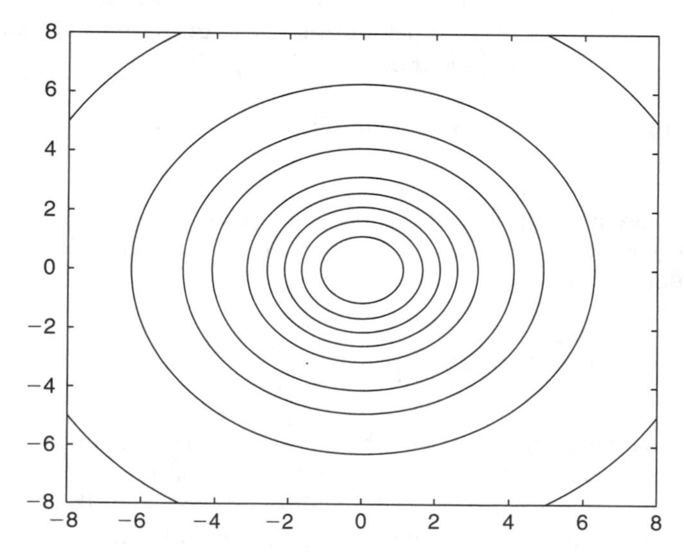

FIGURE 1.4
The **contour** graphics of Example 1.29.

IN-CLASS EXERCISES

Pb. 1.26 Use the **`contour`** function to graphically find the locus of points on the above sinc surface that are 1/2 units above the x–y plane (i.e., the surface intersection with the $z = 1/2$ plane).

Pb. 1.27 Find graphically the point(s) of intersection of the x–y plane and the following two surfaces:

$$z_1 = 3 + x + y$$
$$z_2 = 4 - 2x - 4y$$

Homework Problems

Pb. 1.28 Verify your answers to Pb. 1.27 with which you would obtain analytically for the shape of the intersection curves of the given surfaces with the x–y plane. Also, compute the coordinates of the point(s) of intersection of the two curves obtained.

Pb. 1.29 Plot the surfaces that describe the hyperbolic paraboloid and the elliptic paraboloid . Using the **`view`** command which is available in the graphic window, explore the view of these surfaces from different angles and points.

1.10.4 Contour: A Powerful Tool for Exploring 2-D Geometries

In this section, we shall discuss the manner in which the MATLAB command **`contour`** can be put to use to graphically solve problems in 2-D geometry. We shall illustrate the use of this powerful tool in both Euclidean and non-Euclidean geometries.

The non-Euclidean geometry that we will consider here is the so-called Taxicab geometry (the field of expertise of most taxi drivers in checkerboard laid-out cities). The Taxicab geometry is defined through its metric (i.e., the expression for measuring distance between two points). While the Euclidean distance between two points is determined as the crow flies; in Taxicab geometry the distance is determined as the taxi drives. While not invented here — Minkowski introduced this metric in the nineteenth century — it became the staple of urban planning in New York. (In Manhattan, north of Wall Street the city is criss-crossed by avenues that run south to north and by streets that run east to west, Broadway being the important exception.)

In Euclidean geometry, the distance between points A and B is given by

$$d_E(A, B) = \sqrt{(x_A - x_B)^2 + (y_A - y_B)^2}$$

In Taxicab geometry, the distance between two points is instead given by

$$d_T(A, B) = |x_A - x_B| + |y_A - y_B|$$

Curves in 2-D, as we previously discussed, can be represented by equations of the form $f(x, y) = 0$. Consequently, we can graph any 2-D curve by taking the **contour** of the function $z = f(x, y)$ at zero height. We shall use this technique to compare the loci of points satisfying certain relations in both geometries.

Example 1.30

Find graphically for each of the following cases, in both Euclidean geometry and Taxicab geometry the loci of points that satisfy the following properties:

(a) The points, whose distance from the axes origin is 8.
(b) The points which have the sum of their distances to the points $A(-3, -4)$ and $B(4, 3)$ equal to 16.
(c) The points which have the difference of their distances to the points $A(-3, -4)$ and $B(4, 3)$ equal to 2.
(d) The point S such that $d(S, P) + d(S, Q) = d(P, Q)$, where $P(-2, 3)$ and $Q(1, -4)$.

Solution: We show in the adjoining figures the solutions for each of the above cases for both geometries. The solution for the Euclidian geometry is in dashed lines while the corresponding solution for the Taxicab geometry is in solid line. Below, for purpose of illustration, is the program that generated the solution for case (b).

```
x=-10:0.1:10;y=-10:0.1:10; Ax=-3;Ay=-4;Bx=4;By=3;
[X,Y]=meshgrid(x,y);
Z1=(abs(X-Ax)+abs(Y-Ay))+(abs(X-Bx)+abs(Y-By));
contour(X,Y,Z1,[16 16],'r')
axis([-10 10 -10 10]); axis square
set(gca,'Xtick',[-10:1:10],'Ytick',[-10:1:10])
hold on; plot(Ax,Ay,'k*',Bx,By,'ko')
Z2=sqrt((X-Ax).^2+(Y-Ay).^2)+sqrt((X-Bx).^2+(Y-By).^2);
contour(X,Y,Z2,[16 16],'b--')
hold off
%grid      %add this line to see better details
```

The graphs representing each of the above cases (a through d) are shown in Figures 1.5–1.8.

Comment: In Euclidean geometry the loci are:

(a) a circle centered at the origin and whose radius is 8,
(b) an ellipse with A and B as foci,
(c) a hyperbola branch with A and B as foci, and

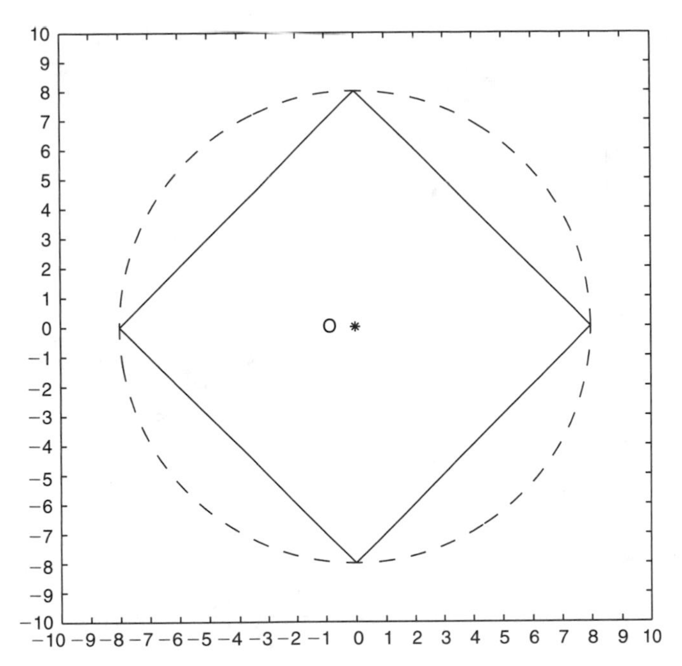

FIGURE 1.5
Graphics of Example 1.30a.

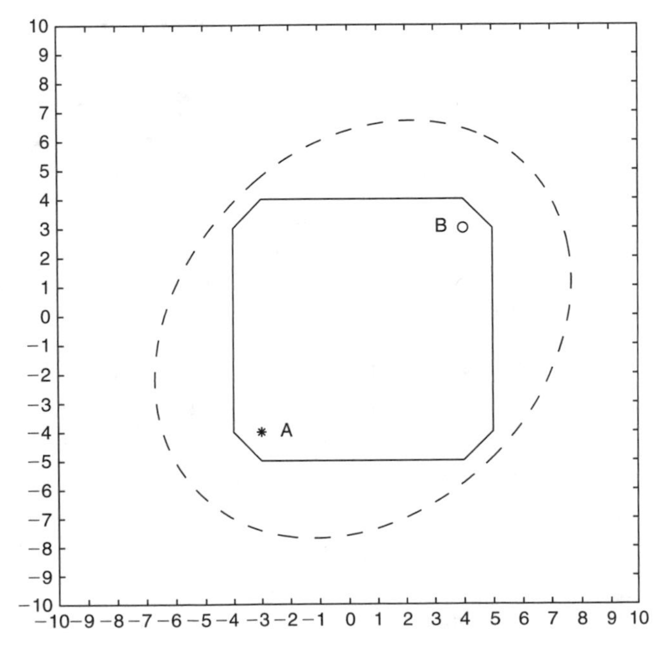

FIGURE 1.6
Graphics of Example 1.30b.

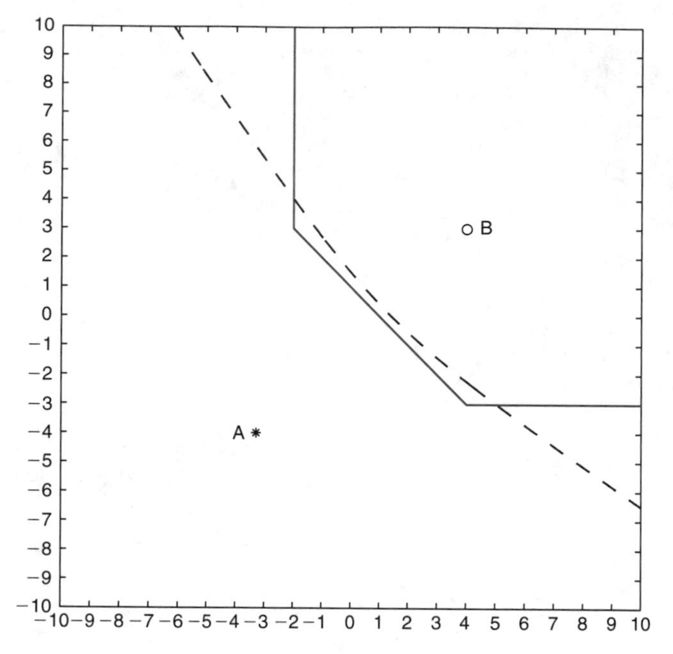

FIGURE 1.7
Graphics of Example 1.30c.

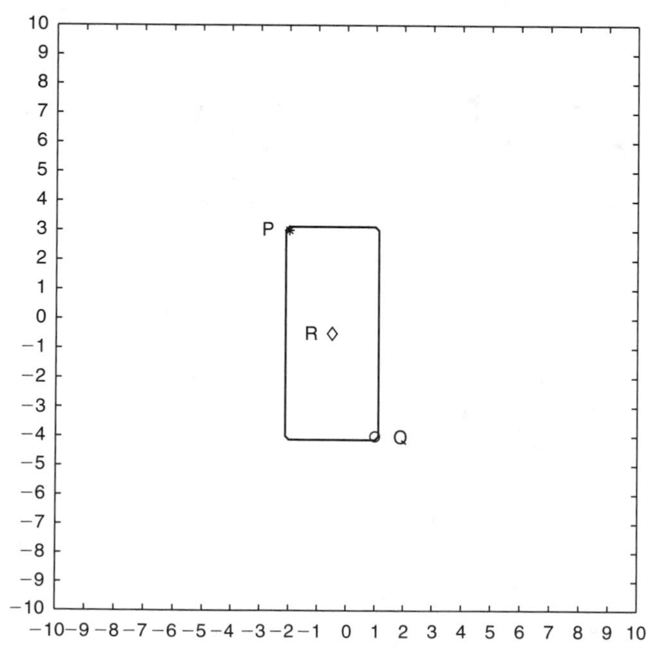

FIGURE 1.8
Graphics of Example 1.30d.

(d) a point $R(-1/2, -1/2)$. The solution for this case, in Taxicab geometry, is the set of all points inside the rectangle shown (not shaded here to show the point R). □

Homework Problems

Pb. 1.30 Find the point(s) P such that

$$d(P, A) + d(P, B) + d(P, C) \quad \text{is minimum}$$

where $A(-7, 5)$, $B(-1, -8)$, and $C(6, 9)$. Solve the problem for both Euclidean and Taxicab geometries.

Pb. 1.31 Benefiting from the solution to Pb. 1.30, solve the following problem with only a ruler, graph paper, and a pencil:

Find the pair of integers (x, y) such that the following quantity is a minimum:

$$S = |x - A_x| + |y - A_y| + |x - B_x| + |y - B_y| + |x - C_x| + |y - C_y|$$

where $A_x = 1, A_y = 1, B_x = 5, B_y = 10, C_x = 10, C_y = 3$.

Also find the value of this minimum. Check your results by computing the 2-D array structure giving the value of S for different integers.

Pb. 1.32 Find the curves generated by the points P such that the sum of the Taxicab distances to points A, B, and C is a given number (these curves are called the trifocal ellipses).

Assume that $A_x = -3, A_y = 4, B_x = 4, B_y = 7, C_x = -5, C_y = 8$.

Pb. 1.33 The taxicab distance between the point $P(x, y)$ and the vertical line $L(x = b)$ is

$$d_T(P, L) = |x - b|$$

Plot the locus of points which satisfy the condition

$$d_T(P, L) = d_T(P, A)$$

where $A_x = -8, A_y = 0$, and $b = 8$.

(This curve is called the taxicab parabola. A is its focus and L its directrix.)

Pb. 1.34 Another non-Euclidean geometry of interest is that associated with the distance

$$d_m(A, B) = \max\left(|A_x - B_x|, |A_y - B_y|\right)$$

Go back and solve the preceding problem using this geometry.

NOTE Often it is easier to visualize the shape of the intersection of a surface with planes parallel to the x–y plane by drawing the contours in 3-D rather than in 2-D. The MATLAB command for this graphic is

```
contour3(X,Y,Z)
```

In effect, **contour3(X,Y,Z)** is the same as **contour(X,Y,Z)**, except that the contours are drawn at their corresponding Z heights.

This command is especially useful when one is examining the surface near an extremum, or a waist, and when it is desirable to quickly visualize the qualitative change in the shape of the contours.

1.11 Animation

A very powerful feature of MATLAB is its ability to render an animation. For example, suppose that we want to visualize the oscillations of an ordinary spring. What are the necessary steps to implement this objective?

1. Determine the parametric equations that describe the curve at a fixed time.
2. Introduce the time dependence in the appropriate curve parameters. In this instance, make the helix pitch to be oscillatory in time.
3. Generate 3-D plots of the curve at different times. Make sure that your axis definition includes all cases.
4. Use the **movie** command to display consecutively the different frames obtained in step 3.

EXAMPLE 1.31

Make a movie showing the dynamics of a helix whose pitch is varying sinusoidaly in time.

Solution: The following *script M-file* implements the above work-plan:

```
th=0:pi/60:32*pi;
a=1;
A=0.25;
w=2*pi/15;
     for t=1:16;
          x=a*cos(th);
          y=a*sin(th);
          z=(1+A*cos(w*(t-1)))*th;
          plot3(x,y,z,'r');
          axis([-2 2 -2 2 0 40*pi]);
          M(:,t)=getframe;
     end
movie(M,15)
```
□

Each of the movie frames at different times are generated within the **`for`** loop. The **`getframe`** function returns a pixel image of the image of the different frames. The last command **`movie(M,n)`** plays the movie n times (15, in this instance).

Homework Problem

Pb. 1.35 Animate in a movie the motion of the tip of the top while in the nutation mode as described in Example 1.28.

1.12 Specialized Plots: Velocity, Gradient, etc.*

In visualizing the motion of a particle, one gains better understanding of the dynamics if in addition to the trajectory, one also sees plotted the velocity of the particle as it moves along its trajectory. Furthermore, in analyzing electrostatics problems, in addition to obtaining the electric potential, it is essential to also visualize the direction and magnitude of the electric field produced. The purpose of this section is to introduce those MATLAB commands and graphics that help us achieve these tasks.

1.12.1 Velocity Plots from Parametric Equations

As we have seen in Section 1.8, the parametric equations of a point determine its curve. If the parameter is time, the curve is the trajectory of the particle.

The velocity of a particle is the time derivative of its position vectors (more on that in Chapter 7). This means that the x-component of the velocity is obtained by taking the derivative of $x(t)$ with respect to time, and in similar fashion we obtain the other components of the velocity, i.e.,

$$v_x = \frac{dx(t)}{dt}, \quad v_y = \frac{dy(t)}{dt}, \quad v_z = \frac{dz(t)}{dt}$$

To compute the derivative in MATLAB, we use the command **`diff`**. This command when applied to an array X gives a new array of length one shorter than X, and its elements are respectively: $[X(2) - X(1) \;\; X(3) - X(2) \;\; \ldots \;\; X(n) - X(n-1)]$. Therefore, from the definition of the derivative, we can approximate numerically its value by taking the ratio of the new array and dt the increment in the time array. (In Chapter 4, we discuss in more detail how this should be done to achieve a desired accuracy.) To obtain a velocity array with the same dimension as the array for the corresponding coordinate, we expand the time interval by an extra dt, compute the position and velocity arrays, and then drop the last element of the expanded position array.

* The asterisk indicates more advanced material that may be skipped in a first reading.

Having now obtained, both the position and velocity arrays, the MATLAB graphic command **quiver3** allows us to plot the velocity vectors at the different trajectories points. The syntax for this command is

```
quiver3(X,Y,Z,Vx,Vy,Vz)
```

EXAMPLE 1.32

The motion of an electron moving in the presence of constant electric and magnetic fields parallel to the z-axis is described by the following set of parametric equations:

$$x(t) = 5\cos(t); \quad y(t) = 5\sin(t); \quad z(t) = 0.1\,t^2$$

the initial time $t_{\text{init}} = 0$ and the final time $t_{\text{fin}} = 10$.

Plot the magnitude and direction of the velocity at different points on the electron trajectory.

Solution:

```
tin=0;
tfin=10;
dt=0.25;
t=tin:dt:tfin+dt;
l=length(t);
xe=5*cos(t);
ye=5*sin(t);
ze=0.1*t.^2;
x=xe(1:l-1);
y=ye(1:l-1);
z=ze(1:l-1);
vx=diff(xe)/dt;
vy=diff(ye)/dt;
vz=diff(ze)/dt;
quiver3(x,y,z,vx,vy,vz)
axis([-6 6 -6 6 0 10])
xlabel('x-axis')
ylabel('y-axis')
zlabel('z-axis')
```

The resulting curve is shown in Figure 1.9. □

1.12.2 Gradient of a Potential

In analyzing electrostatic problems, or other time-independent classical field problems, one computes the field scalar potential at all points of space. The electric field itself is obtained by taking (minus the gradient of the potential).

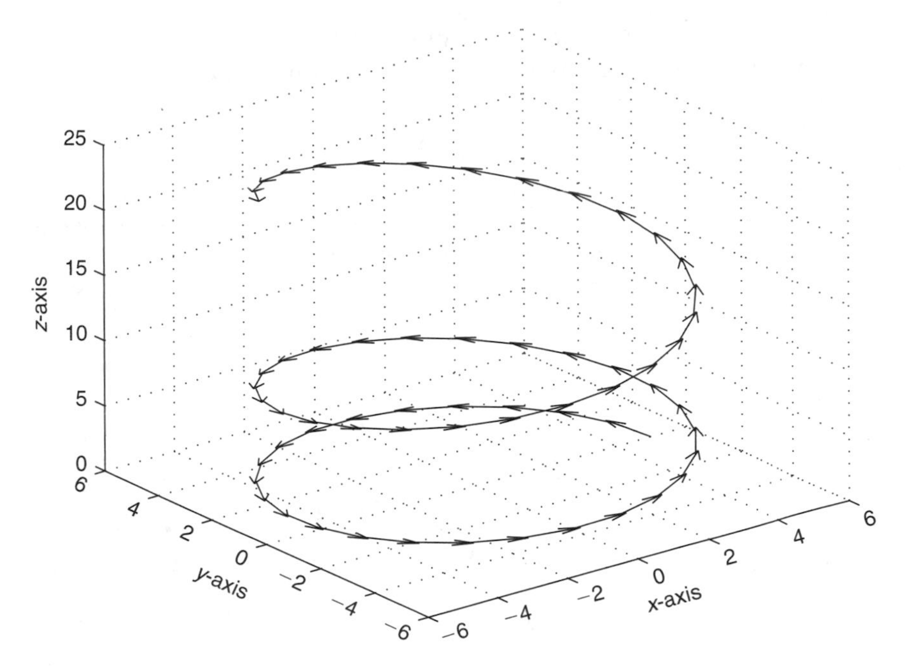

FIGURE 1.9
Graphics of Example 1.32.

The gradient of the scalar potential is a vector. In the case of an electrostatic problem, the electric field is obtained through:

$$\vec{E} = -\vec{\nabla}\Phi = -\left\{\frac{\partial\Phi}{\partial x}\hat{e}_x + \frac{\partial\Phi}{\partial y}\hat{e}_y + \frac{\partial\Phi}{\partial z}\hat{e}_z\right\}$$

where Φ is the electrostatic potential, and the $\hat{e}_i$ are the unit vectors in the different directions. To simplify, we shall consider here the case, where the potential is only a function of the variables x and y alone.

The electric potential can be plotted through the techniques covered in Section 1.10, i.e., we define a **meshgrid** in the x–y plane, and plot the field potential (as the third component of a 3-D plot) through any of the **mesh, surf,** or **contour** commands. Now, the gradient is called through the command

```
[DX, DY]=gradient(Z,dx,dy)
```

where **DX** and **DY** are the x- and y-components of the gradient, respectively, and **dx** and **dy** are the increments in the x- and y-directions, respectively, over which the gradient is approximated. When the dx and dy are omitted, the default is to assume them both to be 1.

Once the arrays approximating the gradient are obtained, we can plot the value and direction of the gradient through the command:

```
quiver(X,Y,DX,DY)
```

EXAMPLE 1.33

The electric potential in the region bounded by $0 \leq x \leq a$ and $0 \leq y$, and such that

$$\Phi(x=0,y)=\Phi(x=a,y)=0 \quad \text{and} \quad \Phi(x,y=0)=V_0$$

is given by

$$\Phi(x,y)=\frac{2V_0}{\pi}\tan^{-1}\left(\frac{\sin(\pi x/a)}{\sinh(\pi y/a)}\right)$$

Show the magnitude and direction of the electric field, assuming $V_0 = 10, a = 3$.

Solution: *Be careful to stay away from the apparent singularity*. The following program solves this problem:

```
V0=10;
a=3;
x=0:0.3:3;
y=eps:0.3:3+eps;
[X,Y]=meshgrid(x,y);
minusV=-(2*V0/pi)*atan(sin(pi*X/a)./sinh(pi*Y/a));
[Ex,Ey]=gradient(minusV,0.3);
quiver(X,Y,Ex,Ey)
axis([0 3 0 3])
axis square
```

The resulting graph is shown in Figure 1.10.

NOTE The electric field will be everywhere perpendicular to the contour lines if these were plotted on the same graph (be careful to increase the finesse of the meshgrid in the plot of the contours, in case you desire a continuous representation of these curves). □

1.13 Printing and Saving Work in MATLAB

1.13.1 Printing a Figure

Use the MATLAB `print` function to print a displayed figure directly to your printer. Notice that the printed figure does not take up the entire page. This is because the default orientation of the graph is in portrait mode. To change

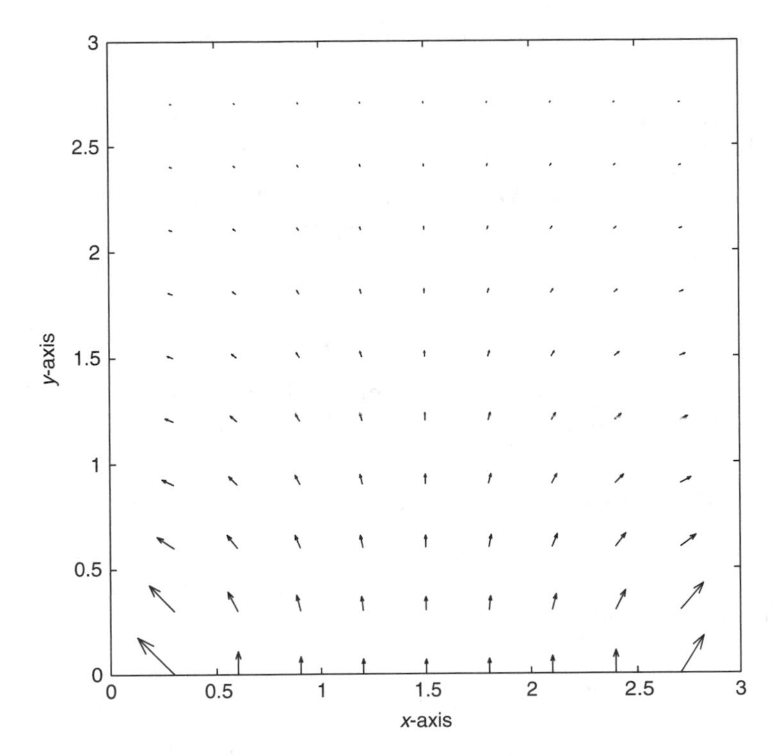

FIGURE 1.10
Graphics of Example 1.33.

Homework Problems

Pb. 1.36 The parametric equations for the position of a particle are given in the interval $0 \le t \le 4\pi$ by

$$x = 3\cos(t)$$
$$y = 6\sin(2t)$$
$$z = 0.1t^2$$

Plot the magnitude and direction of the velocity at different points of the particle trajectory.

Pb. 1.37 The distribution of voltage on a metal plate is given by

$$\Phi(x, y) = 25 - x^2 - 2y^2$$

over the domain $-5 \le x \le 5$ and $-5 \le y \le 5$:

a. At the point $P(1, -2)$,
- In what direction does the voltage increases most rapidly?
- What is the magnitude of this maximum increase?

b. Plot the electric field over the domain of definition of the potential.

these settings, try the following commands while printing an already generated graphic window:

```
orient landscape   %full horizontal layout
orient tall        %full vertical layout
```

In addition to these basic commands, the **Page Setup** in the figure **File** pull-down menu gives you all the layout flexibility that you desire. Explore it and use it to advantage.

1.13.2 Printing a Program File (*Script M-file*)

From the **File** pull-down menu, highlight the **Open** command. Select the *M-file* that you want to print. Go to the **File** pull-down menu, and select **Print**.

1.13.3 Converting a MATLAB Graphic into an Image File

Go to the **File** pull-down menu and select **Save as**. This window allows you to save your graph in most commonly digital images used formats. Select the format that your publisher (or end user) designates.

1.13.4 Saving Values of Variables

The value of a certain variable x can be saved through the command

```
save 'user volume:x'
```

1.13.5 Loading the Value of a Variable

Use

```
load 'user volume:x'
```

1.13.6 Saving a MATLAB Session

Often, you may like to save the contents of a session, i.e., you would like to leave your work temporarily but you desire to return to it some time later and have your computer in the same state that you left it. You can use the MATLAB **save** function

```
save 'user volume:workspace'
```

This will save your session in a folder called **workspace.mat**.

1.13.7 Loading `workspace.mat`

Enter MATLAB and use the MATLAB load function

```
load 'user volume:workspace'.
```

1.14 MATLAB Commands Review

`acos`	Inverse cosine.
`acosh`	Inverse hyperbolic cosine.
`acot`	Inverse cotangent.
`acoth`	Inverse hyperbolic cotangent.
`asec`	Inverse secant.
`asech`	Inverse hyperbolic secant.
`asin`	Inverse sine.
`asinh`	Inverse hyperbolic sine.
`atan`	Inverse tangent.
`atanh`	Inverse hyperbolic tangent.
`axis`	Sets the axis limits for both 2-D and 3-D plots. Axis supports the arguments equal and square, which makes the current graphs aspect ratio 1.
`bar`	Plots the bar chart of an array.
`ceil`	Round a number to the next integer toward $+\infty$.
`clear`	Clears all variables from the workspace.
`clf`	Clears figure.
`contour`	Plots contour lines of a surface.
`contour3`	Same as contour but drawn in 3-D at corresponding heights.
`cos`	Cosine.
`cosh`	Hyperbolic cosine.
`cot`	Cotangent.
`coth`	Hyperbolic cotangent.
`diff`	gives for an array X, the new arrary: $[X(2) - X(1)\ X(3) - X(2) \ldots X(n) - X(n-1)]$.
`disp`	Displays the contents of an array.
`else`	Delineates an alternate set of commands.
`elseif`	Conditionally executes an alternate set of commands.
`end`	Terminates **for**, **if**, and **while** statements.
`eps`	Specifies the accuracy of a floating point calculation.
`erf`	Error function.
`exp`	Exponential.
`figure`	Creates a new figure window.
`fill3`	Fills a 3-D polygon with a color.
`find`	Computes an array where the successive elements indicate the positions in the original array where the values are nonzero.
`fix`	Rounds a number to the next integer toward 0.
`floor`	Rounds a number to the next integer toward $-\infty$.
`for`	Runs a sequence of commands a given number of times.
`format`	Controls the screen output display.
`getframe`	Returns the pixel image of a movie frame.
`ginput`	Read off a 2-D graph the coordinates of a point with the mouse.

`gradient`	Approximates the gradient of a function.
`grid`	Create grid lines at the tick positions.
`gtext`	Places a text in a graph at a point specified by the mouse.
`help`	Online help.
`hist`	Bins the elements of an array in separate containers.
`hold on(off)`	Holds the plot axis with existing graphics on, so that multiple traces can be plotted on the same graph (release the hold).
`if`	Conditionally executes a set of commands.
`Inf`	Infinity.
`length`	Gives the number of elements in an array.
`line`	Creates a line.
`linspace`	Generates an array with a specified number of points between two values.
`load`	Loads data or variable values from previous sessions into current MATLAB session.
`log`	Natural logarithm.
`log10`	Base 10 logarithm.
`loglog`	Plots a log–log 2-D graph.
`max`	Finds the value of the largest element of an array.
`mean`	Finds the mean of the elements of an array.
`mesh`	Plots a mesh surface of a surface stored in a matrix.
`meshc`	The same as mesh, but also plots in the same figure the contour plot.
`meshgrid`	Makes a 2-D array of coordinate squares suitable for plotting surface meshes.
`min`	Finds the value of the smallest element of an array.
`movie`	Plays the movie described by a matrix **M**.
`NaN`	Indicates an undefined numerical result.
`orient`	Orients the current graph to your needs.
`pi`	The number π.
`pie`	Plots pie charts.
`plot`	Plots points or 2-D arrays in a 2-D graph.
`plot3`	Plots points or arrays in a 3-D graph.
`pol2cart`	Polar to Cartesian conversion.
`polar`	Plots a polar plot on a polar grid.
`print`	Prints a figure to the default printer.
`quit`	Exits the MATLAB program.
`quiver, quiver3`	Plot velocity vectors as arrows at $P(x, y)$ or $P(x, y, z)$.
`rand`	Generates an array with elements randomly chosen from the uniform distribution over the interval [0, 1].
`randn`	Generates an array with elements randomly chosen from the normal distribution function with zero mean and standard deviation 1.
`round`	Rounds a number to the nearest integer.

save	Saves MATLAB variables.
sec	Secant.
sech	Hyperbolic secant.
semilogx	Plots a semi logarithmic 2-D graph, with x in logarithmic scale.
semilogy	Plots a semi logarithmic 2-D graph, with y in logarithmic scale.
set	Specifies the properties of an object.
sin	Sine.
sinh	Hyperbolic sine.
sort	Sorts the elements of an array in ascending order.
sqrt	Squareroot.
std	Finds the standard deviation of the elements of an array.
stem	Plots the data sequence as stems from the x-axis terminated with circles for the data value.
subplot	Partitions the graphics window into subwindows.
sum	Sums the elements of an array.
surf	Renders a 3-D colored surface.
surfc	The same as surf, but also plots in the same figure the contour plot.
tan	Tangent.
tanh	Hyperbolic tangent.
view	Views 3-D graphics from different perspectives.
while	Repeats a set of commands indefinitely until a certain condition(s) is met.
who	Lists all variables in the workspace.
xlabel, ylabel	Labels the appropriate axes with text and title.
(x>=x1)	Boolean function that is equal to 1 (0) when the condition inside the parenthesis is (is not) satisfied.
zlabel, title	
zoom	Zooms in and out of a 2-D plot.

Symbols

`==`	Equal to	`&`	AND
`~=`	Not Equal to	`|`	OR
`<`	Less than	`~`	NOT
`<=`	Less or Equal to	`;`	Suppress display
`>`	Greater than	`...`	Continue line
`>=`	Greater or Equal	`:`	Array generation
`>>`	Prompt	`+,-,*,/,^`	Scalar operations
`%`	Commenting	`+,-,.*,./,.^`	Array operations

2

Difference Equations

This chapter introduces difference equations and examines some simple but important cases of their applications. We develop simple algorithms for their numerical solutions and apply these techniques to the solution of some problems of interest to the engineering professional. In particular, it illustrates each type of difference equation that is of widespread interest.

2.1 Simple Linear Forms

The following components are needed to define and solve a difference equation:

1. An ordered array defining an index for the sequence of elements.
2. An equation connecting the value of an element having a certain index with the values of some of the elements having lower indices (the order of the equation being defined by the number of lower indices terms appearing in the difference equation).
3. A sufficient number of the values of the elements at the lowest indices to act as seeds in the recursive generation of the higher indexed elements.

For example, the Fibonacci numbers are defined as follows:

1. The ordered array is the set of positive integers.
2. The defining difference equation is of second order and is given by

$$F(k+2) = F(k+1) + F(k). \tag{2.1}$$

3. The initial conditions are $F(1) = F(2) = 1$ (note that the required number of initial conditions should be the same as the order of the equation).

From the above, it is then straightforward to compute the first few Fibonacci numbers:

$$1, 1, 2, 3, 5, 8, 13, 21, 34, 55, 89, \ldots$$

Example 2.1

Write a program for finding the first 20 Fibonacci numbers.

Solution: The following program fulfills this task:

```
N=18;
F(1)=1;
F(2)=1;
        for k=1:N
                F(k+2)=F(k)+F(k+1);
        end
F
```

It should be noted that the value of the different elements of the sequence depends on the values of the initial conditions, as illustrated in Pb. 2.1, which follows. □

In-Class Exercises

Pb. 2.1 Find the first 20 elements of the sequence that obeys the same recursion relation as that of the Fibonacci numbers, but with the following initial conditions:

$$F(1) = 0.5 \quad \text{and} \quad F(2) = 1$$

Pb. 2.2 Find the first 20 elements of the sequence generated by the following difference equation:

$$F(k + 3) = F(k) + F(k + 1) + F(k + 2)$$

with the following boundary conditions:

$$F(1) = 1, \quad F(2) = 2, \quad \text{and} \quad F(3) = 3$$

Why do we need to specify three initial conditions?

2.2 Amortization

In this application of difference equations, we examine simple problems of finance that are of major importance to every engineer, at both the personal and professional levels. When the purchase of any capital equipment or real estate is made on credit, the assumed debt is normally paid for by means of a process known as amortization. Under this plan, a debt is repaid in a sequence of periodic payments where a portion of each payment reduces the outstanding principal, while the remaining portion is for interest on the loan.

Suppose that the original debt to be paid is C and that interest charges are compounded at the rate r per payment period. Let $y(k)$ be the outstanding principal after the kth payment, and $u(k)$ the amount of the kth payment.

After the kth payment period, the outstanding debt increased by the interest due on the previous principal $y(k-1)$, and decreased by the amount of payment $u(k)$, this relation can be written in the following difference equation form:

$$y(k) = (1 + r)\, y(k - 1) - u(k) \tag{2.2}$$

We can simplify the problem and assume here that the bank wants its money back in equal amounts over N periods (this can be in days, weeks, months, or years; note, however, that the unit used here should be the same as used for the assignment of the value of the interest rate r). Therefore, let

$$u(k) = p \quad \text{for } k = 1, 2, 3, \ldots, N \tag{2.3}$$

Now, using Eq. (2.2), let us iterate the first few terms of the difference equation:

$$y(1) = (1 + r)y(0) - p = (1 + r)C - p \tag{2.4}$$

Since C is the original capital borrowed:
at $k = 2$, using Eq. (2.2) and Eq. (2.4), we obtain

$$y(2) = (1 + r)y(1) - p = (1 + r)^2C - p(1 + r) - p \tag{2.5}$$

at $k = 3$, using Eq. (2.2) through Eq. (2.5), we obtain

$$y(3) = (1 + r)y(2) - p = (1 + r)^3C - p(1 + r)^2 - p(1 + r) - p \tag{2.6}$$

$\vdots$

and for an arbitrary k, we can write, by induction, the general expression

$$y(k) = (1 + r)^k C - p\sum_{i=0}^{k-1} (1 + r)^i \tag{2.7}$$

Using the expression for the sum of a geometric series, from Appendix D, the expression for $y(k)$ then reduces to

$$y(k) = (1+r)^k C - p\left[\frac{(1+r)^k - 1}{r}\right] \tag{2.8}$$

At $k = N$, the debt is paid off and the bank is owed no further payment; therefore,

$$y(N) = 0 = (1+r)^N C - p\left[\frac{(1+r)^N - 1}{r}\right] \tag{2.9}$$

From this equation, we can determine the amount of each of the (equal) payments:

$$p = \frac{r(1+r)^N}{(1+r)^N - 1} C \tag{2.10}$$

Question: What percentage of the first payment is going into retiring the principal?

In-Class Exercises

Pb. 2.3 A home insurance company advertises that it will return to a customer every cent of all premiums that he/she paid over a period of 10 years, if the customer does not file a single claim during that period.

Assume that (i) the yearly premium for a particular customer is \$1000.00 payable on January 1 of each year of the contract; and (ii) the insurance company invests the premiums in a financial instrument with guaranteed returns of 10% per year, payable to it on the close of business on December 31 of every year. Compute the insurance company's profits from this contract; assuming that the customer did not make any claim and before the company's administrative and sales costs for securing and managing this account are factored in. The bonus to the customer is paid on the close of business on December 31 of the 10th year.

Pb. 2.4 Use the same reasoning as for the amortization problem to write the difference equation for an individual's savings plan. Let $y(k)$ be the savings balance on the first day of the kth year and $u(k)$ the amount of deposit made in the kth year.

Write a MATLAB program to compute $y(k)$ if the sequence $u(k)$ and the interest rate r are given. Specialize to the case where you deposit an amount that increases by the rate of inflation i. Compute and plot the total value of the savings as a function of k if the deposit in the first year is \$1000, the yearly interest rate is 6%, and the yearly rate of inflation is 3%. (Hint: For simplicity, assume that the deposits are made on December 31 of each year, and that the balance statement is issued on January 1 of each year.)

Homework Problem

Pb. 2.5 In this problem, we shall explore the algorithm used by financial analysts for common stock valuation (i.e., determining what would be a fair price for a stock).

The basic assumption made in the simplest model is that the common stock has an infinite lifetime, and that the present value of a stock is the sum of all present values of its future dividends. Each of the future dividends present value can be computed from its future value discounted by the risk-adjusted cost of capital (i.e., the investor's required rate of return) and the number of years until the actual payment:

$$PV(d(i)) = \frac{d(i)}{(1+k)^i}$$

where PV refers to the present value of a quantity, $d(i)$ is the dividend in the ith year, and k the investor's required rate of return.

The present value of the stock is then

$$\text{Stock value} = \sum_{i=1}^{\infty} \frac{d(i)}{(1+k)^i}$$

The one stage (or single mode) for a dividend assumes that the dividend's growth over the lifetime of the stock is constant, and that the growth rate is g (where $g < k$), i.e.,

$$d(i+1) = (1+g)\, d(i)$$

giving

$$d(i) = (1+g)^i\, d_0$$

where d_0 is the present year dividend declared by the company. Combining this expression with the above expression for the stock value, we obtain

$$\text{Stock value} = \sum_{i=1}^{\infty} d_0 \frac{(1+g)^i}{(1+k)^i}$$

This is the expression for a geometric series. Using the expression for the sum of a geometric series given in Appendix D, this gives

$$\text{Stock value} = \frac{d_0(1+g)}{k-g}$$

(cont'd.)

Homework Problem *(cont'd.)*

Therefore, if this year's declared dividend is \$3.00, and the investor's required rate of return and the dividend growth are 16 and 10% respectively, then the stock value is

$$\frac{3.00(1+0.1)}{(0.16-0.10)} = \$55.00$$

If the quoted price of the stock is below this level then the stock is a good value.

Single-product technology companies do not follow the single-mode model. Typically, in the years immediately following a technological breakthrough and the introduction of a new product, the dividend growth can be much higher than when the technology matures, and if the introduction of new products is absent the dividend's growth rate will decrease or even can become negative in the company's declining profits and aging phase.

Your task is to value the share of this single-product company, assuming that the company lifetime is 20 years, this year's dividend is \$1.00, the investor's required rate of return is 16% and the dividend growth rate is as follows:

$$g = \begin{cases} 25\% & \text{years 1 to 5} \\ 10\% & \text{years 6 to 14} \\ -6\% & \text{years 15 to 20} \end{cases}$$

Assume that at the end of 20 years, the company has no salvage value.

2.3 An Iterative Geometric Construct: The Koch Curve

In your previous studies of 2-D geometry, you would have encountered classical geometric objects such as the circle, the triangle, the square, and different polygons. These shapes only approximate the shapes that you observe in nature (e.g., the shapes of clouds, mountain ranges, rivers, and coastlines). In a successful effort to address the limitations of classical geometry, mathematicians have developed, over the last century and more intensely over the last three decades, a new geometry called fractal geometry. This geometry defines the geometrical object through an iterative transformation applied an infinite number of times on an initial simple geometrical object. We illustrate this new concept in geometry by considering the Koch curve (see Figure 2.1).

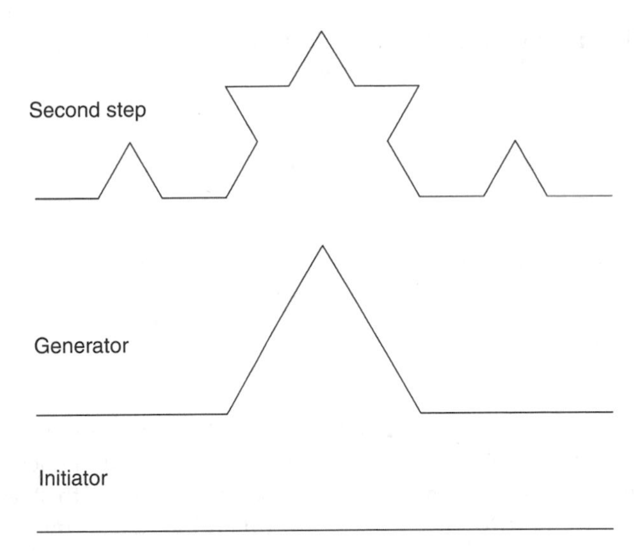

FIGURE 2.1
The first few steps in the construction of the Koch curve.

The Koch curve has the following simple geometrical construction. Begin with a straight line of length L. This initial object is called the initiator. Now partition it into three equal parts. Then replace the middle line segment by an equilateral triangle (the segment you removed is its base). This completes the basic construction, which transformed the line segment into four noncollinear smaller parts. This constructional prescription is called the generator. We now repeat the transformation, taking each of the resulting line segments, partitioning them into three equal parts, removing the middle section, and so on. This process is repeated indefinitely. The first two steps of this construction is shown in Figure 2.1. It is interesting to observe that the Koch curve is an example of a curve where there is no way to fit a tangent to any of its points. In a sense, it is an example of a curve that is made out of corners everywhere.

The detailed study of these objects is covered in courses in fractal geometry, chaos, dynamic systems, etc. We limit ourselves here to the simple problems of determining the number of segments, the length of each segment, the length of the curve, and the area bounded by the curve and the horizontal axis, following the kth step:

1. After the first step, we are left with a curve made up of four line segments of equal length; after the second step, we have (4×4) segments, and the number of segments after k steps, is

$$n(k) = 4^k \tag{2.11}$$

2. If the initiator had length L, the length of the segment after the first step is $L/3$, $L/(3)^2$, after the second step and after k steps:

$$s(k) = L/(3)^k \tag{2.12}$$

3. Combining the results of Eq. (2.11) and Eq. (2.12), we deduce that the length of the curve after k steps:

$$P(k) = L \times \left(\frac{4}{3}\right)^k \tag{2.13}$$

4. The number of vertices in this curve, denoted by $u(k)$, is equal to the number of segments plus one:

$$u(k) = 4^k + 1 \tag{2.14}$$

5. The area enclosed by the Koch curve and the horizontal line can be deduced from solving a difference equation: the area enclosed after the kth step is equal to the area enclosed in the $(k-1)$th step plus the number of the added triangles multiplied by their individual area:

$$\text{Number of new triangles} = \left(\frac{u(k) - u(k-1)}{3}\right) \tag{2.15}$$

$$\text{Area of the new equilateral triangle} = \frac{\sqrt{3}}{4} s^2(k) = \frac{\sqrt{3}}{4}\left(\frac{1}{3}\right)^{2k} L^2 \tag{2.16}$$

from which the difference equation for the area can be deduced:

$$\begin{aligned} A(k) &= A(k-1) + \left[\frac{u(k) - u(k-1)}{3}\right] \frac{\sqrt{3}}{4} \frac{L^2}{3^{2k}} \\ &= A(k-1) + \frac{\sqrt{3}}{24}\left(\frac{2}{3}\right)^{2k-1} L^2 \end{aligned} \tag{2.17}$$

The initial condition for this difference equation is

$$A(1) = \frac{\sqrt{3}}{4} \frac{L^2}{9} \tag{2.18}$$

Clearly, the solution of the above difference equation is the sum of a geometric series, and can therefore be written analytically. For $k \to \infty$, this area has the limit:

$$A(k \to \infty) = \frac{\sqrt{3}}{20} L^2 \tag{2.19}$$

However, if you did not notice the relationship of the above difference equation with the sum of a geometric series, you can still solve this equation numerically, using the following routine and assuming $L = 1$:

```
N=25;
A=zeros(N,1); %preallocating size of array speeds
              % computation
m=1:N;
A(1)=(sqrt(3)/24)*(2/3);
  for k=2:N
    A(k)=A(k-1)+(sqrt(3)/24)*((2/3)^(2*k-1));
  end
stem(m,A,'*')
```

The above plot shows the value of the area on the first 20 iterations of the function, and as can be verified, the numerical limit of this area has the same value as the analytical expression given in Eq. (2.19).

Before leaving the Koch curve, we note that although the area of the curve goes to a finite limit as the index increases, the value of the length of the curve [Eq. (2.13)] continues to increase. This is a feature not encountered in the classical geometric objects with which you are most familiar.

2.4 Solution of Linear Constant Coefficients Difference Equations

In Section 2.1, we explored the general numerical techniques for solving difference equations. In this section, we consider, some special techniques for obtaining the analytical solutions for the class of linear constant coefficients difference equations. The related physical problem is to determine, for a linear system, the output $y(k)$, $k > 0$, given a specific input $u(k)$ and a specific set of initial conditions. We discuss, at this stage, the so-called direct method.

The general expression for this class of difference equation is given by

$$\sum_{j=0}^{N} a_j y(k-j) = \sum_{m=0}^{M} b_m u(k-m) \tag{2.20}$$

The direct method assumes that the total solution of a linear difference equation is the sum of two parts — the homogeneous solution and the particular solution:

$$y(k) = y_{\text{homog.}}(k) + y_{\text{partic.}}(k) \tag{2.21}$$

The homogeneous solution is independent of the input $u(k)$, and the RHS of the difference equation is equated to zero; that is,

$$\sum_{j=0}^{N} a_j y(k-j) = 0 \tag{2.22}$$

2.4.1 Homogeneous Solution

Assume that the solution is of the form

$$y_{\text{homog.}}(k) = \lambda^k \tag{2.23}$$

Substituting into the homogeneous equation, we obtain the following algebraic equation:

$$\sum_{j=0}^{N} a_j \lambda^{k-j} = 0 \tag{2.24}$$

or

$$\lambda^{k-N}(a_0\lambda^N + a_1\lambda^{N-1} + a_2\lambda^{N-2} + \cdots + a_{N-1}\lambda + a_N) = 0 \tag{2.25}$$

The polynomial in parentheses is called the characteristic polynomial of the system. The roots can be obtained analytically for all polynomials up to order 4; otherwise, they are obtained numerically. In MATLAB, they can be obtained graphically when they are all real, or through the `roots` command in the most general case. We introduce this command in Chapter 5. In all the following examples in this chapter, we restrict ourselves to cases for which the roots can be obtained analytically.

If we assume that the roots are all distinct, the general solution to the homogeneous difference equation is

$$y_{\text{homog.}}(k) = C_1\lambda_1^k + C_2\lambda_2^k + \cdots + C_N\lambda_N^k \tag{2.26}$$

where $\lambda_1, \lambda_2, \lambda_3, \ldots, \lambda_N$ are the roots of the characteristic polynomial.

Example 2.2

Find the homogeneous solution of the difference equation

$$y(k) - 3y(k-1) - 4y(k-2) = 0$$

Solution: The characteristic polynomial associated with this equation leads to the quadratic equation

$$\lambda^2 - 3\lambda - 4 = 0$$

The roots of this equation are -1 and 4, respectively. Therefore, the solution of the homogeneous equation is

$$y_{\text{homog.}}(k) = C_1(-1)^k + C_2(4)^k$$

The constants C_1 and C_2 are determined from the initial conditions $y(1)$ and $y(2)$. Substituting, we obtain

$$C_1 = -\frac{4}{5}y(1) + \frac{y(2)}{5} \quad \text{and} \quad C_2 = \frac{y(1)+y(2)}{20}$$

NOTE If the characteristic polynomial has roots of multiplicity m, then the portion of the homogeneous solution corresponding to that root can be written, instead of $C_1\lambda^k$, as

$$C_1^{(1)}\lambda^k + C_1^{(2)}k\lambda^k + \cdots + C_1^{(m)}k^{m-1}\lambda^k \quad \square$$

This can be verified by taking the example of a second-order difference equation. We show that when the roots of the characteristic polynomial are degenerate, then $y(k) = k\lambda^k$ is also a solution of the homogeneous difference equation. Substituting this solution into the difference equation of the form

$$ay(k) + by(k-1) + cy(k-2) = 0$$

we obtain

$$a(k\lambda^k) + b((k-1)\lambda^{k-1}) + c((k-2)\lambda^{k-2}) = \{\lambda^{k-2}[(a\lambda^2 + b\lambda + c)k - (b\lambda + 2c)]\}$$

The first term in the square bracket is zero because this is the characteristic polynomial for this difference equation, while the second term is zero, because in the case of degeneracy, $\lambda = -b/(2a)$ and $b^2 - 4ac = 0$.

In-Class Exercises

Pb. 2.6 Find the homogeneous solution of the following second-order difference equation:

$$y(k) = 3y(k-1) - 2y(k-2)$$

with the initial conditions $y(0) = 1$ and $y(1) = 2$. Then check your results numerically.

Pb. 2.7 Find the homogeneous solution of the following second-order difference equation:

$$y(k) = 4y(k-1) - 4y(k-2)$$

with the initial conditions $y(1) = 6$ and $y(2) = 20$. Check your results numerically.

2.4.2 Particular Solution

The particular solution depends on the form of the input signal. The following table summarizes the form of the particular solution of a linear equation for some simple input functions:

Input signal	Particular solution
A (constant)	B (constant)
AM^k	BM^k
Ak^M	$B_0k^M + B_1k^{M-1} + \cdots + B_M$
$\{A\cos(\omega_0 k), A\sin(\omega_0 k)\}$	$B_1\cos(\omega_0 k) + B_2\sin(\omega_0 k)$

For more complicated input signals, the z-transform technique provides the simplest solution method. This technique is discussed in great detail in courses on linear systems. It is briefly discussed in Appendix C.

IN-CLASS EXERCISE

Pb. 2.8 Find the particular solution of the following second-order difference equation:

$$y(k) - 3y(k-1) + 2y(k-2) = (3)^k \quad \text{for } k > 0$$

2.4.3 General Solution

The general solution of a linear difference equation is the sum of its homogeneous solution and its particular solution, with the constants adjusted, so as to satisfy the initial conditions. We illustrate this general prescription with an example.

EXAMPLE 2.3

Find the complete solution of the first-order difference equation:

$$y(k+1) + y(k) = k$$

with the initial condition $y(0) = 0$.

Solution: First, solve the homogeneous equation $y(k+1) + y(k) = 0$. The characteristic polynomial is $\lambda + 1 = 0$; therefore,

$$y_{\text{homog.}} = C(-1)^k$$

The particular solution can be obtained from the above table. Noting that the input signal has the functional form k^M, with $M = 1$, then the particular solution is of the form

$$y_{\text{partic.}} = B_0\,k + B_1 \tag{2.27}$$

Substituting back into the original equation, and grouping the different powers of k, we deduce that

$$B_0 = 1/2 \quad \text{and} \quad B_1 = -1/4$$

The complete solution of the difference equation is then

$$y(k) = C(-1)^k + \frac{2k-1}{4}$$

The constant C is determined from the initial condition

$$y(0) = 0 = C(-1)^0 + \frac{(-1)}{4}$$

as 1/4. □

In-Class Exercises

Pb. 2.9 Use the following program to model Example 2.3:

```
N=19;
y(1)=0;
        for k=1:N
                y(k+1)=k-y(k);
        end
y
```

Verify the closed-form answer.

Pb. 2.10 Find, for $k \geq 2$, the general solution of the second-order difference equation

$$y(k) - 3y(k-1) - 4y(k-2) = 4^k + 2 \times 4^{k-1}$$

with the initial conditions $y(0) = 1$ and $y(1) = 9$. (Hint: When the functional form of the homogeneous and particular solutions are the same, use the same functional form for the solutions as in the case of multiple roots for the characteristic polynomial.)

Answer: $y(k) = \left[-\frac{1}{25}(-1)^k + \frac{26}{25}(4)^k\right] + \left[\frac{6}{5}k4^k\right]$

Homework Problems

Pb. 2.11 Given the general geometric series $y(k)$, where

$$y(k) = 1 + a + a^2 + \cdots + a^k$$

show that $y(k)$ obeys the first-order equation

$$y(k) = y(k-1) + a^k$$

Pb. 2.12 Show that the response of the system

$$y(k) = (1-a)u(k) + ay(k-1)$$

to a step signal of amplitude c, that is, $u(k) = c$ for all positive k, is given by

$$y(k) = c(1 - a^{k+1}) \quad \text{for } k = 0, 1, 2, \ldots$$

where the initial condition $y(-1) = 0$.

Pb. 2.13 Given the first-order difference equation

$$y(k) = u(k) + y(k-1) \quad \text{for } k = 0, 1, 2, \ldots$$

with the input signal $u(k) = k$ and the initial condition $y(-1) = 0$, verify that its solution also satisfies the second-order difference equation

$$y(k) = 2y(k-1) - y(k-2) + 1$$

with the initial conditions $y(0) = 0$ and $y(-1) = 0$.

Pb. 2.14 Verify that the response of the system governed by the first-order difference equation

$$y(k) = bu(k) + ay(k-1)$$

to the alternating input: $u(k) = (-1)^k$ for $k = 0, 1, 2, 3, \ldots$ is given by

$$y(k) = \frac{b}{1+a}[(-1)^k + a^{k+1}] \quad \text{for } k = 0, 1, 2, 3, \ldots$$

if the initial condition is $y(-1) = 0$.

Pb. 2.15 The impulse response of a system is the output from this system when excited by an input signal $\delta(k)$ that is zero everywhere, except at $k = 0$, where it is equal to 1. Using this definition and the general form

(cont'd.)

Homework Problems *(cont'd.)*

of the solution of a difference equation, write the output of a linear system described by

$$y(k) - 3y(k-1) - 4y(k-2) = \delta(k) + 2\delta(k-1)$$

The initial conditions are $y(-2) = y(-1) = 0$.

Answer: $y(k) = \left[-\frac{1}{5}(-1)^k + \frac{6}{5}(4)^k\right] \quad \text{for } k > 0$

Pb. 2.16 The expression for the national income for a political entity is given by

$$y(k) = c(k) + i(k) + g(k)$$

where c is consumer expenditure, i the induced private investment, g the government expenditure, and k the accounting period, typically corresponding to a particular quarter. Samuelson theory, introduced to many engineers in Cadzow's classic Discrete-Time Systems (see reference list), assumes the following properties for the above three components of the national income:

1. Consumer expenditure in any period k is proportional to the national income at the previous period:

$$c(k) = ay(k-1)$$

2. Induced private investment in any period k is proportional to the increase in consumer expenditure from the preceding period:

$$i(k) = b[c(k) - c(k-1)] = ab[y(k-1) - y(k-2)]$$

3. Government expenditure is the same for all accounting periods:

$$g(k) = g$$

Combining the above equations, the national income obeys the second-order difference equation:

$$y(k) = g + a(1+b)\,y(k-1) - aby(k-2) \quad \text{for } k = 1, 2, 3, \ldots$$

The initial conditions $y(-1)$ and $y(0)$ are to be specified.

Plot the national income for the first 40 quarters of a new national entity, assuming that

$a = 1/6$, $b = 1$, $g = \$20{,}000{,}000$, $y(-1) = \$20{,}000{,}000$, $y(0) = \$30{,}000{,}000$.

How would the national income curve change if the marginal propensity to consume (i.e., the constant a) is decreased to $1/8$?

2.5 Convolution-Summation of a First-Order System with Constant Coefficients

The amortization problem in Section 2.2 was solved by obtaining the present output, $y(k)$, as a linear combination of the present and all past inputs, $(u(k), u(k-1), u(k-2), \ldots)$. This solution technique is referred to as the convolution-summation representation:

$$y(k) = \sum_{i=0}^{\infty} w(i)u(k-i) \tag{2.28}$$

where $w(i)$ is the weighting function (or weight). Usually, the infinite sum is reduced to a finite sum because the inputs with negative indexes are usually assumed to be zeros.

On the other hand, in the difference equation formulation of this class of problems, the present output $y(k)$ is expressed as a linear combination of the present and m most recent inputs and of the n most recent outputs, specifically:

$$\begin{aligned} y(k) &= b_0u(k) + b_1u(k-1) + \cdots + b_mu(k-m) \\ &\quad -a_1y(k-1) - a_2y(k-2) - \cdots - a_ny(k-n) \end{aligned} \tag{2.29}$$

where, of course, n is the order of the difference equation. Elementary techniques for solving this class of equations were introduced in Section 2.4. However, the most powerful technique to directly solve the linear difference equation with constant coefficients is, as pointed out earlier, the z-transform technique.

Each of the above formulations of the input–output problem has distinct advantages in different circumstances. The direct difference equation formulation is the most amenable to numerical computations because of lower computer memory requirements, while the convolution-summation technique has the advantage of being suitable for developing mathematical proofs and finding general features for the difference equation.

Relating the parameters of the two formulations of this problem is usually cumbersome without the z-transform technique. However, for first-order difference equations, this task is rather simple.

Example 2.4

Relate, for a first-order difference equation with constant coefficients, the sets $\{a_n\}$ and $\{b_n\}$ with $\{w_n\}$.

Solution: The first-order difference equation is given by

$$y(k) = b_0u(k) + b_1u(k-1) - a_1y(k-1)$$

where $u(k) = 0$ for all k negative. From the difference equation and the initial conditions, we can directly write

$$y(0) = b_0 u(0)$$

$$\text{for } k = 1, \quad \begin{cases} y(1) = b_0 u(1) + b_1 u(0) - a_1 y(0) \\ \quad\;\; = b_0 u(1) + b_1 u(0) - a_1 b_0 u(0) \\ \quad\;\; = b_0 u(1) + (b_1 - a_1 b_0) u(0) \end{cases}$$

Similarly,

$$y(2) = b_0 u(2) + (b_1 - a_1 b_0) u(1) - a_1 (b_1 - a_1 b_0) u(0)$$

$$y(3) = b_0 u(3) + (b_1 - a_1 b_0) u(2) - a_1 (b_1 - a_1 b_0) u(1) + a_1^2 (b_1 - a_1 b_0) u(0)$$

or, more generally, if

$$y(k) = w(0) u(k) + w(1) u(k-1) + \cdots + w(k) u(0)$$

then

$$w(0) = b_0$$

$$w(i) = (-a_1)^{i-1} (b_1 - a_1 b_0) \quad \text{for } i = 1, 2, 3, \ldots \quad \square$$

In-Class Exercises

Pb. 2.17 Using the convolution-summation technique, find the closed-form solution for

$$y(k) = u(k) - \frac{1}{3} u(k-1) + \frac{1}{2} y(k-1)$$

$$\text{and the input function given by } \begin{cases} u(k) = 0 & \text{for } k \text{ negative} \\ u(k) = 1 & \text{otherwise} \end{cases}$$

Compare your analytical answer with the numerical solution.

Pb. 2.18 Show that the resultant weight functions for two systems are, respectively,

$$w(k) = w_1(k) + w_2(k) \qquad \text{if connected in parallel}$$

$$w(k) = \sum_{i=0}^{k} w_2(i) w_1(k-i) \qquad \text{if connected in cascade}$$

2.6 General First-Order Linear Difference Equations*

Thus far, we have considered difference equations with constant coefficients. Now we consider first-order difference equations with arbitrary functions as coefficients:

$$y(k+1) + A(k)\,y(k) = B(k) \tag{2.30}$$

The homogeneous equation corresponding to this form satisfies the following equation:

$$l(k+1) + A(k)\,l(k) = 0 \tag{2.31}$$

Its expression can be easily found:

$$\begin{aligned} l(k+1) &= -A(k)l(k) = A(k)A(k-1)l(k-1) = \ \dots \ = \\ &= (-1)^{k+1}A(k)A(k-1)\dots A(0)l(0) = \left\{\prod_{i=0}^{k}[-A(i)]\right\} l(0) \end{aligned} \tag{2.32}$$

Assuming that the general solution is of the form

$$y(k) = l(k)\,v(k) \tag{2.33}$$

let us find $v(k)$. Substituting the above trial solution in the difference equation, we obtain

$$l(k+1)\,v(k+1) + A(k)\,l(k)\,v(k) = B(k) \tag{2.34}$$

Further, assuming that

$$v(k+1) = v(k) + \Delta v(k) \tag{2.35}$$

substituting in the difference equation, and recalling that $l(k)$ is the solution of the homogeneous equation, we obtain

$$\Delta v(k) = \frac{B(k)}{l(k+1)} \tag{2.36}$$

Summing this over the variable k from 0 to k, we deduce that

$$v(k+1) = \sum_{j=0}^{k} \frac{B(j)}{l(j+1)} + C \tag{2.37}$$

where C is a constant.

---

* The asterisk indicates more advanced material that may be skipped in a first reading.

EXAMPLE 2.5

Find the general solution of the following first-order difference equation:

$$y(k+1) - k^2 y(k) = 0$$

with $y(1) = 1$.

Solution:

$$\begin{aligned} y(k+1) &= k^2 y(k) = k^2(k-1)^2 y(k-1) = k^2(k-1)^2(k-2)^2 y(k-2) \\ &= k^2(k-1)^2(k-2)^2(k-3)^2 y(k-3) = \ldots \\ &= k^2(k-1)^2(k-2)^2(k-3)^2 \cdots (2)^2(1)^2 y(1) = (k!)^2 \quad \square \end{aligned}$$

EXAMPLE 2.6

Find the general solution of the following first-order difference equation:

$$(k+1)\, y(k+1) - k y(k) = k^2$$

with $y(1) = 1$.

Solution: Reducing this equation to the standard form, we have

$$A(k) = -\frac{k}{k+1} \quad \text{and} \quad B(k) = \frac{k^2}{k+1}$$

The homogeneous solution is given by

$$l(k+1) = \frac{k!}{(k+1)!} = \frac{1}{(k+1)}$$

The particular solution is given by

$$v(k+1) = \sum_{j=1}^{k} \frac{j^2}{(j+1)}(j+1) + C = \sum_{j=1}^{k} j^2 + C = \frac{(k+1)(2k+1)k}{6} + C$$

where we used the expression for the sum of the square of integers (see Appendix D).

The general solution is then

$$y(k+1) = \frac{(2k+1)k}{6} + \frac{C}{(k+1)}$$

From the initial condition $y(1) = 1$, we deduce that $C = 1$. $\square$

IN-CLASS EXERCISE

Pb. 2.19 Find the general solutions for the following difference equations, assuming that $y(1) = 1$.

a. $y(k + 1) - 3ky(k) = 0$.
b. $y(k + 1) - ky(k) = 1$.

2.7 Nonlinear Difference Equations

In this and the following section, we explore a number of nonlinear difference equations that exhibit some general features typical of certain classes of solutions and observe other instances with novel qualitative features. Our exploration is purely experimental, in the sense that we restrict our treatment to guided computer runs. The underlying theories of most of the models presented are the subject of more advanced courses; however, many educators, including this author, believe that there is virtue in exposing students qualitatively early on to these fascinating and generally new developments in mathematics.

2.7.1 Computing Irrational Numbers

In this model, we want to exhibit an example of a nonlinear difference equation whose solution is a sequence that approaches a specific limit, irrespective, within reasonable constraints, of the initial condition imposed on it. This type of difference equation has been used to compute a class of irrational numbers. For example, a well-defined approximation for computing $\sqrt{A}$ is the feedback process:

$$y(k+1) = \frac{1}{2}\left[y(k) + \frac{A}{y(k)}\right] \tag{2.38}$$

The main features of the above equation are explored in the following exercise.

IN-CLASS EXERCISE

Pb. 2.20 Using the difference equation given by Eq. (2.38):

a. Write down a routine to compute $\sqrt{2}$. As an initial guess, take the initial value to be successively 1, 1.5, and 2; even consider 5, 10, and 20. What is the limit of each of the obtained sequences?
b. How many iterations are required to obtain the limit accurate to four digits for each of the above initial conditions?
c. Would any of the above properties be different for a different choice of A.

Now, having established that the above sequence goes to a limit, let us prove that this limit is indeed $\sqrt{A}$. To prove the above assertion, let this limit be denoted by y_{lim}; that is, for large k, both $y(k)$ and $y(k+1) \Rightarrow y_{\text{lim}}$, and the above difference equation goes in the limit to

$$y_{\text{lim}} = \frac{1}{2}\left[y_{\text{lim}} + \frac{A}{y_{\text{lim}}} \right] \tag{2.39}$$

Solving this equation, we obtain

$$y_{\text{lim}} = \sqrt{A} \tag{2.40}$$

It should be noted that the above derivation is meaningful only when a limit exists and is in the domain of definition of the sequence (in this case, the real numbers). In Section 2.7.2, we encounter a sequence where, for some values of the parameters, there is no limit.

2.7.2 The Logistic Equation

Section 2.7.1 illustrated the case in which the solution of a nonlinear difference equation converges to a single limit for large values of the iteration index. In this subsection, we consider the case in which a succession of iterates (called orbits) bifurcate, yielding orbits of period length 2, 4, 8, 16, *ad infinitum*, ending in what is called a "chaotic" orbit of infinite period length. We illustrate the prototype for this class of difference equations by exploring the logistic difference equation.

The logistic equation was introduced by Verhulst to model the growth of populations limited by finite resources (the name logistic was coined by the French army under Napoleon when this equation was used for the planning of "logement" of troops in camps). In more modern settings of ecology, the above model is used to simulate a population growth model. Specifically, in an ecological or growth process, the normalized measure $y(k+1)$ of the next generation of a specie (the number of animals, for example) is a linear function of the present measure $y(k)$, that is,

$$y(k+1) = r\, y(k) \tag{2.41}$$

where r is the growth parameter. If unchecked, the growth of the specie follows a geometric series, which for $r > 1$ grows to infinity. But growth is often limited by finite resources. In other words, the larger the $y(k)$, the smaller the growth factor. The simplest way to model this decline in the growth factor is to replace r by $r(1 - y(k))$, so that as $y(k)$ approaches the theoretical limit

(1 in this case), the effective growth factor goes to zero. The difference equation goes to

$$y(k+1) = r(1 - y(k))y(k) \tag{2.42}$$

which is the standard form for the logistic equation.

In the next series of exercises, we explore the solution of Eq. (2.42) as we vary the value of r. We find that qualitatively different classes of solutions may appear for different values of r.

We start by writing the simple subroutine that models Eq. (2.42):

```
N=127; r=?; y(1)=?;  %Enter the values of r and y(1)
m=1:N+1;
  for k=1:N
    y(k+1)= r*(1-y(k))*y(k);
  end
plot(m, y,'*')
y
```

Remember to key in the values of r and $y(1)$ for each of the specific cases under consideration.

In-Class Exercises

In the following two problems, we take in the logistic equation $r > 1$ and $y(1) < 1$.

Pb. 2.21 Consider the case that $1 < r < 3$ and $y(1) = 0.5$.

a. Show that by running the above program for different values of r and $y(1)$ the iteration of the logistic equation leads to the limit

$$y(N \gg 1) = \left(\frac{r-1}{r}\right).$$

b. Does the value of this limit change if the value of $y(1)$ is modified, while r is kept fixed?

Pb. 2.22 Find the iterates of the logistic equation for the following values of r: 3.1, 3.236068, 3.3, 3.498561699, 3.566667, and 3.569946, assuming the following three initial conditions:

$$y(1) = 0.2, \quad y(1) = 0.5, \quad \text{and} \quad y(1) = 0.7$$

In particular, specify for each case:

a. The period of the orbit for large N, and the values of each of the iterates.
b. Whether the orbit is superstable (i.e., the periodicity is present for all values of N).

This section provided a quick glimpse of two types of nonlinear difference equations, one of which may not necessarily converge to one value. We discovered that a great number of classes of solutions may exist for different values of the equation's parameters. In Section 2.8 we generalize to 2-D. Section 2.8 illustrates nonlinear difference equations in 2-D geometry. The study of these equations has led in the last few decades to various mathematical discoveries in the branches of mathematics called symbolic dynamical theory, fractal geometry, and chaos theory, which have far-reaching implications in many fields of engineering. The interested student/reader is encouraged to consult the publications quoted in the References section of this book for a deeper understanding of this subject.

2.8 Fractals and Computer Art

In Section 2.4, we introduced a fractal type having *a priori* well-defined and apparent spatial symmetries, namely, the Koch curve. In Section 2.7, we discovered that a certain type of 1-D nonlinear difference equation may lead, for a certain range of parameters, to a sequence that may have different orbits. Section 2.8.1 explores examples of 2-D fractals, generated by coupled difference equations, whose solution morphology can also be quite distinct solely due to a minor change in one of the parameters of the difference equations. Section 2.8.2 illustrates another possible feature observed in some types of fractals. We show how the 2-D orbit representing the solution of a particular nonlinear difference equation can also be substantially changed through a minor variation in the initial conditions of the equation.

2.8.1 Mira's Model

The coordinates of the points on the Mira curve are generated iteratively through the following system of nonlinear difference equations:

$$\begin{aligned} x(k+1) &= b\,y(k) + F(x(k)) \\ y(k+1) &= -x(k) + F((x(k+1))) \end{aligned} \tag{2.43}$$

where

$$F(x) = ax + \frac{2(1-a)x^2}{1+x^2} \tag{2.44}$$

We illustrate the different morphologies of the solutions in two different cases, and leave other cases as exercises for your fun and exploration.

Case 1 Here, $a = -0.99$, and we consider the cases $b = 1$ and $b = 0.98$. The starting point coordinates are (4, 0) (see Figure 2.2). This case can be viewed by editing and executing the following *script M-file*:

```
a=-0.99;b1=1;b2=0.98;
x1(1)=4;y1(1)=0;x2(1)=4;y2(1)=0;
  for n=1:12000
    x1(n+1)=b1*y1(n)+a*x1(n)+2*(1-a)*...
                (x1(n))^2/(1+(x1(n)^2));
    y1(n+1)=-x1(n)+a*x1(n+1)+2*(1-a)*...
                (x1(n+1)^2)/(1+(x1(n+1)^2));
    x2(n+1)=b2*y2(n)+a*x2(n)+2*(1-a)*...
                (x2(n))^2/(1+(x2(n)^2));
    y2(n+1)=-x2(n)+a*x2(n+1)+2*(1-a)*...
                (x2(n+1)^2)/(1+(x2(n+1)^2));
  end
subplot(2,1,1); plot(x1,y1,'.')
title('a=-0.99 b=1')
subplot(2,1,2); plot(x2,y2,'.')
title('a=-0.99 b=0.98')
```

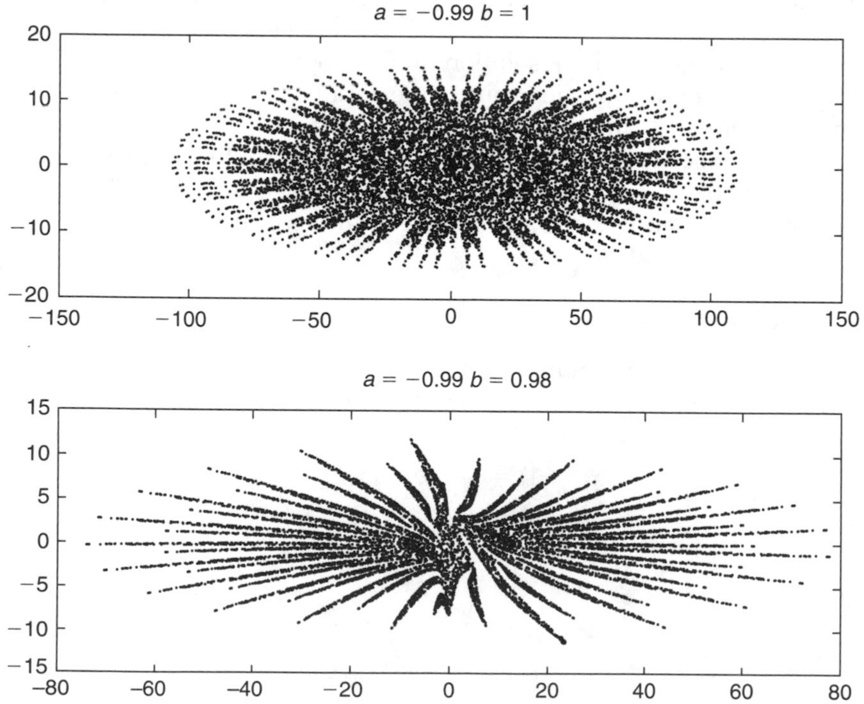

FIGURE 2.2
Plot of the Mira curve for $a = 0.99$. The starting point coordinates are (4, 0). Top panel: $b = 1$, bottom panel: $b = 0.98$.

Case 2 Here, $a = 0.7$, and we consider the cases $b = 1$ and $b = 0.9998$. The starting point coordinates are (0, 12.1) (see Figure 2.3).

IN-CLASS EXERCISE

Pb. 2.23 Manifest the computer artist inside yourself. Generate new geometrical morphologies, in Mira's model, by new choices of the parameters ($-1 < a < 1$ and $b \approx 1$) and of the starting point. You can start with:

a	b_1	b_2	(x_1, y_1)
−0.48	1	0.93	(4, 0)
−0.25	1	0.99	(3, 0)
0.1	1	0.99	(3, 0)
0.5	1	0.9998	(3, 0)
0.99	1	0.9998	(0, 12)

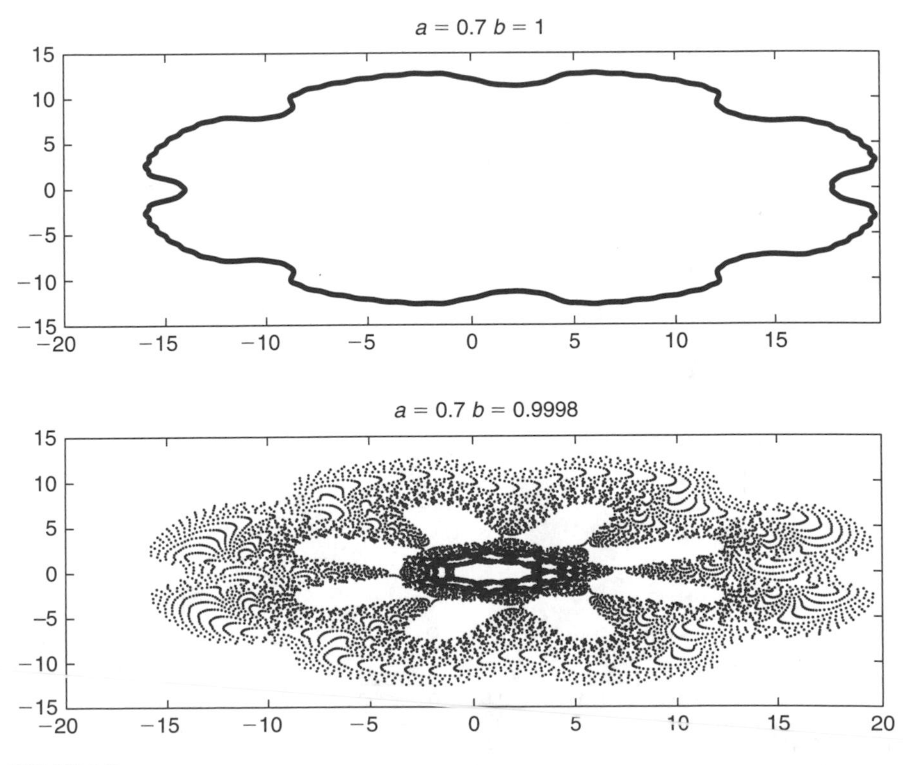

FIGURE 2.3
Plot of the Mira curve for $a = 0.7$. The starting point coordinates are (0, 12.1). Top panel: $b = 1$, bottom panel: $b = 0.9998$.

2.8.2 Hénon's Model

The coordinates of the Hénon's orbits are generated iteratively through the following system of nonlinear difference equations:

$$\begin{aligned} x(k+1) &= ax(k+1) - b(y(k) - (x(k))^2) \\ y(k+1) &= bx(k+1) + a(y(k) - (x(k))^2) \end{aligned} \tag{2.45}$$

where $|a| \leq 1$ and $b = \sqrt{1 - a^2}$.

Executing the following *script M-file* illustrates the possibility of generating two distinct orbits if the starting points of the iteration are slightly different (here, $a = 0.24$), and the starting points are slightly different from each other. For the two cases initial point coordinates are given, respectively, by (0.5696, 0.1622) and (0.5650, 0.1650) (see Figure 2.4).

```
a=0.24;
b=0.9708;
x1(1)=0.5696;y1(1)=0.1622;
x2(1)=0.5650;y2(1)=0.1650;
```

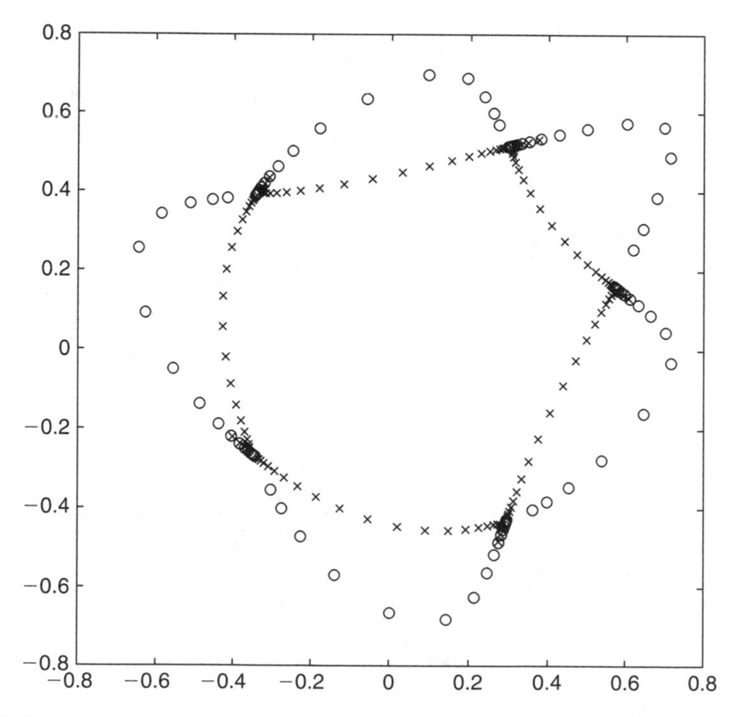

FIGURE 2.4
Plot of two Hénon orbits having the same $a = 0.25$ but different starting points. (o) corresponds to the orbit with starting point (0.5696, 0.1622); (x) corresponds to the orbit with starting point (0.5650, 0.1650).

```
for n=1:120
  x1(n+1)=a*x1(n)-b*(y1(n)-(x1(n))^2);
  y1(n+1)=b*x1(n)+a*(y1(n)-(x1(n))^2);
  x2(n+1)=a*x2(n)-b*(y2(n)-(x2(n))^2);
  y2(n+1)=b*x2(n)+a*(y2(n)-(x2(n))^2);
end
plot(x1,y1,'ro',x2,y2,'bx')
```

2.8.2.1 Demonstration

Different orbits for Hénon's model can be plotted if different starting points are randomly chosen. Executing the following *script M-file* illustrates the $a = 0.24$ case, with random initial conditions (see Figure 2.5).

```
a=0.24;
b=sqrt(1-a^2);
rx=rand(1,40);
ry=rand(1,40);
   for n=1:1500
           for m=1:40
                  x(1,m)=-0.99+2*rx(m);
                  y(1,m)=-0.99+2*ry(m);
```

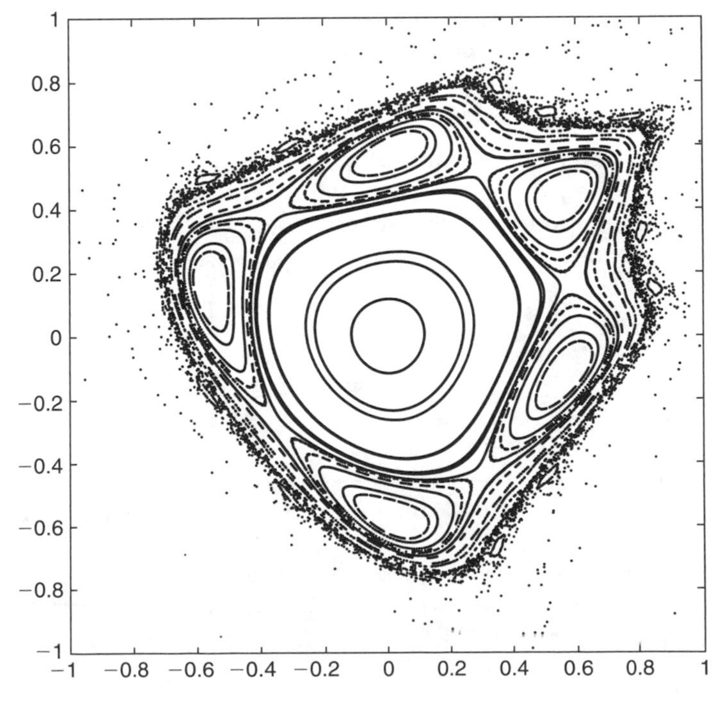

FIGURE 2.5
Plot of multiple Hénon orbits having the same $a = 0.25$ but random starting points.

```
                x(n+1,m)=a*x(n,m)-b*(y(n,m)-(x(n,m))^2);
                y(n+1,m)=b*x(n,m)+a*(y(n,m)-(x(n,m))^2);
        end
    end
plot(x,y,'r.')
axis([-1 1 -1 1])
axis square
```

2.9 Generation of Special Functions from Their Recursion Relations*

In this section, we go back to more classical mathematics. We consider the case of the special functions of mathematical physics. In this case, we need to define the iterated quantities by two indices: the order of the function and the value of the argument of the function.

In many electrical engineering problems, it is convenient to use a class of polynomials called the orthogonal polynomials. For example, in filter design, the set of Chebyshev polynomials are of particular interest.

The Chebyshev polynomials can be defined through recursion relations, which are similar to difference equations and relate the value of a polynomial of a certain order at a particular point to the values of the polynomials of lower orders at the same point. These are defined by the following recursion relation:

$$T_k(x) = 2xT_{k-1}(x) - T_{k-2}(x) \tag{2.46}$$

Now, instead of giving two values for the initial conditions as we would have in difference equations, we need to give the explicit functions for two of the lower-order polynomials. For example, the first- and second-order Chebyshev polynomials are

$$T_1(x) = x \tag{2.47}$$

$$T_2(x) = 2x^2 - 1 \tag{2.48}$$

Example 2.7

Plot over the interval $0 \leq x \leq 1$, the fifth-order Chebyshev polynomial.

Solution: The strategy to solve this problem is to build an array to represent the x-interval, and then use the difference equation routine to find the value of the Chebyshev polynomial at each value of the array, remembering that the indexing should always be a positive integer.

---

* The asterisk indicates more advanced material that may be skipped in a first reading.

The following program implements the above strategy:

```
N=5;
x1=1:101;
x=(x1-1)/100;
T(1,x1)=x;
T(2,x1)=2*x.^2-1;
  for k=3:N
    T(k,x1)=2.*x.*T(k-1,x1)-T(k-2,x1);
  end
y=T(N,x1);
plot(x,y) □
```

In-Class Exercise

Pb. 2.24 By comparing their plots, verify that the above definition for the Chebyshev polynomial gives the same graph as that obtained from the closed-form expression:

$$T_N(x) = \cos(N\cos^{-1}(x)) \quad \text{for } 0 \le x \le 1$$

In addition to the Chebyshev polynomials, you will encounter other orthogonal polynomials in your engineering studies. In particular, the solutions of a number of problems in electromagnetic theory and in quantum mechanics (QM) call on the Legendre, Hermite, Laguerre polynomials, etc. In the following exercises, we explore, in a preliminary manner, some of these polynomials. We also explore another important type of the special functions, the spherical Bessel function.

Homework Problems

Pb. 2.25 Plot the function y defined, in each case:

a. Legendre polynomials:

$$\begin{cases} (m+2)P_{m+2}(x) = (2m+3)xP_{m+1}(x) - (m+1)P_m(x) \\ P_1(x) = x \quad \text{and} \quad P_2(x) = \dfrac{1}{2}(3x^2 - 1) \end{cases}$$

For $0 \le x \le 1$, plot $y = P_5(x)$.

These polynomials describe the electric field distribution from a nonspherical charge distribution. *(cont'd.)*

***Homework Problems** (cont'd.)*

b. Hermite polynomials: $\begin{cases} H_{m+2}(x) = 2xH_{m+1}(x) - 2(m+1)H_m(x) \\ H_1(x) = 2x \quad \text{and} \quad H_2(x) = 4x^2 - 2 \end{cases}$

For $0 \le x \le 6$, plot $y = A_5H_5(x)\exp(-x^2/2)$, where $A_m = (2^m m!\sqrt{\pi})^{-1/2}$

The function y describes the QM wave-function of the harmonic oscillator.

c. Laguerre polynomials:

$$\begin{cases} L_{m+2}(x) = [(3+2m-x)L_{m+1}(x) - (m+1)L_m(x)]/(m+2) \\ L_1(x) = 1 - x \quad \text{and} \quad L_2(x) = (1 - 2x + x^2/2) \end{cases}$$

For $0 \le x \le 20$, plot $y = \exp(-x/2)L_5(x)$.

The Laguerre polynomials figure in the solutions of the QM problem of atoms and molecules.

Pb. 2.26 The recursion relations can, in addition to defining orthogonal polynomials, also define some special functions of mathematical physics. For example, the spherical Bessel functions that play an important role in defining the modes of spherical cavities in electrodynamics and scattering amplitudes in both classical and quantum physics are defined by the following recursion relation:

$$j_{m+2}(x) = \left(\frac{3+2m}{x}\right) j_{m+1}(x) - j_m(x)$$

With

$$j_1(x) = \frac{\sin(x)}{x^2} - \frac{\cos(x)}{x} \quad \text{and} \quad j_2(x) = \left[\frac{3}{x^3} - \frac{1}{x}\right]\sin(x) - \frac{3\cos(x)}{x^2}$$

Plot $j_4(x)$ over the interval $0 < x < 30$.

3

Elementary Functions and Some of Their Uses

The purpose of this chapter is to illustrate and build some practice in the use of elementary functions in selected basic electrical engineering problems. We also construct some simple signal functions that you will encounter in future engineering analysis and design problems.

NOTE It is essential to review Appendix A at the end of this book in case you want to refresh your memory on the particular elementary functions covered in the different chapter sections.

3.1 Function Files

To analyze and graph functions using MATLAB, we have to be able to construct functions that can be called from within the MATLAB environment. In MATLAB, functions are made and stored in *function M-files*. We have already used one kind of *M-file* (script file) to store various executable commands in a routine. *Function M-files* differ from *script M-files* in that they have designated input(s) and output(s).

The following is an example of a function. Type and save the following function in a file named **aline.m**:

```
function y=aline(x)
% (x,y) is a point on a line that has slope 3
% and y-intercept=-5
y=3*x-5;
```

NOTES

1. The word **function** at the beginning of the file makes it a *function M-file* rather than a *script M-file*.

2. The function name, **aline**, that appears in the first line of this file should match the name that we assign to this file name when saving it (i.e., **aline.m**).

Having created a *function M-file* in your user volume, move to the command window to learn how to call this function. There are two basic ways to use a function file:

1. To evaluate the function for a specified value **x=x1**, enter **aline(x1)** to get the function value at this point; that is, $y_1 = 3x_1 - 5$.
2. To plot $y_1 = 3x_1 - 5$ for a range of x values, say $[-2, 7]$, enter:

```
fplot(@aline,[-2,7])
```

Above, using a *function M-file*, we showed a method to plot the defined function **aline** on the interval $(-2, 7)$ using the **fplot** command. An alternative method is, of course, to use arrays, in the manner specified in Chapter 1. Specifically, we could have plotted the **'aline'** function in the following alternate method:

```
x=-2:.01:7;
y=3*x-5;
plot(x,y)
```

To compare the two methods, we note that:

1. **plot** requires a user-supplied x-array (abscissa points) and a constructed y-array (ordinate points), while **fplot** only requires the name of the function file, defined previously and stored in a *function M-file* and the endpoints of the interval.
2. The **fplot** automatically creates a sampled domain that is used to plot the function, taking into account the type of function being plotted and using enough points to make the display appear continuous. On the other hand, **plot** requires that you choose the array length yourself.

Both methods, therefore, have their own advantages and it depends on the particular problem whether to use **plot** or **fplot**.

We are now in position to explore the use of some of the most familiar functions.

NOTE The above example illustrates a function with one input and one output. The construction of a *function M-file* of a function having n inputs and m outputs starts with:

```
function [y1,y2,...,ym]=funname(x1,x2,...,xn)
```

3.2 Examples with Affine Functions

The equation of an affine function is given by

$$y(x) = ax + b \tag{3.1}$$

IN-CLASS EXERCISES

Pb. 3.1 Generate four *function M-files* for the following four functions:

$$y_1(x) = 3x + 2;\ y_2(x) = 3x + 5;\ y_3(x) = -\frac{x}{3} + 3;\ y_4(x) = -\frac{x}{3} + 4$$

Pb. 3.2 Sketch the functions of Pb. 3.1 on the interval $-5 < x < 5$. What can you say about the angle between each of the two lines' pairs. (Did you remember to make your aspect ratio = 1?)

Pb. 3.3 Read off the graphs the coordinates of the points of intersection of the lines in Pb. 3.1. (Review the syntax of the `zoom` and `ginput` commands for a more accurate reading of the coordinates of a point.)

Pb. 3.4 Write a *function M-file* for the line passing through a given point and intersecting another given line at a given angle.

3.2.1 Application to a Simple Circuit

The purpose of this application is to show that:

1. The solution to a simple circuit problem can be viewed as the simultaneous solution of two affine equations, or, equivalently, as the intersection of two straight lines.
2. The variations in the circuit performance can be studied through a knowledge of the affine functions, relating the voltages and the current.

Consider the simple circuit shown in Figure 3.1. In the terminology of the circuit engineer, the voltage source V_s is called the input to the circuit, and the current I and the voltage V are called the circuit outputs. Thus, this is an example of a system with one input and two outputs. As you may have studied in high school physics courses, all of circuit analysis with resistors as

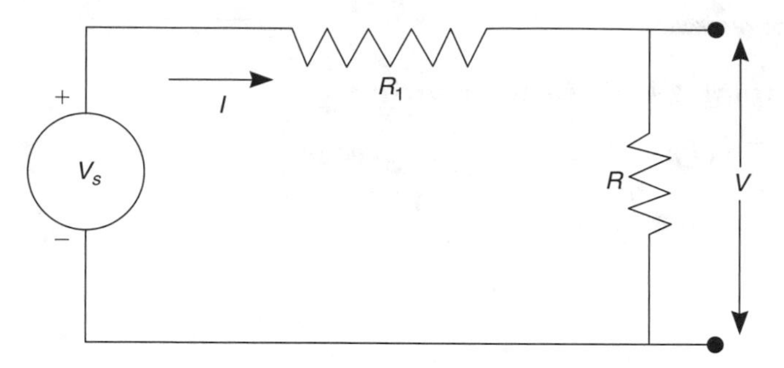

FIGURE 3.1
A simple resistor circuit.

elements can be accomplished using Kirchoff's current law, Kirchoff's voltage law, and Ohm's law.

- Kirchoff's voltage law: The sum of all voltage drops around a closed loop is balanced by the sum of all voltage sources around the same loop.
- Kirchoff's current law: The algebraic sum of all currents entering (exiting) a circuit node must be zero. (Assign the + sign to those currents that are entering the node, and the − sign to those current exiting the node.)
- Ohm's law: The ratio of the voltage drop across a resistor to the current passing through the resistor is a constant, defined as the resistance of the element; that is, $R = V/I$.

The quantities we are looking for include (1) the current I through the circuit, and (2) the voltage V across the load resistor R.

Using Kirchoff's voltage law and Ohm's law for resistance R_1, we obtain

$$V_s = V + V_1 = V + IR_1 \tag{3.2}$$

while applying Ohm's law for the load resistor gives

$$V = IR \tag{3.3}$$

These two equations can be rewritten in the form of affine functions of I as functions of V:

$$L_1\text{: } I = \frac{(V_s - V)}{R_1} \tag{3.4}$$

$$L_2\text{: } I = \frac{V}{R} \tag{3.5}$$

If we know the value of V_s, R, and R_1, then Eq. (3.4) and Eq. (3.5) can be represented as lines drawn on a plane with ordinate I and abscissa V.

Suppose we are interested in finding the value of the current I and the voltage V when $R_1 = 100\,\Omega$, $R = 100\,\Omega$, and $V_s = 5\,\text{V}$. To solve this problem graphically, we plot each of the L_1 and L_2 functions on the same graph and find their point of intersection.

The functions L_1 and L_2 are programmed as follows:

```
function I=L1(V)
R1=100;
R=100;
Vs=5;
I=(Vs-V)/R1;

function I=L2(V)
R1=100;
R=100;
Vs=5;
I=V/R;
```

Because the voltage V is smaller than the source potential, due to losses in the resistor, a suitable domain for V would be [0, 5]. We now plot the two lines on the same graph:

```
fplot(@L1,[0,5])
hold on
fplot(@L2,[0,5])
hold off
```

IN-CLASS EXERCISE

Pb. 3.5 Verify that the two lines L_1 and L_2 intersect at the point: ($I = 0.025$, $V = 2.5$).

In the above analysis, we had to declare the numerical values of the parameters R_1 and R in the definition of each of the two functions. This can, at best, be tedious if you are dealing with more than two *function M-files* or two parameters; or worse, can lead to errors if you overlook changing the values of the parameters in any of the relevant *function M-files* when you decide to modify them. To avoid these types of problems, it is a good practice to call all functions from a single *script M-file* and link the parameters' values together so that you only need to edit the calling *script M-file*. To link the values of parameters to all functions in use, you can use the MATLAB **global** command. To see how this works, rewrite the above *function M-files* as follows:

```
function I=L1(V)
global R1 R     % global statement
Vs=5;
I=(Vs-V)/R1;
```

```
function I=L2(V)
global R1 R                % global statement
Vs=5;
I=V/R;
```

The calling *script M-file* now reads:

```
global R1 R              %global statement
R1=100;                  %set global resistance values
R=100;
V=0:.01:5;               %set the voltage range
I1=L1(V);                %evaluate I1
I2=L2(V);                %evaluate I2
plot(V,I1,V,I2,'-')      %plot the two curves
```

IN-CLASS EXERCISE

Pb. 3.6 In the above *script M-file*, we used arrays and the **plot** command. Rewrite this script file such that you make use of the **fplot** command.

3.2.2 Further Consideration of Figure 3.1

Calculating the circuit values for fixed resistor values is important, but we can also ask about the behavior of the circuit as we vary the resistor values. Suppose we keep $R_1 = 100\,\Omega$ and $V_s = 5\,\text{V}$ fixed, but vary the value that R can take. To this end, an analytic solution would be useful because it would give us the circuit responses for a range of values of the circuit parameters R_1, R, V_s. However, a plot of the lines L_1 and L_2 for different values of R can also provide a great deal of qualitative information regarding how the simultaneous solution to L_1 and L_2 changes as the value of R changes.

The following problem serves to give you a better qualitative idea as to how the circuit outputs vary as different values are chosen for the resistor R.

IN-CLASS EXERCISE

Pb. 3.7 This problem still refers to the circuit of Figure 3.1.

a. Redraw the lines L_1 and L_2, using the previous values for the circuit parameters.
b. Holding the graph for the case $R = 100\,\Omega$, sketch L_1 and L_2 again for $R = 50\,\Omega$ and $R = 500\,\Omega$. How do the values of the voltage and the current change as R increases and decreases?
c. Determine the largest values of the current and voltage that can exist in this circuit when R varies over nonnegative values.
d. The usual nomenclature for the circuit conditions is as follows: the circuit is called an open circuit when $R = \infty$, while it is called a short circuit when $R = 0$. What are the (V, I) solutions for these two cases? Can you generalize your statement?

Now, to validate the graphical results obtained in Pb. 3.7, let us solve analytically the L_1 and L_2 system. Solving this system of two linear equations in two unknowns gives, for the current and the voltage, the following expressions:

$$V(R) = \left(\frac{R}{R + R_1} \right) V_s \tag{3.6}$$

$$I(R) = \left(\frac{1}{R + R_1} \right) V_s \tag{3.7}$$

Note that the above analytic expressions for V and I are neither linear nor affine functions in the value of the resistance.

IN-CLASS EXERCISE

Pb. 3.8 This problem still refers to the circuit of Figure 3.1.

a. Keeping the values of V_s and R_1 fixed, sketch the functions $V(R)$ and $I(R)$ for this circuit, and verify that the solutions you found previously in Pb. 3.7, for the various values of R, agree with those found here.

b. Given that the power lost in a resistive element is the product of the voltage across the resistor multiplied by the current through the resistor, plot the power through the variable resistor as a function of R.

c. Determine the value of R such that the power lost in this resistor is maximized.

d. Find, in general, the relation between R and R_1 that ensures that the power lost in the load resistance is maximized. (This general result is called Thevenin's theorem.)

3.3 Examples with Quadratic Functions

A quadratic function is of the form

$$y(x) = ax^2 + bx + c \tag{3.8}$$

PREPARATORY EXERCISES

Pb. 3.9 Find the coordinates of the vertex of the parabola described by Eq. (3.8) as functions of the a, b, and c parameters.

Pb. 3.10 If $a = 1$, show that the quadratic Eq. (3.8) can be factored as

$$y(x) = (x - x_+)(x - x_-)$$

where $x_\pm$ are the roots of the quadratic equation. Further, show that, for arbitrary a, the product of the roots is $\frac{c}{a}$ and their sum is $\frac{-b}{a}$.

IN-CLASS EXERCISES

Pb. 3.11 Develop a *function M-file* that inputs the two real roots of a second-degree equation and returns the value of this function for an arbitrary x. Is this function unique?

Pb. 3.12 In your elementary mechanics course, you learned that the trajectory of a projectile in a gravitational field (oriented in the $-y$ direction) with an initial velocity $v_{0,x}$ in the x-direction and $v_{0,y}$ in the y-direction satisfies the following parametric equations:

$$x = v_{0,x} t \quad \text{and} \quad y = -\frac{1}{2} g t^2 + v_{0,y} t$$

where t is time and the origin of the axis was chosen to correspond to the position of the particle at $t = 0$ and $g = 9.8\,\text{m}\,\text{s}^{-2}$.

a. By eliminating the time t, show that the projectile trajectory $y(x)$ is a parabola.

b. Noting that the components of the initial velocity can be written as function of the projectile initial speed and its angle of inclination:

$$v_{0,y} = v_0 \sin(\phi) \quad \text{and} \quad v_{0,x} = v_0 \cos(\phi)$$

show that, for a given initial speed, the maximum range for the projectile is achieved when the inclination angle of the initial velocity is 45°.

c. Plot the range for a fixed inclination angle as a function of the initial speed.

3.4 Examples with Polynomial Functions

As pointed out in Appendix A, a polynomial function is an expression of the form

$$p(x) = a_n x^n + a_{n-1} x^{n-1} + \cdots + a_1 x + a_0 \tag{3.9}$$

where $a_n \neq 0$ for an nth-degree polynomial. In MATLAB, we can represent the polynomial function as an array:

$$p = [a_n \; a_{n-1} \; \cdots \; a_0] \tag{3.10}$$

NOTES

1. MATLAB does not provide a direct function for adding polynomials unless they are of the same order, in which case the array addition operation works. In cases where the order of the two polynomials are not the same, the row vector corresponding to the lower-order polynomial has to be padded with zeros on the left, so that it will have the same length as that of the row vector corresponding to the higher-order polynomial, at which point array addition will give the sum of the polynomials.
2. Polynomial multiplication of p_1 and p_2 is supported by the function

   ```
   p3=conv(p1,p2)
   ```

   where length(p_3) = length(p_1) + length(p_2) − 1.
3. Polynomial division is accomplished with the command:

   ```
   [q,r]=deconv(p1,p2)
   ```

   where q is the quotient polynomial and r the remainder term.
4. The derivative of a polynomial is obtained through the command:

   ```
   pder=polyder(p)
   ```

   where length(*pder*) = length(p) − 1.
5. The integral of a polynomial with an integration constant c is obtained through the command:

   ```
   pint_c=polyint(p,c)
   ```

   where length(*pint_c*) = length(p) + 1.

The following printout is self-explanatory:

Entering

```
p1=[1 4 2 6]
length(p1)
p2=[5 7 3 2 5 6]
length(p2)
pprod=conv(p1,p2)
length(pprod)
[pquot,r]=deconv(p2,p1)
derp2=polyder(p2)
length(derp2)
intp1_2=polyint(p1,2)
length(intp1_2)
```

which returns

```
p1 =
        1       4       2       6
length(p1)
        4
p2 =
        5       7       3       2       5       6
length(p2)
        6
pprod =
        5      27      41      58      61      48      46      42      36
length(pprod)
        9
pquot =
        5     -13      45
r =
        0       0       0    -182      -7    -264
derp2 =
       25      28       9       4       5
length(derp2)
        5
intp1_2 =
       0.2500    1.3333    1.0000    6.0000    2.0000
length(intp1_2)
       5
```

6. In Example 3.1, we show how we can construct a function that, starting with the row vector representation of a polynomial, can evaluate this polynomial at a particular value. MATLAB has a built-in function that can do the same, specifically:

```
polyval(p,x)
```

EXAMPLE 3.1

You are given the array of coefficients of the polynomial. Write a *function M-file* for evaluating this polynomial using array operations. Assume that x is a scalar. Let $p = [1\ 3\ 2\ 1\ 0\ 3]$:

Solution:

```
function y=polfct(x)
p=[1 3 2 1 0 3];
L=length(p);
v=x.^[(L-1):-1:0];
y=sum(p.*v);
```

□

IN-CLASS EXERCISES

Pb. 3.13 Show that, for the polynomial p defined by Eq. (3.9), the product of the roots is $(-1)^n \frac{a_0}{a_n}$, and the sum of the roots is $-\frac{a_{n-1}}{a_n}$.

Pb. 3.14 Find graphically the real roots of the polynomial $p = [1\ 3\ 2\ 1\ 0\ 3]$.

3.5 Examples with the Trigonometric Functions

A time-dependent cosine function of the form:

$$x = a\cos(\omega t + \phi) \tag{3.11}$$

appears often in many applications of electrical engineering: a is called the amplitude, ω the angular frequency, and ϕ the phase. Note that we do not have to have a separate discussion of the sine function because the sine function, as shown in Appendix A, differs from the cosine function by a constant phase. Therefore, by suitably changing only the value of the phase parameter, it is possible to transform the sine function into a cosine function.

In the following example, we examine the period of the different powers of the cosine function; your preparatory task is to predict analytically the relationship between the periods of the two curves given in Example 3.2 and then verify your answer numerically.

EXAMPLE 3.2

Plot simultaneously, $x_1(t) = \cos^3(t)$ and $x_2 = \cos(t)$ on $t \in [0, 6\pi]$.

Solution: To implement this task, edit and execute the following *script M-file*:

```
t=0:.2:6*pi;              % t-array
a=1;w=1;                  % desired parameters
x1=a*(cos(w*t)).^3;       % x1-array constructed
x2=a*cos(w*t);            % x2-array constructed
plot(t,x1,t,x2,'--')
```
□

IN-CLASS EXERCISES

Pb. 3.15 Determine the phase relation between the sine and cosine functions of the same argument.

Pb. 3.16 The meaning of amplitude, angular frequency, and phase can be better understood using MATLAB to obtain graphs of the cosine function for a family of a values, ω values, and ϕ values.

a. With $\omega = 1$ and $\phi = \pi/3$, plot the cosine curves corresponding to $a = 1{:}0.1{:}2$.
b. With $a = 1$ and $\omega = 1$, plot the cosine curves corresponding to $\phi = 0{:}\pi/10{:}\pi$.
c. With $a = 1$ and $\phi = \pi/4$, plot the cosine curves corresponding to $\omega = 1{:}0.1{:}2$.

Homework Problem

Pb. 3.17 Find the period of the function obtained by summing the following three cosine functions:

$$x_1 = 3\cos(t/3 + \pi/3),\ x_2 = \cos(t + \pi),\ x_3 = \frac{1}{3}\cos\left(\frac{3}{2}(t + \pi)\right)$$

Verify your result graphically.

3.6 Examples with the Logarithmic Function

3.6.1 Ideal Coaxial Capacitor

An ideal capacitor can be loosely defined as two metallic plates separated by an insulator. If a potential is established between the plates, for example, through the means of connecting the two plates to the different terminals of a battery, the plates will be charged by equal and opposite charges, with the battery serving as a pump to move the charges around. The capacitance of a capacitor is defined as the ratio of the magnitude of the charge accumulated on either of the plates divided by the potential difference across the plates.

Using Gauss' law of electrostatics, it can be shown that the capacitance per unit length of an infinitely long coaxial cable is

$$\frac{C}{l} = \frac{2\pi\varepsilon}{\ln(b/a)} \tag{3.12}$$

where a and b are the radius of the internal and external conductors, respectively, and ε is the permittivity of the dielectric material sandwiched between the conductors. (The permittivity of vacuum is approximately $\varepsilon_0 = 8.85 \times 10^{-12}$, while that of oil, polystyrene, glass, quartz, bakelite, and mica are, respectively, 2.1, 2.6, 4.5–10, 3.8–5, 5, and 5.4–6 larger.)

IN-CLASS EXERCISE

Pb. 3.18 Find the ratio of the capacitance of two coaxial cables with the same dielectric material for, respectively, $b/a = 5$ and 50.

3.6.2 The Decibel Scale

In the SI units used by electrical engineers, the unit of power is the Watt. However, in a number of applications, it is convenient to express the power as a ratio of its value to a reference value. Because the value of this ratio can vary over several orders of magnitude, it is often more convenient to represent this ratio on a logarithmic scale, called the decibel scale:

$$G[\text{dB}] = 10\log\left(\frac{P}{P_{ref}}\right) \tag{3.13}$$

where the function log is the logarithm to base 10. The table below converts the power ratio to its value in decibels (dB):

P/P_{ref} (10^n)	dB values (10 n)
4	6
2	3
1	0
0.5	−3
0.25	−6
0.1	−10
10^{-3}	−30

IN-CLASS EXERCISE

Pb. 3.19 In a measurement of two power values, P_1 and P_2, it was determined that:

$$G_1 = 9\,\text{dB} \quad \text{and} \quad G_2 = -11\,\text{dB}$$

Using the above table, determine the value of the ratio P_1/P_2.

3.6.3 Entropy

Given a random variable X (such as the number of spots on the face of a thrown die) whose possible outcomes are $x_1, x_2, x_3, \ldots$, and such that the probability for each outcome is, respectively, $p(x_1), p(x_2), p(x_3), \ldots$; then the entropy for this system described by the outcome of one random variable is defined by

$$H(X) = -\sum_{i=1}^{N} p(x_i)\log_2(p(x_i)) \tag{3.14}$$

where N is the number of possible outcomes, and the logarithm is to base 2.

The entropy is a measure of the uncertainty in the value of the random variable. In information theory, it will be shown that the entropy, so defined, is the number of bits, on average, required to describe the random variable X.

In-Class Exercises

Pb. 3.20 In each of the following cases, find the entropy:

a. $N = 32$ and $p(x_i) = \dfrac{1}{32}$ for all i

b. $N = 8$ and $p = \left[\dfrac{1}{2}, \dfrac{1}{4}, \dfrac{1}{8}, \dfrac{1}{16}, \dfrac{1}{64}, \dfrac{1}{64}, \dfrac{1}{64}, \dfrac{1}{64}\right]$

c. $N = 4$ and $p = \left[\dfrac{1}{2}, \dfrac{1}{4}, \dfrac{1}{8}, \dfrac{1}{8}\right]$

d. $N = 4$ and $p = \left[\dfrac{1}{2}, \dfrac{1}{4}, \dfrac{1}{4}, 0\right]$

Pb. 3.21 Assume that you have two dices, one red and the other blue. Tabulate all possible outcomes that you can obtain by throwing these dices together. Now assume that all you care about is the sum of spots on the two dices. Find the entropy of the outcome.

Homework Problem

Pb. 3.22 A so-called A-law compander (compressor followed by an expander) uses a compressor that relates output to input voltages by:

$$y = \pm\frac{A|x|}{1+\log(A)} \qquad \text{for } |x| \leq 1/A$$

$$y = \pm\frac{1+\log(A|x|)}{1+\log(A)} \qquad \text{for } \frac{1}{A} \leq |x| \leq 1$$

Here, the + sign applies when x is positive and the − sign when x is negative. $x = v_i/V$ and $y = v_o/V$, where v_i and v_o are the input and output voltages, respectively. The range of allowable voltages is $-V$ to V. The parameter A determines the degree of compression.

For a value of $A = 87.6$, plot y versus x in the interval $[-1, 1]$.

3.7 Examples with the Exponential Function

Take a few minutes to review the section on the exponential function in Appendix A before proceeding further. (Recall that exp(1) = e.)

IN-CLASS EXERCISES

Pb. 3.23 Plot the function $y(x) = (x^{13} + x^9 + x^5 + x^2 + 1)\exp(-4x)$ over the interval [0, 10].

Pb. 3.24 Plot the function $y(x) = \cos(5x)\exp(-x/2)$ over the interval [0, 10].

Pb. 3.25 From the results of Pb. 3.23 and Pb. 3.24, what can you deduce about the behavior of a function at ∞ if one of its factors is an exponentially decreasing function of *x*, while the other factor is a polynomial or trigonometric function of *x*? What modification to the curve is observed if the degree of the polynomial is increased?

3.7.1 Application to a Simple *RC* Circuit

The solution giving the voltage across the capacitor in Figure 3.2 following the closing of the switch can be written in the following form:

$$V_c(t) = V_c(0)\exp\left[-\frac{t}{RC}\right] + V_s\left[1 - \exp\left[-\frac{t}{RC}\right]\right] \tag{3.15}$$

$V_c(t)$ is called the time response of the *RC* circuit, or the circuit output resulting from the constant input V_s. The time constant *RC* of the circuit has the units of seconds and, as you will observe in the present analysis and other problems

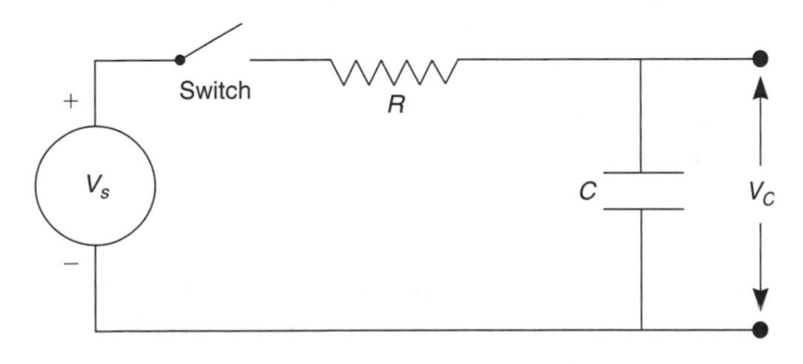

FIGURE 3.2
The circuit used in charging a capacitor.

in subsequent chapters, its ratio to the characteristic time of a given input potential determines qualitatively the output of the system.

IN-CLASS EXERCISE

Pb. 3.26 A circuit designer can produce outputs of various shapes by selecting specific values for the circuit time constant *RC*. In the following simulations, you can examine the influence of this time constant on the response of the circuit of Figure 3.2.

Using $V_c(0) = 3$ V, $V_s = 10$ V (capacitor charging process), and $RC = 1$ s:

a. Sketch a graph of $V_c(t)$. What is the asymptotic value of the solution? How long does it take the capacitor voltage to reach the value of 9 V?

b. Produce an *M-file* that will plot several curves of $V_c(t)$ corresponding to:
  (i) $RC = 1$
  (ii) $RC = 5$
  (iii) $RC = 10$

  Which of these time constants results in the fastest approach of $V_c(t)$ toward V_s?

c. Repeat the above simulations for the case $V_s = 0$ (capacitor discharge)?

d. What would you expect to occur if $V_c(0) = V_s$?

Homework Problem

Pb. 3.27 The Fermi-Dirac distribution, which gives the average population of electrons in a state with energy ε, neglecting the electron spin for the moment, is given by:

$$f(\varepsilon) = \frac{1}{\exp[(\varepsilon - \mu)/\Theta] + 1}$$

where μ is the Fermi (or chemical) potential and Θ is proportional to the absolute (or Kelvin) temperature.

a. Plot the function $f(\varepsilon)$ as function of ε, for the following cases:
  (i) $\mu = 1$ and $\Theta = 0.002$
  (ii) $\mu = 1$ and $\Theta = 0.02$
  (iii) $\mu = 1$ and $\Theta = 0.2$
  (iv) $\mu = 1$ and $\Theta = 2$

b. What is the value of $f(\varepsilon)$ when $\varepsilon = \mu$?

c. Determine the condition under which we can approximate the Fermi-Dirac distribution function by:

$$f(\varepsilon) \approx \exp[(\mu - \varepsilon)/\Theta]$$

3.8 Examples with the Hyperbolic Functions and Their Inverses

3.8.1 Application to the Capacitance of Two Parallel Wires

The capacitance per unit length of two parallel wires, each of radius a and having their axis separated by distance D, is given by

$$\frac{C}{l} = \frac{\pi\varepsilon_0}{\cosh^{-1}\left(\frac{D}{2a}\right)} \tag{3.16}$$

where ε_0 is the permittivity of air (taken to be that of vacuum) = 8.854×10^{-12} Farad/m.

Question: Write this expression in a different form using the logarithmic function.

IN-CLASS EXERCISES

Pb. 3.28 Find the capacitance per unit length of two wires of radii 1 cm separated by a distance of 1 m. Express your answer using the most appropriate of the following subunits:

mF = 10^{-3}F (milli-Farad); μF = 10^{-6}F (micro-Farad);
nF = 10^{-9}F (nano-Farad); pF = 10^{-12}F (pico-Farad);
fF = 10^{-15}F (femto-Farad); aF = 10^{-18}F (atto-Farad);

Pb. 3.29 Assume that you have two capacitors, one consisting of a coaxial cable (radii a and b) and the other of two parallel wires, separated by a distance D. Further assume that the radius of the wires is equal to the radius of the inner cylinder of the coaxial cable. Plot the ratio D/a as a function of b/a, if we desire the two geometrical configurations for the capacitor to end up having the same value for the capacitance. $\left(\text{Take } \frac{\varepsilon}{\varepsilon_0} = 2.6.\right)$

3.9 Commonly Used Signal Processing Functions

In studying signals and systems, you will also encounter, *inter alia*, the following functions (or variation thereof), in addition to the functions discussed previously in this chapter:

- Unit step function
- Unit slope ramp function
- Unit area rectangle pulse
- Unit slope right angle triangle function

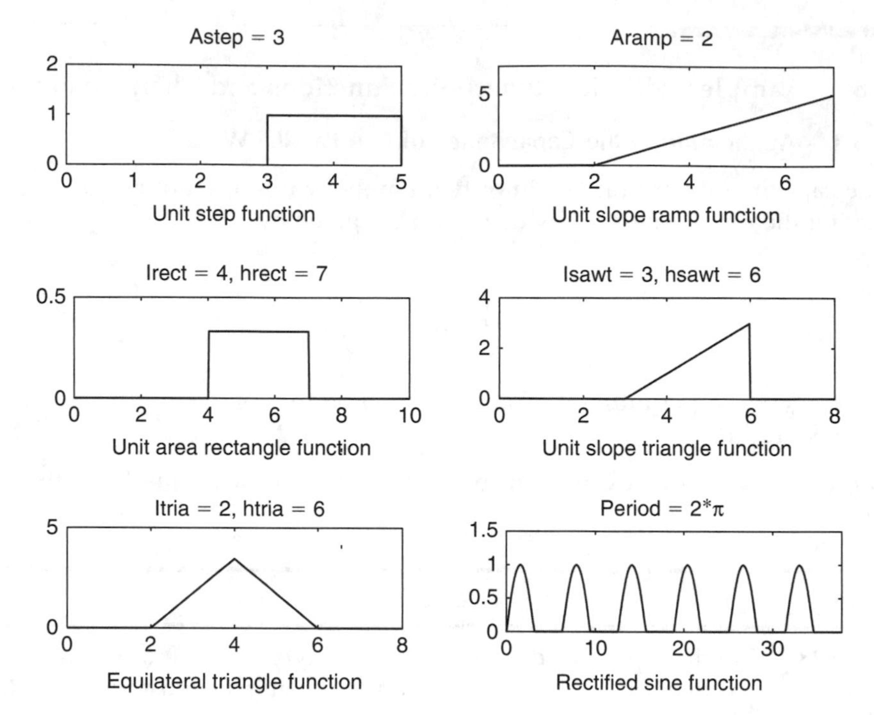

FIGURE 3.3
Various useful signal processing functions.

- Equilateral triangle function
- Periodic traces

These functions are plotted in Figure 3.3, and the corresponding *function M-files* are (x is everywhere a scalar):

A. Unit step function

```
function y=stepf(x)
global astep
  if x<astep
    y=0;
  else
    y=1;
  end
```

B. Unit slope ramp function

```
function y=rampf(x)
global aramp
  if x<aramp
    y=0;
  else
    y=x-aramp;
  end
```

C. Unit area rectangle function

```
function y=rectf(x)
 global lrect hrect
  if x<lrect
    y=0
  elseif lrect<x & x<hrect
    y=1/(hrect-lrect);
  else
    y=0;
  end
```

D. Unit slope right angle triangle function

```
function y=sawtf(x)
global lsawt hsawt
  if x<lsawt
    y=0;
  elseif lsawt<x & x<hsawt
    y=x-lsawt;
  else
    y=0;
  end
```

E. Equilateral triangle function

```
function y=triaf(x)
global ltria htria
  if x<ltria
    y=0;
  elseif ltria<x & x<(ltria+htria)/2
    y=sqrt(3)*(x-ltria);
  elseif (ltria+htria)/2<=x & x<htria
    y=sqrt(3)*(-x+htira);
  else
    y=0
  end
```

F. Periodic functions

It is often necessary to represent a periodic signal train where the elementary representation on one cycle can easily be written. The technique is to use the modulo arithmetic to map the whole of the x-axis over a finite domain. This is, of course, possible because the function is periodic. For example, consider the rectified sine function train. Its *function M-file* is

```
function y=psinef(x)
s=rem(x,2*pi)
  if s>0 & s=<pi
     y=sin(s);
  elseif s>pi & s=<2*pi
     y=0;
```

```
else
   y=0
end
```

In-Class Exercises

Pb. 3.30 In the above definition of all the special shape functions, we used the if-else-end form. Write each of the *function M-files* to define these same functions using only Boolean expressions.

Pb. 3.31 An adder is a device that adds the input signals to give an output signal equal to the sum of the inputs. Using the functions previously obtained in this section, write the *function M-file* for the signal in Figure 3.4.

Pb. 3.32 A multiplier is a device that multiplies two inputs. Find the product of the inputs given in Figure 3.5 and Figure 3.6.

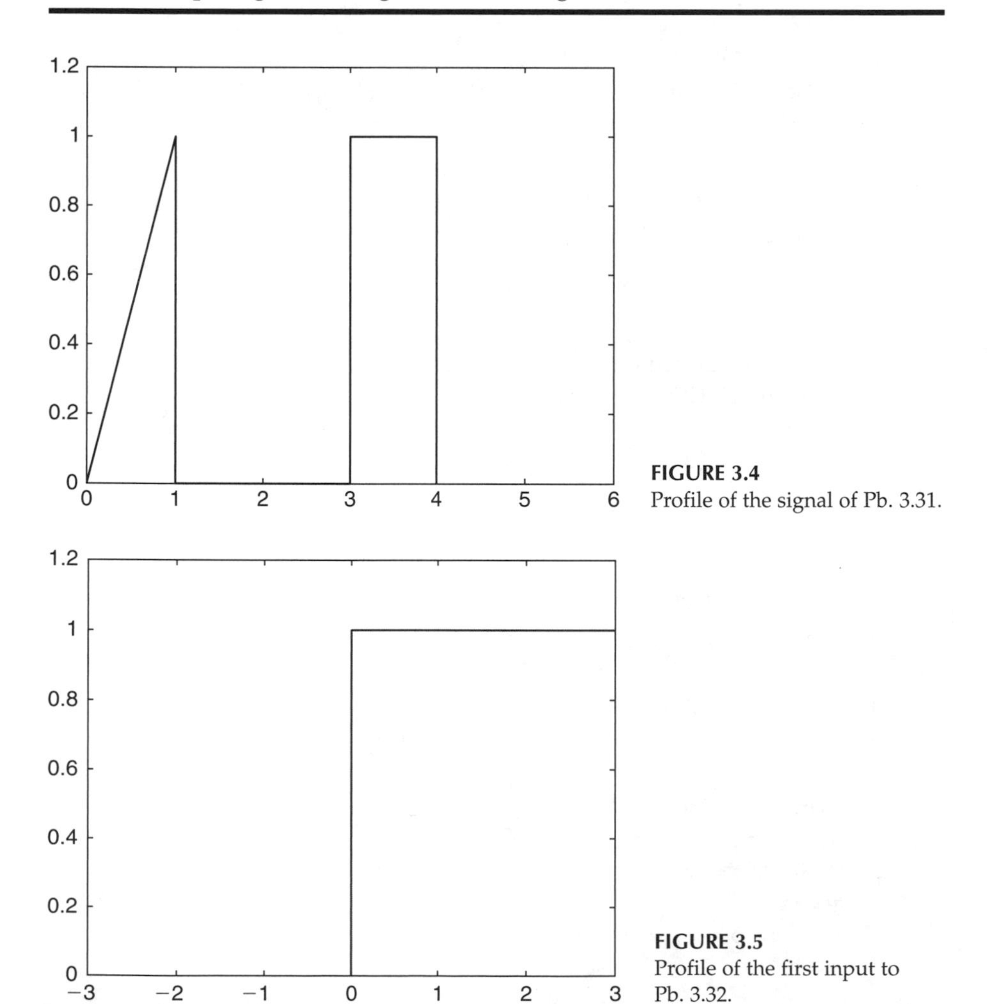

FIGURE 3.4
Profile of the signal of Pb. 3.31.

FIGURE 3.5
Profile of the first input to Pb. 3.32.

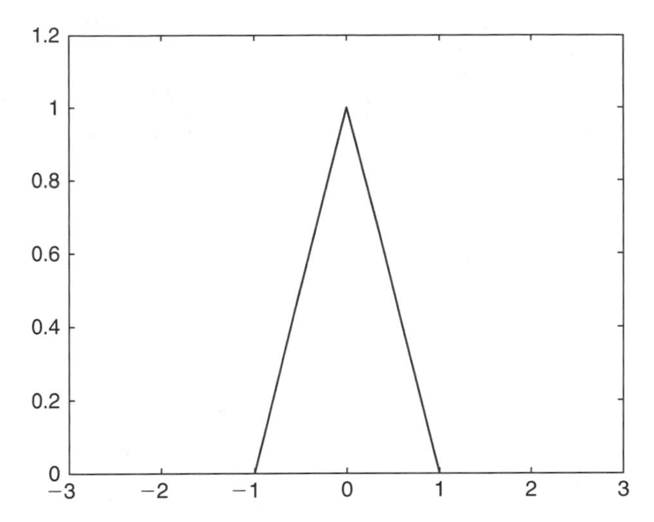

FIGURE 3.6
Profile of the second input to Pb. 3.32.

Homework Problems

The first three problems in this set are a brief introduction to the different analog modulation schemes of communication theory.

Pb. 3.33 In DSB-AM (double-sideband amplitude modulation), the amplitude of the modulated signal is proportional to the message signal, which means that the time domain representation of the modulated signal is given by

$$u_{\text{DSB}}(t) = A_c m(t)\cos(2\pi f_c t)$$

where the carrier-wave shape is

$$c(t) = A_c \cos(2\pi f_c t)$$

and the message signal is $m(t)$.

For a message signal given by

$$m(t) = \begin{cases} 1 & 0 \leq t \leq t_0/3 \\ -3 & t_0/3 < t \leq 2t_0/3 \\ 0 & \text{otherwise} \end{cases}$$

a. Write the expression for the modulated signal using the unit area rectangle and the trigonometric functions.

b. Plot the modulated signal as function of time. (Let $f_c = 200$ and $t_0 = 0.01$.)

(cont'd.)

Homework Problems (cont'd.)

Pb. 3.34 In conventional AM, $m(t)$ in the DSB-AM expression for the modulated signal is replaced by $[1 + am_n(t)]$, where $m_n(t)$ is the normalized message signal (i.e., $m_n(t) = \dfrac{m(t)}{\max(m(t))}$ and a is the index of modulation $(0 \le a \le 1)$. The modulated signal expression is then given by

$$u_{AM}(t) = A_c[1 + am_n(t)]\cos(2\pi f_c t)$$

For the same message as that of Pb. 3.33 and the same carrier frequency, and assuming the modulation index $a = 0.85$:

a. Write the expression for the modulated signal.
b. Plot the modulated signal.

Pb. 3.35 The angle modulation scheme, which includes frequency modulation (FM) and phase modulation (PM), has the modulated signal given by

$$u_{PM}(t) = A_c \cos(2\pi f_c t + k_p m(t))$$
$$u_{FM}(t) = A_c \cos\left(2\pi f_c t + 2\pi k_f \int_{-\infty}^{t} m(\tau)d\tau\right)$$

Assuming the same message as in Pb. 3.33:

a. Write the expression for the modulated signal in both schemes.
b. Plot the modulated signal in both schemes. Let $k_p = k_f = 100$.

Pb. 3.36 If $f(x) = f(-x)$ for all x, then the graph of $f(x)$ is symmetric with respect to the y-axis, and the function $f(x)$ is called an even function. If $f(x) = -f(-x)$ for all x, the graph of $f(x)$ is antisymmetric with respect to the origin, and we call such a function an odd function.

a. Show that any function can be written as the sum of an odd function and an even function. List as many even and odd functions as you can.
b. State the conditions that be true for a polynomial to be even, or to be odd.
c. Show that the product of two even functions is even; the product of two odd functions is even; and the product of an odd and even function is odd.
d. Replace in **c** above the word product by quotient and deduce the parity of the resulting function.
e. Deduce from the above results that the sign/parity of a function follows algebraic rules.

(cont'd.)

Homework Problems *(cont'd.)*

f. Find the even and odd parts of the following functions:

(i) $f(x) = x^7 + 3x^4 + 6x + 2$

(ii) $f(x) = (\sin(x) + 3)\sinh^2(x)\exp(-x^2)$

Pb. 3.37 Decompose the signal shown in Figure 3.7 into its even and odd parts.

Pb. 3.38 Plot the function y defined by

$$y(x) = \begin{cases} x^2 + 4x + 4 & \text{for } -2 \le x < -1 \\ 0.16x^2 - 0.48x & \text{for } -1 < x < 1.5 \\ 0 & \text{elsewhere} \end{cases}$$

and find its even and odd parts.

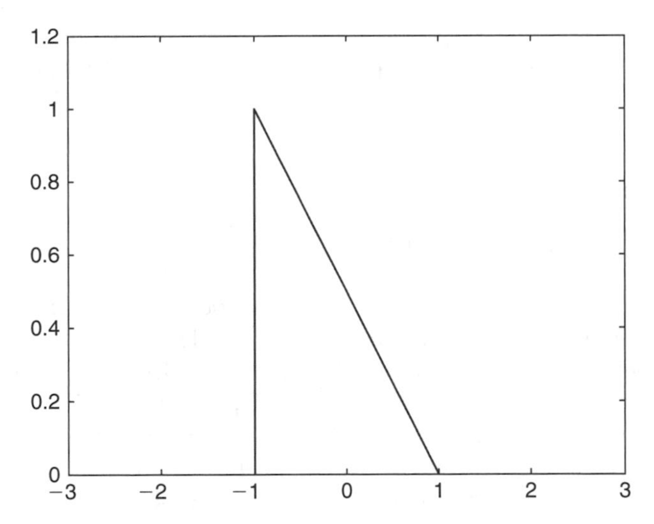

FIGURE 3.7
Profile of the signal of Pb. 3.37.

3.10 Animation of a Moving Rectangular Pulse

You might often want to plot the time development of one of the above signal processing functions if its defining parameters are changing in time. Take, for example, a theatrical spotlight of constant intensity density across its cross-section, but assume that its position varies with time. The light spot size can be represented by a rectangular pulse (e.g., of width 2 m and height 1 m) that is moving to the right with a constant speed of 1 m/s. Assume that the center of the spot is originally at $x = 1$ m, and that its final position is at $x = 8$ m. We want to write a program that will illustrate its time development, and then play the resulting movie.

To illustrate the use of other commands not often utilized in this chapter, we can, instead of the **if-else-end** syntax used in the previous section, use the Boolean syntax, and define the array by the **linspace** command.

Edit and execute the following *script M-file*:

```
lrect=0;hrect=2;
x=linspace(0,10,200);
t=linspace(0,8,40);
  for m=1:40
     y=(x>=lrect+t(m)).*(x<=hrect+t(m));
     plot(x,y,'r')
     axis([-2 12 0 1.2]);
     M(:,m)=getframe;
  end
movie(M,3)
```

Question: How would you modify the above program if the speed of the beam spot is not 1?

3.11 Use of the Function Handle

A function handle (@) is a MATLAB data type that contains information used in referencing a function. The function handle stores all the information MATLAB needs to evaluate the function from any location in the MATLAB environment. So far, we made trivial use of it, when we called a function of one variable with the **fplot** command. In the examples below, we illustrate its broader use, using in some examples problems already solved by other techniques. In the following chapters, you will discover other uses of the function handle that simplify many programs, in particular, passing a function reference to another function.

Example 3.3

Use the function handle to define and plot the unit area rectangle function of Section 3.9(C), without using global variables statements.

Solution: First, construct the function file:

```
function y=rectf2(x,lrect,hrect)
y=(1/(hrect-lrect)).*(lrect<x).*(x<hrect);
```

Second, go to the command window and instruct MATLAB to plot this function for the values used in panel(3) of Figure 3.3 (i.e., **lrect = 4 and hrect = 7**):

Enter:

```
>>fplot(@(x)rectf2(x,4,7),[0 10])
```

Instead of the **fplot** function we could have used the **ezplot** command (see Appendix C). □

We can use the anonymous function technique used above to also create anonymous functions having multiple input arguments.

EXAMPLE 3.4

Use the command **ezcontour** and the function handle to obtain the contour lines of the paraboloid $z = x^2 + y^2 + 3$ over the domain $-1 \leq x \leq 1$ and $-1 \leq y \leq 1$.

Solution: First, construct the function file:

```
function z=parab2(x,y,k)
z=x.^2+y.^2+k;
```

Then, enter in the command window:

```
>>ezcontour(@(x,y)parab2(x,y,3),[-1 1],[-1 1])
```

(The syntax for the **ezcontour** command can be found in Appendix C.)

This problem could as well also have been solved by directly entering the function in the **ezcontour** command, if an *M-file* did not already exist for the paraboloid, specifically:

```
>> ezcontour(@(x,y)x.^2+y.^2+3,[-1 1], [-1 1])
```
□

The function handle can also be used for calling one function within another; specifically, one anonymous function can call another to implement function composition.

EXAMPLE 3.5

Given the functions,

$$f(x) = \sin(x)\cos(x^2) \quad \text{and} \quad g(x) = x^3 + x^2$$

construct the function $h(x) = g(f(x))$ and plot it over the interval $0 \leq x \leq 3$.

Solution:

```
>>f=@(x) sin(x).*cos(x.^2);
g=@(x) x.^3+x.^2;
h=@(x) g(f(x));
fplot(h,[0 3])
```
□

3.12 MATLAB Commands Review

`conv`	Computes the product of two polynomials.
`deconv`	Finds the quotient and the remainder of two polynomials.
`ezcontour`	Generates the contours of an expression in two variables.
`ezplot`	Generates a 2-D plot of a function in one variable.
`function`	The header of a *function M-file.*
`fplot`	Plot a specified function over a specified interval.
`global`	Allows variables to share their values in multiple programs.
`polyder`	Finds the derivative of a polynomial.
`polyint`	Finds the integral of a polynomial.
`polyval`	Computes the value of a polynomial at a specific point.
`@`	Function handle.

4

Differentiation, Integration, and Solutions of Ordinary Differential Equations

The first seven sections of this chapter discuss numerical techniques for the basic operations of calculus. We describe the basic methods for numerically finding the value of the limit of an indeterminate form, the value of a derivative, the value of a convergent infinite sum, and the value of a definite integral. Using an improved form of the differentiator, we also present first-order iterator techniques for solving ordinary first- and second-order linear differential equations. The Runge–Kutta technique for solving ordinary differential equations (ODE) is briefly discussed. The mode of use of some of the MATLAB® packages to perform each of the previous tasks is also described in each instance of interest. In Section 4.8, we find analytically the integral equation equivalent to an ODE with given boundary conditions. The numerical technique for solving integral equations is postponed to Chapter 8.

4.1 Limits of Indeterminate Forms

A fundamental result from elementary calculus states that

- if the quotient $u(x)/v(x)$ has an indeterminate form of the $0/0$ kind, or
- if the quotient $u(x)/v(x)$ has an indeterminate form of the ∞/∞ kind,

then the standard technique for solving these problems is through the use of L'Hopital's rule, which states that

if
$$\lim_{x \to x_0} \frac{u'(x)}{v'(x)} = C \tag{4.1}$$

then
$$\lim_{x \to x_0} \frac{u(x)}{v(x)} = C \tag{4.2}$$

In this section, we discuss a simple algorithm to obtain numerically this limit using MATLAB. The method consists of the following steps:

1. Construct a sequence of points whose limit is x_0. In the examples below, consider the sequence $\left\{x_n = x_0 - \left(\frac{1}{2}\right)^n\right\}$. Recall in this regard that as $n \to \infty$, the nth power of any number whose magnitude is smaller than one goes to zero.
2. Construct the sequence of function values corresponding to the x-sequence, and find its limit.

Example 4.1

Compute numerically $\lim_{x \to 0} \frac{\sin(x)}{x}$.

Solution: Enter the following instructions in your MATLAB command window:

```
N=20; n=1:N;
x0=0;
dxn=-(1/2).^n;
xn=x0+dxn;
yn=sin(xn)./xn;
plot(xn,yn,'o')
```

The limit of the **yn** sequence is clearly equal to 1. The deviation of the sequence of the **yn** from the value of the limit can be obtained by entering:

```
figure
dyn=yn-1;
semilogy(n,dyn,'o')
```

The last command plots the curve with the ordinate y expressed logarithmically. This mode of display is the most convenient in this case because the ordinate spans many decades of values. □

In-Class Exercises

Find the limits of the following functions at the indicated points:

Pb. 4.1 $\frac{(x^2 - 2x - 3)}{(x - 3)}$ at $x \to 3$

Pb. 4.2 $\left(\dfrac{1+\sin(x)}{x} - \dfrac{1}{\sin(x)}\right)$ at $x \to 0$

Pb. 4.3 $(x\cot(x))$ at $x \to 0$

Pb. 4.4 $\dfrac{(1-\cos(2x))}{x^2}$ at $x \to 0$

Pb. 4.5 $\sin(2x)\cot(3x)$ at $x \to 0$

4.2 Derivative of a Function

DEFINITION The derivative of a certain function at a particular point is defined as

$$f'(x_0) = \lim_{x \to x_0} \frac{f(x) - f(x_0)}{x - x_0} \tag{4.3}$$

Numerically, the derivative is computed at the point x_0 as follows:

1. Construct an x-sequence that approaches x_0.
2. Compute a sequence of the function values corresponding to the x-sequence.
3. Evaluate the sequence of the ratio, appearing in the definition of the derivative in Eq. (4.3).
4. Read off the limit of this ratio sequence. This will be the value of the derivative at the point x_0.

Example 4.2

Find numerically the derivative of the function $\ln(1 + x)$ at $x = 0$.

Solution: Edit and execute the following *script M-file*:

```
N=20;n=1:N;
x0=0;
dxn=(1/2).^[1:N];
xn=x0+dxn;
yn=log(1+xn);
dyn=yn-log(1+x0);
deryn=dyn./dxn;
plot(n,deryn, 'o')
```

The limit of the **deryn**'s sequence is clearly equal to 1, the value of this function derivative at 0.

NOTE The choice of **N** should always be such that **dxn** is larger than the machine precision; that is, **N** < 53, since $(1/2)^{53} \approx 10^{-16}$. □

In-Class Exercises

Find numerically, to one part per 10,000 accuracy, the derivatives of the following functions at the indicated points:

Pb. 4.6 $x^4(\cos^3(x) - \sin(2x))$ at $x \to \pi$

Pb. 4.7 $\dfrac{\exp(x^2+3)}{(2+\cos^2(x))}$ at $x \to 0$

Pb. 4.8 $\dfrac{(1+\sin^2(x))}{(2-\cos^3(x))}$ at $x \to \pi/2$

Pb. 4.9 $\ln\left(\dfrac{x-1/2}{x+1}\right)$ at $x \to 1$

Pb. 4.10 $\tan^{-1}(x^2+3)$ at $x \to 0$

Example 4.3

Plot the derivative of the function $x^2\sin(x)$ over the interval $0 \leqslant x \leqslant 2\pi$.

Solution: Edit and execute the following *script M-file*:

```
dx=10^(-4);
x=0:dx:2*pi+dx;
df=diff(sin(x).*x.^2)/dx;
plot(0:dx:2*pi,df)
```

where we remind the reader that **diff** is a MATLAB command, which when acting on an array X, gives the new array [X(2) − X(1) X(3) − X(2) … X(n) − X(n − 1)], whose length is one unit shorter than the array X.

The accuracy of the above algorithm depends on the choice of **dx**. Ideally, the smaller it is, the more accurate the result. However, using any computer, we should always choose a **dx** that is larger than the machine precision, while still much smaller than the value of the variation of **x** over which the function changes appreciably.

For a systematic method to choose an upper limit on **dx**, you might want to follow these simple steps:

1. Plot the function on the given interval and identify the point where the derivative is largest.
2. Compute the derivative at that point using the sequence method of Example 4.2, and determine the **dx** that would satisfy the desired tolerance; then go ahead and use this value of **dx** in the above routine to evaluate the derivative throughout the given interval. □

IN-CLASS EXERCISES

Plot the derivatives of the following functions on the indicated intervals:

Pb. 4.11 $\ln\left(\dfrac{x-1}{x+1}\right)$ on $2 < x < 3$

Pb. 4.12 $\ln\left(\dfrac{1+\sqrt{1+x^2}}{x}\right)$ on $1 < x < 2$

Pb. 4.13 $\ln(\tanh(x/2))$ on $1 < x < 5$

Pb. 4.14 $\tan^{-1}(\sinh(x))$ on $0 < x < 10$

Pb. 4.15 $\ln(\csc(x) + \tan(x))$ on $0 < x < \pi/2$

4.3 Infinite Sums

An infinite series is denoted by the symbol $\sum_{n=1}^{\infty} a_n$. It is important not to confuse the series with the sequence $\{a_n\}$. The sequence is a list of terms while the series is a sum of these terms. A sequence is convergent if the term a_n approaches a finite limit; however, convergence of a series requires that the sequence of partial sums $S_N = \sum_{n=1}^{N} a_n$ approaches a finite limit. There are cases where the sequence may approach zero, while the series is divergent. The classical example is that of the sequence $\left\{\frac{1}{n}\right\}$; this sequence approaches the limit zero, while the corresponding series is divergent.

In any numerical calculation, we cannot perform the operation of adding an infinite number of terms. We can only add a finite number of terms. The infinite sum of a convergent series is the limit of the partial sums S_N.

You will study in your calculus course the different tests for checking the convergence of a series. We summarize below the most useful of these tests.

- The Ratio Test, which is very useful for series with terms that contain factorials and nth power of a constant, states that:

$$\text{for } a_n > 0, \text{ the series } \sum_{n=1}^{\infty} a_n \text{ is convergent if } \lim_{n\to\infty}\left(\frac{a_{n+1}}{a_n}\right) < 1$$

- The Root Test stipulates that for $a_n > 0$, the series $\sum_{n=1}^{\infty} a_n$ is convergent if

$$\lim_{n\to\infty} (a_n)^{1/n} < 1$$

- For an alternating series, the series is convergent if it satisfies the conditions that

$$\lim_{n\to\infty} |a_n| = 0 \quad \text{and} \quad |a_{n+1}| < |a_n|$$

Now look at the numerical routines for evaluating the limit of the partial sums when they exist.

Example 4.4

Compute the sum of the geometrical series $S_N = \sum_{n=1}^{N} \left(\frac{1}{2}\right)^n$.

Solution: Edit and execute the following *script M-file*:

```
for N=1:20
  n=N:-1:1;
  fn=(1/2).^n;
  Sn(N)=sum(fn);
end
NN=1:20;
plot(NN,Sn,'o')
```

You will observe that this partial sum converges to 1.

NOTE The above summation was performed backwards because this scheme will ensure a more accurate result as it keeps all the significant digits of the smallest term of the sum. □

In-Class Exercises

Compute the following infinite sums, accurate to one part in 10,000:

Pb. 4.16 $\displaystyle\sum_{k=1}^{\infty} \frac{1}{(2k-1)2^{2k-1}}$

Pb. 4.17 $\displaystyle\sum_{k=1}^{\infty} \frac{\sin(2k-1)}{(2k-1)^2}$

Pb. 4.18 $\displaystyle\sum_{k=1}^{\infty} \frac{\cos(k)}{k^4}$

Pb. 4.19 $\sum_{k=1}^{\infty} \frac{\sin(k/2)}{k^3}$

Pb. 4.20 $\sum_{k=1}^{\infty} \frac{1}{2^k} \sin(k)$

4.4 Numerical Integration

The algorithm for integration discussed in this section is the second simplest available (the trapezoid rule being the simplest, beyond the trivial, is given at the end of this section as a problem). It has been generalized to become more accurate and efficient through other approximations, including Simpson's rule, the Newton–Cotes rule, the Gaussian–Laguerre rule, etc. Simpson's rule is derived in Section 4.6, while other advanced techniques are left to more advanced numerical methods courses.

Here, we perform numerical integration through the means of a Rieman sum: we subdivide the interval of integration into many subintervals. Then we take the area of each strip to be the value of the function at the midpoint of the subinterval multiplied by the length of the subinterval, and we add the strip areas to obtain the value of the integral. This technique is referred to as the midpoint rule.

We can justify the above algorithm by recalling the mean value theorem of calculus, which states that

$$\int_a^b f(x)dx = (b-a)f(c) \tag{4.4}$$

where $c \in [a, b]$. Thus, if we divide the interval of integration into narrow subintervals, then the total integral can be written as the sum of the integrals over the subintervals, and we approximate the location of c in a particular subinterval by the midpoint between its boundaries.

Example 4.5

Use the above algorithm to compute the value of the definite integral of the function $\sin(x)$ from 0 to π.

Solution: Edit and execute the following program:

```
dx=pi/200;
x=0:dx:pi-dx;
xshift=x+dx/2;
yshift=sin(xshift);
Int=dx*sum(yshift)
```

You get for the above integral a result that is within 1/1000 error from the analytical result. □

In-Class Exercises

Find numerically, to a 1/10,000 accuracy, the values of the following definite integrals:

Pb. 4.21 $\int_0^{\infty} \frac{1}{x^2+1} dx$

Pb. 4.22 $\int_0^{\infty} \exp(-x^2)\cos(2x)dx$

Pb. 4.23 $\int_0^{\pi/2} \sin^6(x)\cos^7(x)dx$

Pb. 4.24 $\int_0^{\pi} \frac{2}{1+\cos^2(x)} dx$

Example 4.6

Plot the value of the indefinite integral $\int_0^x f(x)dx$ as a function of x, where $f(x)$ is the function $\sin(x)$ over the interval $[0, \pi]$.

Solution: We solve this problem for the general function $f(x)$ by noting that

$$\int_0^x f(x)dx \approx \int_0^{x-\Delta x} f(x)dx + f(x - \Delta x + \Delta x/2)\Delta x \tag{4.5}$$

where we are dividing the x-interval into subintervals and discretizing x to correspond to the coordinates of the boundaries of these subintervals. An array $\{x_k\}$ represents these discrete points, and the above equation is then reduced to a difference equation:

$$\text{Integral}(x_k) = \text{Integral}(x_{k-1}) + f(\text{Shifted}(x_{k-1}))\Delta x \tag{4.6}$$

where

$$\text{Shifted}(x_{k-1}) = x_{k-1} + \Delta x/2 \tag{4.7}$$

and the initial condition is $\text{Integral}(x_1) = 0$.

The above algorithm can then be programmed, for the above specific function, as follows:

```
a=0;
b=pi;
dx=0.001;
x=a:dx:b-dx;
N=length(x);
xshift=x+dx/2;
yshift=sin(xshift);
Int=zeros(1,N+1);
Int(1)=0;
  for k=2:N+1
    Int(k)=Int(k-1)+yshift(k-1)*dx;
  end
plot([x b],Int)
```

It may be useful to remind the reader, at this point, that the algorithm in Example 4.6 can be generalized to any arbitrary function. However, it should be noted that the key to the numerical calculation accuracy is a good choice for the increment **dx**. A very rough prescription for the estimation of this quantity, for an oscillating function, can be obtained as follows:

1. Plot the function inside the integral (i.e., the integrand) over the desired interval domain.
2. Verify that the function does not blow-out (i.e., goes to infinity) anywhere inside this interval.
3. Choose **dx** conservatively, such that at least 30 subintervals are included in any period of oscillation of the function (see Section 6.8 for more details). □

In-Class Exercises

Plot the following indefinite integrals as function of x over the indicated interval:

Pb. 4.25 $\int_0^x \left(\frac{\cos(x)}{\sqrt{1+\sin(x)}} \right) dx \qquad 0 \le x \le \pi/2$

Pb. 4.26 $\int_1^x \frac{(1+x^{2/3})^6}{x^{1/3}} dx \qquad 1 \le x \le 8$

Pb. 4.27 $\int_0^x \left[\frac{(x+2)}{(x^2+2x+4)^2} \right] dx \qquad 0 \le x \le 1$

Pb. 4.28 $\int_0^x x^2 \sin(x^3)dx \qquad 0 \le x \le \pi/2$

Pb. 4.29 $\int_0^x \sqrt{\tan(x)} \sec^2(x)dx \quad 0 \le x \le \pi/4$

Homework Problem

Pb. 4.30 Another simpler algorithm than the midpoint rule for evaluating a definite integral is the Trapezoid rule: the area of the slice is approximated by the area of the trapezoid with vertices having the following coordinates: $(x(k), 0)$; $(x(k+1), 0)$; $(x(k+1), y(k+1))$; $(x(k), y(k))$; giving for this trapezoid area the value:

$$\frac{1}{2}[x(k+1) - x(k)][y(k+1) + y(k)] = \frac{\Delta x}{2}[y(k+1) + y(k)]$$

thus leading to the following iterative expression for the Trapezoid integrator:

$$I_T(k+1) = I_T(k) + \frac{\Delta x}{2}[y(k+1) + y(k)]$$

The initial condition is $I_T(1) = 0$.

a. Evaluate the integrals of Pb. 4.25 through Pb. 4.29 using the Trapezoid rule.
b. Compare for the same values of Δx, the accuracy of the Trapezoid rule with that of the midpoint rule.
c. Give a geometrical interpretation for the difference in accuracy obtained using the two integration schemes.

NOTE MATLAB has a built-in command for evaluating the integral by the Trapezoid rule. If the sequence of the sampling points and of the function values are given, `trapz(x,y)` gives the desired integral.

4.5 A Better Numerical Differentiator

In Section 4.2, for the numerical differentiator, we used the simple expression

$$d(k) = \frac{1}{\Delta x}(y(k) - y(k-1)) \tag{4.8}$$

Our goal in this section is to find a more accurate expression for the differentiator. We shall use the difference equation for the Trapezoid rule to derive this improved differentiator, which we shall denote by $D(k)$.

The derivation of the difference equation for $D(k)$ hinges on the basic observation that differentiating the integral of a function gives back the original function. We say that the numerical differentiator is the inverse of the numerical integrator. We shall use the convolution-summation representation of the solution of a difference equation to find the iterative expression for $D(k)$.

Denoting the weighting sequence representations of the identity operation, the numerical integrator, and the numerical differentiator by $\{w\}$, $\{w_1\}$, and $\{w_2\}$, respectively, and using the notation and results of Section 2.5, we have for the identity operation the following weights:

$$w(0) = 1 \tag{4.9a}$$

$$w(i) = 0 \quad \text{for } i = 1, 2, 3, \ldots \tag{4.9b}$$

The Trapezoid numerical integrator, as given in Pb. 4.25, is a first-order system with the following parameters:

$$b_0^{(1)} = \frac{\Delta x}{2} \tag{4.10a}$$

$$b_1^{(1)} = \frac{\Delta x}{2} \tag{4.10b}$$

$$a_1^{(1)} = -1 \tag{4.10c}$$

giving for its weight sequence, as per Example 2.4, the values:

$$w_1(0) = \frac{\Delta x}{2} \tag{4.11a}$$

$$w_1(i) = \Delta x \quad \text{for } i = 1, 2, 3, \ldots \tag{4.11b}$$

The improved numerical differentiator's weight sequence can now be directly obtained by noting, as noted above, that if we successively cascade integration with differentiation, we are back to the original function. Using the results of Pb. 2.18, we can write

$$w(k) = \sum_{i=0}^{k} w_2(i) w_1(k-i) \tag{4.12}$$

Combining the above values for $w(k)$ and $w_1(k)$, we can deduce the following equalities:

$$w(0) = 1 = \frac{\Delta x}{2} w_2(0) \tag{4.13a}$$

$$w(1) = 0 = \Delta x \left[\frac{1}{2} w_2(1) + w_2(0) \right] \tag{4.13b}$$

$$w(2) = 0 = \Delta x \left[\frac{1}{2} w_2(2) + w_2(1) + w_2(0) \right] \tag{4.13c}$$

etc ....

from which we can directly deduce the following expressions for the weighting sequence $\{w_2\}$:

$$w_2(0) = \frac{2}{\Delta x} \tag{4.14a}$$

$$w_2(i) = \frac{4}{\Delta x}(-1)^i \quad \text{for } i = 1, 2, 3, \ldots \tag{4.14b}$$

From these weights we can compute, as per the results of Example 2.4, the parameters of the difference equation for the improved numerical differentiator, namely,

$$b_0^{(2)} = \frac{2}{\Delta x} \tag{4.15a}$$

$$b_1^{(2)} = -\frac{2}{\Delta x} \tag{4.15b}$$

$$a_1^{(2)} = 1 \tag{4.15c}$$

giving for $D(k)$ the following defining difference equation:

$$D(k) = \frac{2}{\Delta x}[y(k) - y(k-1)] - D(k-1) \tag{4.16}$$

In Pb. 4.32 and in other cases, you can verify that indeed this is an improved numerical differentiator. We shall, later in the chapter, use the above expression for $D(k)$ in the numerical solution of ordinary differential equations.

IN-CLASS EXERCISES

Pb. 4.31 Find the inverse system corresponding to the discrete system governed by the difference equation:

$$y(k) = u(k) - \frac{1}{2}u(k-1) + \frac{1}{3}y(k-1)$$

Pb. 4.32 Compute numerically the derivative of the function

$$y = x^3 + 2x^2 + 5 \quad \text{in the interval } 0 \leq x \leq 1$$

using the difference equations for both $d(k)$ and $D(k)$ for different values of Δx. Comparing the numerical results with the analytic results, compute the errors in both methods.

4.5.1 Application

In this application, we make use of the improved differentiator and corresponding integrator (Trapezoid rule) for modeling FM modulation and demodulation. The goal is to show that we retrieve back a good copy of the original message, using the first-order iterators, thus validating the use of these expressions in other communication engineering problems, where reliable numerical algorithms for differentiation and integration are needed in the simulation of different modulation-demodulation schemes.

As pointed out in Pb. 3.35, the FM modulated signal is given by

$$u_{FM}(t) = A_c \cos\left(2\pi f_c t + 2\pi k_f \int_{-\infty}^{t} m(\tau)d\tau\right) \tag{4.17}$$

The following *script M-file* details the steps in the FM modulation, if the signal in some normalized unit is given by the expression

$$m(t) = \text{sinc}(10\,t) \tag{4.18}$$

Assuming that in the same units, we have $f_c = k_f = 25$.

The second part of the program follows the demodulation process: the phase of the modulated signal is unwrapped, and the demodulated signal is obtained by differentiating this phase, while subtracting the carrier phase, which is linear in time.

```
fc=25;kf=25;tlowb=-1;tupb=1;
t=tlowb:0.0001:tupb;
p=length(t);
dt=(tupb-tlowb)/(p-1);
```

```
m=sinc(10*t);
subplot(2,2,1)
plot(t,m)
title('Message')

intm=zeros(1,p);
  for k=1:p-1
     intm(k+1)=intm(k)+0.5*dt*(m(k+1)+m(k));
  end
subplot(2,2,2)
plot(t,intm)
title('Modulation Phase')

uc=exp(j*(2*pi*fc*t+2*pi*kf*intm));
u=real(uc);
phase=unwrap(angle(uc))-2*pi*fc*t;
subplot(2,2,3)
plot(t,u)
axis([-0.15 0.15 -1 1])
title('Modulated Signal')

Dphase(1)=0;
  for k=1:p-1
    Dphase(k+1)=(2/dt)*(phase(k+1)-phase(k))- ...
                    Dphase(k);
  end
md=Dphase/(2*pi*kf);
subplot(2,2,4)
plot(t,md)
title('Reconstructed Message')
```

As can be observed by examining Figure 4.1, the results of the simulation are very good, giving confidence in the expressions of the iterators used.

4.6 A Better Numerical Integrator: Simpson's Rule

Prior to discussing Simpson's rule for integration, we shall derive, for a simple case, an important geometrical result.

THEOREM The area of a parabolic segment is equal to 2/3 of the area of the circumscribed parallelogram.

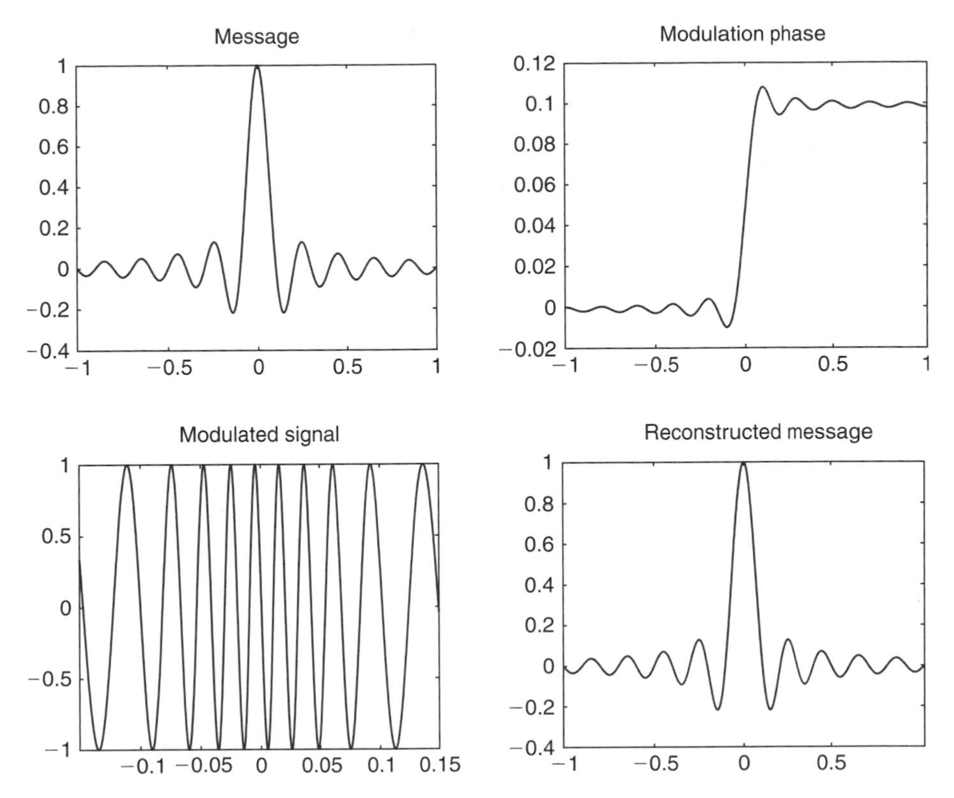

FIGURE 4.1
Simulation of the modulation and demodulation of an FM signal.

PROOF We prove this general theorem in a specialized case, for the purpose of making the derivation simple; however, the result is true for the most general case. Referring to Figure 4.2, we want to show that the area bounded by the x-axis and the parabola is equal to 2/3 the area of the rectangle ABCD. Now the details:

The parabola in Figure 4.2 is described by the equation

$$y = ax^2 + b \tag{4.19}$$

It intersects the x-axis at points $(-(-b/a)^{1/2}, 0)$ and $((-b/a)^{1/2}, 0)$, and the y-axis at point $(0, b)$. The area bounded by the x-axis and the parabola is then simply the following integral:

$$\int_{-(-b/a)^{1/2}}^{(-b/a)^{1/2}} (ax^2 + b)dx = \frac{4}{3}\frac{b^{3/2}}{(-a)^{1/2}} \tag{4.20}$$

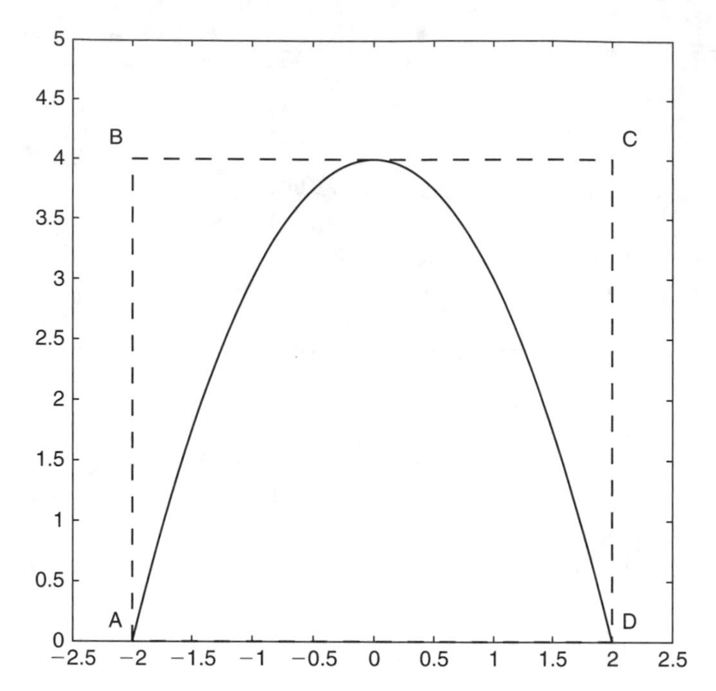

FIGURE 4.2
A parabolic segment and its circumscribed parallelogram.

The area of the rectangle ABCD is $b(2(-b/a)^{1/2}) = \frac{2b^{3/2}}{(-a)^{1/2}}$, which establishes the theorem.

Simpson's algorithm: We shall assume that the interval of integration is sampled at an odd number of points ($2N + 1$), so that we have an even number of intervals. The algorithm groups the intervals in pairs.

Referring to Figure 4.3, points A, H, and G are the first three points in the sampled x-interval. The assumption underlying Simpson's rule is that the curve passing through points B, D, and F, on the curve of the integrand, can have their locations approximated by a parabola. The line CDE is tangent to this parabola at the point D.

Under the above approximation, the value of the integral of the y-function between points A and G is then simply the sum of the area of the trapezoid ABFG plus 2/3 the area of the parallelogram BCEF, namely,

$$\text{Area of the first two slices} = \Delta x\,(y(1) + y(3)) + \frac{4\Delta x}{3}\left(y(2) - \left(\frac{y(1) + y(3)}{2}\right)\right)$$

$$= \frac{\Delta x}{3}(y(1) + 4y(2) + y(3)) \tag{4.21}$$

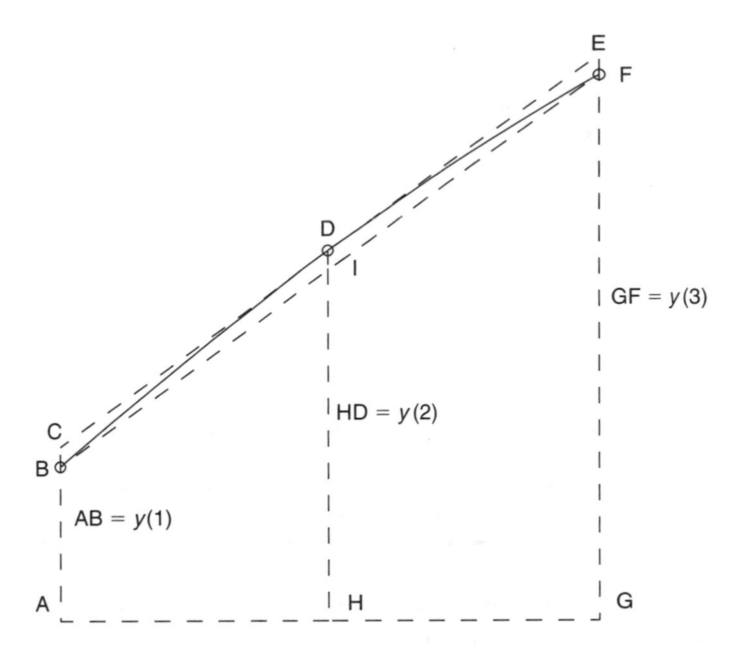

FIGURE 4.3
The first two slices in the Simpson's rule construct. AH = HG = Δx.

In a similar fashion, we can find the area of the third and fourth slices,

$$\text{Area of the third and fourth slices} = \frac{\Delta x}{3}(y(3) + 4y(4) + y(5)) \tag{4.22}$$

Continuing for each successive pair of slices, we obtain for the total integral, or total area of all slices, the expression

$$\text{Total area of all slices} = \frac{\Delta x}{3}\begin{pmatrix} y(1) + 4y(2) + 2y(3) + \cdots + 2y(2N-1) \\ + 4y(2N) + y(2N+1) \end{pmatrix} \tag{4.23}$$

that is, the weights are equal to 1 for the first and last elements, equal to 4 for even elements, and equal to 2 for odd elements.

Example 4.7

Using Simpson's rule, compute the integral of $\sin(x)$ over the interval $0 \le x \le \pi$.

Solution: Edit and execute the following *script M-file*:

```
a=0;b=pi;N=4;
x=linspace(a,b,2*N+1);
y=sin(x);
```

```
for k=1:2*N+1
      if k==1 | k==2*N+1
        w(k)=1;
      elseif rem(k,2)==0
        w(k)=4;
      else
        w(k)=2;
      end
end
Intsimp=((b-a)/(3*(length(x)-1)))*sum(y.*w)
```

Now compare the above answer with the one you obtain if you use the Trapezoid rule, by entering the command: **Inttrapz=trapz(x,y)**.

IN-CLASS EXERCISE

Pb. 4.33 In the above derivation of Simpson's method, we constructed the algorithm by determining the weights sequence. Reformulate this algorithm into an equivalent iterator format.

Homework Problems

In this chapter, we surveyed three numerical techniques for computing the integral of a function. We observed that the different methods lead to different levels of accuracy. In Section 6.9, we shall derive formulas for estimating the accuracy of the different methods discussed here. However, and as noted previously, more accurate techniques than those presented here exist for calculating integrals numerically; many of these are in the MATLAB library and are covered in numerical analysis courses. In particular, familiarize yourself, using the help folder, with the commands **quad** and **quadl**.

Pb. 4.34 The goal of this problem, using the **quadl** command, is to develop a *function M-file* for the Gaussian distribution function of probability theory.

The Gaussian probability density function, previously discussed in Chapter 1, is given by

$$f_X(x) = \frac{1}{(2\pi)^{1/2}\sigma_X} \exp\left[-\frac{(x-a_X)^2}{2\sigma_X^2}\right]$$

where $-\infty < a_X < \infty, 0 < \sigma_X$ are constants, and are respectively equal to the mean and the square root of the variance of x, respectively. *(cont'd.)*

Homework Problems *(cont'd.)*

The Gaussian probability distribution function is defined as

$$F_X(x) \equiv \int_{-\infty}^{x} f_X(\zeta)d\zeta$$

Through a change of variable (specify it!), the Gaussian probability distribution function can be written as a function of the normalized distribution function,

$$F_X(x) = F\left(\frac{x - a_X}{\sigma_X}\right)$$

where

$$F(x) = \frac{1}{\sqrt{2\pi}} \int_{-\infty}^{x} \exp\left(-\frac{\xi^2}{2}\right) d\xi$$

a. Develop the *function M-file* for the normal distribution function.
b. Show that for negative values of x, we have

$$F(-x) = 1 - F(x)$$

c. Plot the normalized distribution function for values of x in the interval $0 \leq x \leq 5$.

Pb. 4.35 The computation of the arc length of a curve can be reduced to a 1-D integration. Specifically, if the curve is described parametrically, then the arc length between the adjacent points $(x(t), y(t), z(t))$ and the point $(x(t + \Delta t), y(t + \Delta t), z(t + \Delta t))$ is given by

$$\Delta s = \sqrt{\left(\frac{dx}{dt}\right)^2 + \left(\frac{dy}{dt}\right)^2 + \left(\frac{dz}{dt}\right)^2} \Delta t$$

giving immediately for the arc length from t_0 to t_1, the expression

$$s = \int_{t_0}^{t_1} \sqrt{\left(\frac{dx}{dt}\right)^2 + \left(\frac{dy}{dt}\right)^2 + \left(\frac{dz}{dt}\right)^2} dt$$

(cont'd.)

Homework Problems *(cont'd.)*

a. Calculate the arc length of the curve described by $x = t^2$ and $y = t^3$ between the points $t = 0$ and $t = 3$.

b. Assuming that a 2-D curve is given in polar coordinates by $r = f(\theta)$, and then noting that

$$x = f(\theta)\cos(\theta) \quad \text{and} \quad y = f(\theta)\sin(\theta)$$

Use the above expression for the arc length (here the parameter is θ) to derive the formula for the arc length in polar coordinates to be

$$s = \int_{\theta_0}^{\theta_1} \sqrt{r^2 + \left(\frac{dr}{d\theta}\right)^2}\, d\theta$$

c. Use the result of (**b**) above to derive the length of the cardioid $r = a(1 + \cos(\theta))$ between the angles 0 and π.

Pb. 4.36 Double and triple integration over rectangular regions can be performed in MATLAB by calling respectively the **`dblquad`** and **`triplequad`** commands, the syntax is respectively:

```
dbl_integ = dblquad(fun,xmin,xmax,ymin,ymax)
triple_integ=triplequad(fun,xmin,xmax,ymin,ymax,
 zmin,zmax)
```

where **`fun`** is a function handle, and **`xmin, xmax, ymin, ymax, zmin, zmax`** are respectively the lower and upper bounds of the integration variables.

a. Find the volume bounded by the x–y plane, the $x = 0$ plane, the $x = 1$ plane, the $y = 0$ plane, the $y = 1$ plane, and the surface $z = 1 + x^2 + y^3$.

b. Find the volume of a hemisphere of radius $R = 2$.

(*Hint*: Integrand should be set equal to zero outside the proper region of integration.)

4.7 Numerical Solutions of Ordinary Differential Equations

Ordinary linear differential equations are of the form:

$$a_n(t)\frac{d^n y}{dt^n} + a_{n-1}(t)\frac{d^{n-1} y}{dt^{n-1}} + \cdots + a_1(t)\frac{dy}{dt} + a_0(t)y = u(t) \tag{4.24}$$

The a's are called the coefficients and $u(t)$ is called the source (or input) term.

Ordinary differential equations show up in many problems of electrical engineering, particularly in circuit problems where, depending on the circuit element, the potential across it may depend on the deposited charge, the current (which is the time derivative of the charge), or the derivative of the current (i.e., the second time derivative of the charge); that is, in the same equation, we may have a function and its first- and second-order derivatives. To focus this discussion, let us start by writing the potential difference across the passive elements of circuit theory. Specifically, the voltage drops across a resistor, capacitor, or inductor are given as follows:

1. The voltage across a resistor is given, using Ohm's law, by

$$V_R(t) = R\, I(t) \tag{4.25}$$

   where R is the resistance measured in Ohms and, I the current.
2. The voltage across a capacitor is proportional to the magnitude of the charge that accumulates on either plate, that is

$$V_C(t) = \frac{Q(t)}{C} = \frac{\int I(t)\,dt}{C} \tag{4.26}$$

   The second equality reflects the relation of the current to the charge. C is the capacitance and, as previously pointed out, is measured in Farads.
3. The voltage across an inductor can be deduced from Lenz's law, which stipulates that the voltage across an inductor is proportional to the time derivative of the current going through it:

$$V_L(t) = L\frac{dI(t)}{dt} \tag{4.27}$$

   where L is the inductance and is measured in Henrys.

From these expressions for the voltage drop across each of the passive elements in a circuit, and using the Kirchoff voltage law, it is then easy to write down the differential equations describing, for example, a series *RC* or an *RLC* circuit.

RC Circuit: Referring to the *RC* circuit diagram in Figure 4.4, the differential equation describing the voltage across the capacitor is given by

$$RC\frac{dV_C}{dt} + V_C = V_s(t) \tag{4.28}$$

RLC Circuit: Referring to the *RLC* circuit in Figure 4.5, the voltage across the capacitor is described by the ODE:

$$LC\frac{d^2V_C}{dt^2} + RC\frac{dV_C}{dt} + V_C = V_s(t) \tag{4.29}$$

Numerically solving these and other types of ODEs will be the subject of the remainder of this section. In Section 4.7.1, we consider first-order iterators to represent the different-order derivatives, apply this algorithm to solve the above types of problems, and conclude by pointing out some of the limitations of this algorithm. In Section 4.7.2, we discuss higher-order iterators,

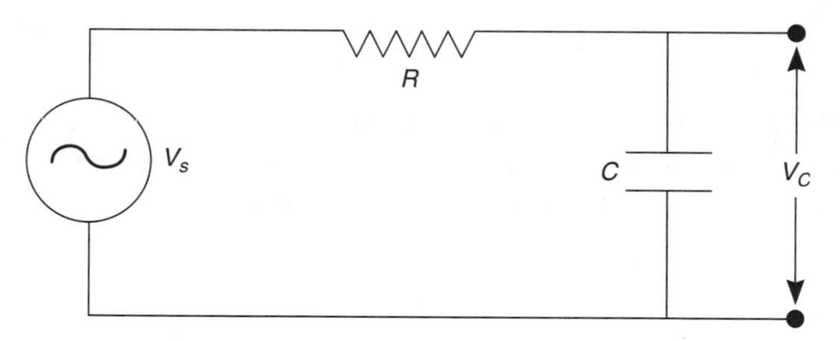

FIGURE 4.4
RC circuit with an ac source.

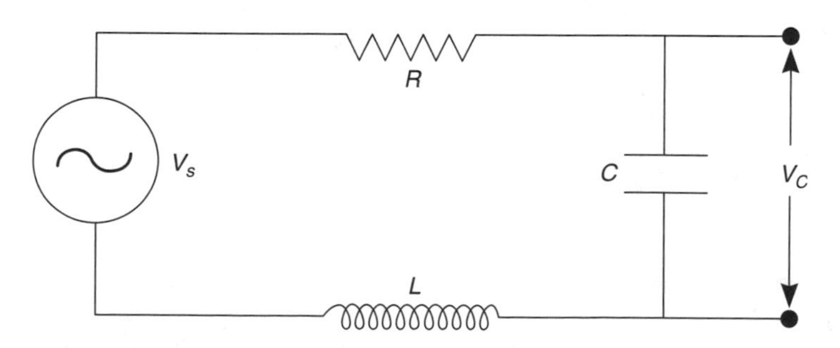

FIGURE 4.5
RLC circuit with an ac source.

particularly the Runge–Kutta technique. In Section 4.7.3, we familiarize ourselves with the use of standard MATLAB solvers for ODEs.

4.7.1 First-Order Iterator

In Section 4.5, we found an improved expression for the numerical differentiator, $D(k)$:

$$D(k) = \frac{2}{\Delta t}[y(k) - y(k-1)] - D(k-1) \tag{4.16}$$

which functionally corresponded to the inverse of the Trapezoid rule for integration. (Note that the independent variable here is t, and not x.)

Applying this first-order differentiator in cascade leads to an expression for the second-order differentiator, namely,

$$\begin{aligned} D2(k) &= \frac{2}{\Delta t}[D(k) - D(k-1)] - D2(k-1) \\ &= \frac{4}{(\Delta t)^2}[y(k) - y(k-1)] - \frac{4}{\Delta t}D(k-1) - D2(k-1) \end{aligned} \tag{4.30}$$

EXAMPLE 4.8

Find the first-order iterative scheme to solve the first-order differential equation given by

$$a(t)\frac{dy}{dt} + b(t)y = u(t) \tag{4.31}$$

with the initial condition $y(t_{in})$ specified.

Solution: Substituting Eq. (4.16) for the numerical differentiator in the differential equation, we deduce the following first-order difference equation for $y(k)$:

$$y(k) = \left[\frac{2a(k)}{\Delta t} + b(k)\right]^{-1}\left[\frac{2a(k)y(k-1)}{\Delta t} + a(k)D(k-1) + u(k)\right] \tag{4.32}$$

to which we should add, in the numerical subroutine, the expression for the first-order differentiator $D(k)$ as given by Eq. (4.16). The initial condition for the function at the origin of time, specify the first elements of the y and D arrays:

$$y(1) = y(t = t_{in})$$

$$D(1) = (1/a(1))[u(1) - b(1)y(1)] \quad \square$$

4.7.1.1 Application

To illustrate the use of the above algorithm, let us solve, over the interval $0 \leq t \leq 6$, for the potential across the capacitor in an RC circuit with an ac source; that is,

$$a\frac{dy}{dt} + y = \sin(2\pi t) \tag{4.33}$$

where $a = RC$ and $y(t = 0) = 0$.

Solution: Edit and execute the following *script M-file*, for $a = 1/(2\pi)$:

```
tin=0;
tfin=6;
t=linspace(tin,tfin,3000);
N=length(t);
y=zeros(1,N);
dt=(tfin-tin)/(N-1);
u=sin(2*pi*t);
a=(1/(2*pi))*ones(1,N);
b=ones(1,N);
y(1)=0;
D(1)=(1/a(1))*(u(1)-b(1)*y(1));
      for k=2:N
            y(k)=((2*a(k)/dt+b(k))^(-1))*...
                    (2*a(k)*y(k-1)/dt+a(k).*D(k-1)+u(k));
            D(k)=(2/dt)*(y(k)-y(k-1))-D(k-1);
      end
plot(t,y,t,u,'--')
```

IN-CLASS EXERCISE

Pb. 4.37 Plot the amplitude of y, and its dephasing from u, as a function of a for large t.

EXAMPLE 4.9

Find the first-order iterative scheme to solve the second-order differential equation given by

$$a(t)\frac{d^2y}{dt^2} + b(t)\frac{dy}{dt} + c(t)y = u(t) \tag{4.34}$$

with initial conditions $y(t = 0)$ and $\left.\frac{dy}{dt}\right|_{t=0}$ given.

Solution: Substituting the above first-order expression of the iterators for the first- and second-order numerical differentiators [respectively Eq. (4.16) and Eq. (4.30), into Eq. (4.34)], we deduce the following iterative equation for $y(k)$:

$$y(k) = \left\{4\frac{a(k)}{(\Delta t)^2} + 2\frac{b(k)}{\Delta t} + c(k)\right\}^{-1} \times$$
$$\left\{y(k-1)\left[4\frac{a(k)}{(\Delta t)^2} + 2\frac{b(k)}{\Delta t}\right] + D(k-1)\left[4\frac{a(k)}{\Delta t} + b(k)\right] + a(k)D2(k-1) + u(k)\right\} \tag{4.35}$$

This difference equation will be supplemented in the ODE numerical solver routine with the iterative equations for $D(k)$ and $D2(k)$, as given respectively by Eq. (4.16) and Eq. (4.30), and with the initial conditions for the function and its derivative. The first elements for the y, D, and $D2$ arrays are given by

$$y(1) = y(t = 0)$$
$$D(1) = \left.\frac{dy}{dt}\right|_{t=0}$$
$$D2(1) = (1/a(1))(-b(1)D(1) - c(1)y(1) + u(1)) \quad \square$$

Application 1

To illustrate the use of the first-order iterator algorithm in solving a second-order ordinary differential equation, let us find, over the interval $0 \leqslant t \leqslant 16\pi$, the voltage across the capacitance in an *RLC* circuit, with an ac voltage source. This reduces to solving the following ODE:

$$a\frac{d^2y}{dt^2} + b\frac{dy}{dt} + cy = \sin(\omega t) \tag{4.36}$$

where $a = LC$, $b = RC$, $c = 1$. Choose in some normalized units, $a = 1$, $b = 3$, $\omega = 1$, and let $y(t = 0) = y'(t = 0) = 0$.

Solution: Edit and execute the following *script M-file*:

```
tin=0;
tfin=16*pi;
t=linspace(tin,tfin,2000);
a=1;
b=3;
c=1;
w=1;
N=length(t);
y=zeros(1,N);
dt=(tfin-tin)/(N-1);
u=sin(w*t);
y(1)=0;
D(1)=0;
D2(1)=(1/a)*(-b*D(1)-c*y(1)+u(1));
     for k=2:N
          y(k)=((4*a/dt^2+2*b/dt+c)^(-1))*...
               (y(k-1)*(4*a/dt^2+2*b/dt)+D(k-1)*...
               (4*a/dt+b)+a*D2(k-1)+u(k));
          D(k)=(2/dt)*(y(k)-y(k-1))-D(k-1);
          D2(k)=(4/dt^2)*(y(k)-y(k-1))-(4/dt)*D(k-1)-...
               D2(k1-1);
     end
plot(t,y,t,u,'--')
```

The dashed curve is the temporal profile of the source term.

IN-CLASS EXERCISE

Pb. 4.38 Plot the amplitude of y and its dephasing from u as function of a for large t, for $0.1 < a < 5$.

Application 2

Solve, over the interval $0 < t < 1$, the following second-order differential equation:

$$(1-t^2)\frac{d^2y}{dt^2} - 2t\frac{dy}{dt} + 20y = 0 \tag{4.37}$$

with the initial conditions: $y(t = 0) = 3/8$ and $y'(t = 0) = 0$.

Then, compare your numerical result with the analytical solution to this problem:

$$y = \frac{1}{8}(35t^4 - 30t^2 + 3) \tag{4.38}$$

Solution: Edit and execute the following *script M-file*:

```
tin=0;
tfin=1;
t=linspace(tin,tfin,2000);
N=length(t);
a=1-t.^2;
b=-2*t;
c=20*ones(1,N);
y=zeros(1,N);
D=zeros(1,N);
dt=(tfin-tin)/(N-1);
u=zeros(1,N);
y(1)=3/8;
D(1)=0;
D2(1)=(1/a(1))*(-b(1)*D(1)-c(1)*y(1)+u(1));
      for k=2:N
        y(k)=((4*a(k)/dt^2+2*b(k)/dt+c(k))^(-1))*...
               (y(k-1)*(4*a(k)/dt^2+2*b(k)/dt)+...
               D(k-1)*(4*a(k)/dt+b(k))+...
               a(k)*D2(k-1)+u(k));
        D(k)=(2/dt)*(y(k)-y(k-1))-D(k-1);
        D2(k)=(4/dt^2)*(y(k)-y(k-1))-...
                (4/dt)*D(k-1)-D2(k-1);
      end
yanal=(35*t.^4-30*t.^2+3)/8;
plot(t,y,t,yanal,'--')
```

As you will observe upon running this program, the numerical solution and the analytical solution agree very well.

NOTE The above ODE is that of the Legendre polynomial of order $l = 4$, encountered earlier in Chapter 2, in Pb. 2.25.

$$(1-t^2)\frac{d^2P_l}{dt^2} - 2t\frac{dP_l}{dt} + l(l+1)P_l = 0 \tag{4.39}$$

where

$$P_l(-t) = (-1)^l P_l(t) \tag{4.40}$$

Homework Problem

Pb. 4.39 Solve, over the interval $0 \leq t \leq 20$, the following first-order differential equation for $a = 2$ and $a = 0.5$:

$$a\frac{dy}{dt} + y = 1$$

where $y(0) = 0$. (Physically, this would correspond to the charging of a capacitor from a dc source connected suddenly to the battery at time zero. Here, y is the voltage across the capacitor, and $a = RC$.)

NOTE The analytic solution to this problem is $y = 1 - \exp(-t/a)$.

4.7.2 Higher-Order Iterators: The Runge–Kutta Method*

In this subsection, we want to explore the possibility that if we sampled the function n times per step, we will obtain a more accurate solution to the ODE than that obtained from the first-order iterator for the same value of Δt.

To focus the discussion, consider the ODE:

$$y'(t) = f(t, y(t)) \tag{4.41}$$

Higher-order ODEs can be reduced, as will be shown at the end of the subsection, to a system of equations having the same functional form as Eq. (4.41). The derivation of a technique using higher-order iterators will be shown below in detail for two evaluations per step. Higher-order recipes can be found in most books on numerical methods for ODE.

The key to the Runge–Kutta method is to properly arrange each of the evaluations in a particular step to depend on the previous evaluations in the same step.

In the second-order model

if
$$k_1 = f(t(n), y(t(n)))(\Delta t) \tag{4.42}$$

then
$$k_2 = f(t(n) + \alpha\Delta t, y(t(n)) + \beta k_1)(\Delta t) \tag{4.43}$$

and
$$y(t(n+1)) = y(t(n)) + ak_1 + bk_2 \tag{4.44}$$

where a, b, α, and β are unknown parameters to be determined. They should be chosen such that Eq. (4.44) is correct to order $(\Delta t)^3$.

* The asterisk indicates more advanced material that may be skipped in a first reading.

To find a, b, α, and β, let us compute $y(t(n+1))$ in two different ways. First, Taylor expanding the function $y(t(n+1))$ to order $(\Delta t)^2$, we obtain

$$y(t(n+1)) = y(t(n)) + \frac{dy(t(n))}{dt}(\Delta t) + \frac{d^2y(t(n))}{dt^2}\frac{(\Delta t)^2}{2} \tag{4.45}$$

Recalling Eq. (4.41) and the total derivative expression of a function in two variables as function of the partial derivatives, we have

$$\frac{dy(t(n))}{dt} = f(t(n), y(t(n))) \tag{4.46}$$

$$\begin{aligned}\frac{d^2y(t(n))}{dt^2} &= \frac{d}{dt}\left(\frac{dy(t(n))}{dt}\right) \\ &= \frac{\partial f(t(n), y(t(n)))}{\partial t} + \frac{\partial f(t(n), y(t(n)))}{\partial y} f(t(n), y(t(n)))\end{aligned} \tag{4.47}$$

Combining Eq. (4.45) to Eq. (4.47), it follows that to second order in (Δt):

$$\begin{aligned}y(t(n+1)) &= y(t(n)) + f(t(n), y(t(n)))(\Delta t) + \\ &+ \left[\frac{\partial f(t(n), y(t(n)))}{\partial t} + \frac{\partial f(t(n), y(t(n)))}{\partial y} f(t(n), y(t(n)))\right]\frac{(\Delta t)^2}{2}\end{aligned} \tag{4.48}$$

Next, let us Taylor expand k_2 to second order in (Δt). This results in

$$\begin{aligned}k_2 &= f(t(n) + \alpha\Delta t, y(t(n)) + \beta k_1)(\Delta t) \\ &= \left[f(t(n), y(t(n))) + \alpha(\Delta t)\frac{\partial f(t(n), y(t(n)))}{\partial t} + (\beta k_1)\frac{\partial f(t(n), y(t(n)))}{\partial y}\right](\Delta t)\end{aligned} \tag{4.49}$$

Combining Eq. (4.42), Eq. (4.44), and Eq. (4.49), we get the other expression for $y(t(n+1))$, correct to second order in (Δt):

$$\begin{aligned}y(t(n+1)) &= y(t(n)) + (a+b)f(t(n), y(t(n)))(\Delta t) + \\ &+ \alpha b\frac{\partial f(t(n), y(t(n)))}{\partial t}(\Delta t)^2 + b\beta\frac{\partial f(t(n), y(t(n)))}{\partial y} f(t(n), y(t(n)))(\Delta t)^2\end{aligned} \tag{4.50}$$

Now, comparing Eq. (4.48) and Eq. (4.50), we obtain the following equalities:

$$a + b = 1; \quad \alpha b = 1/2; \quad b\beta = 1 \tag{4.51}$$

We have three equations in four unknowns; the usual convention is to fix $a = 1/2$, giving for the other quantities:

$$b = 1/2; \quad \alpha = 1; \quad \beta = 1 \tag{4.52}$$

finally leading to the following expressions for the second-order iterator and its parameters:

$$k_1 = f(t(n), y(t(n)))(\Delta t) \tag{4.53a}$$

$$k_2 = f(t(n) + \Delta t, y(t(n)) + k_1)(\Delta t) \tag{4.53b}$$

$$y(t(n+1) = y(t(n)) + \frac{k_1 + k_2}{2} \tag{4.53c}$$

Next, we give, without proof, the famous fourth-order iterator Runge–Kutta expression, one of the most widely used algorithms for solving ODEs in the different fields of science and engineering:

$$k_1 = f(t(n), y(n))(\Delta t) \tag{4.54a}$$

$$k_2 = f(t(n) + \Delta t/2, y(t(n)) + k_1/2)(\Delta t) \tag{4.54b}$$

$$k_3 = f(t(n) + \Delta t/2, y(t(n)) + k_2/2)(\Delta t) \tag{4.54c}$$

$$k_4 = f(t(n) + \Delta t, y(t(n)) + k_3)(\Delta t) \tag{4.54d}$$

$$y(t(n+1) = y(t(n)) + \frac{k_1 + 2k_2 + 2k_3 + k_4}{6} \tag{4.54e}$$

The last point that we need to address before leaving this subsection is what to do in case we have an ODE with higher derivatives than the first. The answer is that we reduce the nth-order ODE to a system of n first-order ODEs.

EXAMPLE 4.10

Reduce the following second-order differential equation into two first-order differential equations:

$$ay'' + by' + cy = \sin(t) \tag{4.55}$$

with the initial conditions $y(t = 0) = 0$ and $y'(t = 0) = 0$

(where, functions with the prime and double primes refer, respectively, to the first and second derivatives of this function).

Solution: Introduce the 2-D array z, and define

$$z(1) = y \tag{4.56a}$$

$$z(2) = y' \tag{4.56b}$$

The system of first-order equations now reads

$$z'(1) = z(2) \tag{4.57a}$$

$$z'(2) = (1/a)(\sin(t) - bz(2) - cz(1)) \tag{4.57b}$$

□

Example 4.11

Using the fourth-order Runge–Kutta iterator, numerically solve the same problem as in Application 1 following Example 4.9.

Solution: Edit and save the following *function M-files*:

```
function zp=zprime(t,z)
a=1; b=3; c=1;
zp(1,1)=z(2,1);
zp(2,1)=(1/a)*(sin(t)-b*z(2,1)-c*z(1,1));
```

The above file specifies the system of ODE that we are trying to solve. □

Next, in another *function M-file*, we edit and save the fourth-order Runge–Kutta algorithm, specifically:

```
function zn=prk4(t,z,dt)
k1=dt*zprime(t,z);
k2=dt*zprime(t+dt/2,z+k1/2);
k3=dt*zprime(t+dt/2,z+k2/2);
k4=dt*zprime(t+dt,z+k3);
zn=z+(k1+2*k2+2*k3+k4)/6;
```

Finally, edit and execute the following *script M-file*:

```
yinit=0;
ypinit=0;
z=[yinit;ypinit];
tinit=0;
tfin=16*pi;
```

```
N=1001;
t=linspace(tinit,tfin,N);
dt=(tfin-tinit)/(N-1);

      for k=1:N-1
        z(:,k+1)=prk4(t(k),z(:,k),dt);
      end

plot(t,z(1,:),t,sin(t),'--')
```

In the above plot, we are comparing the temporal profiles of the voltage difference across the capacitor with that of the source voltage.

4.7.3 MATLAB ODE Solvers

MATLAB has many ODE solvers, **ODE23** and **ODE45** being most commonly used. **ODE23** is based on a pair of second- and third-order Runge–Kutta methods running simultaneously to solve the ODE. The program automatically corrects for the step size if the answers from the two methods have a discrepancy at any stage of the calculation that will lead to a larger error than the allowed tolerance.

To use this solver, we start by creating a *function M-file* that includes the system of equations under consideration. This function is then called from the command window with the **ODE23** or **ODE45** command.

Example 4.12

Using the MATLAB ODE solver, find the voltage across the capacitor in the *RLC* circuit of Example 4.11, and compare it to the source potential time profile.

Solution: Edit and save the following *function M-file*:

```
function zdot=Example_412(t,z)
a=1;
b=3;
c=1;
zdot(1)=z(2);
zdot(2)=(1/a)*(sin(t)-b*z(2)-c*z(1));
zdot=[zdot(1);zdot(2)];
```

Next, edit and execute the following *script M-file*:

```
tspan=[0 16*pi];
zin=[0;0];
[t,z]=ode23(@Example_412,tspan,zin);
plot(t,z(:,1),t,sin(t), '--')
xlabel('Normalized Time')
```

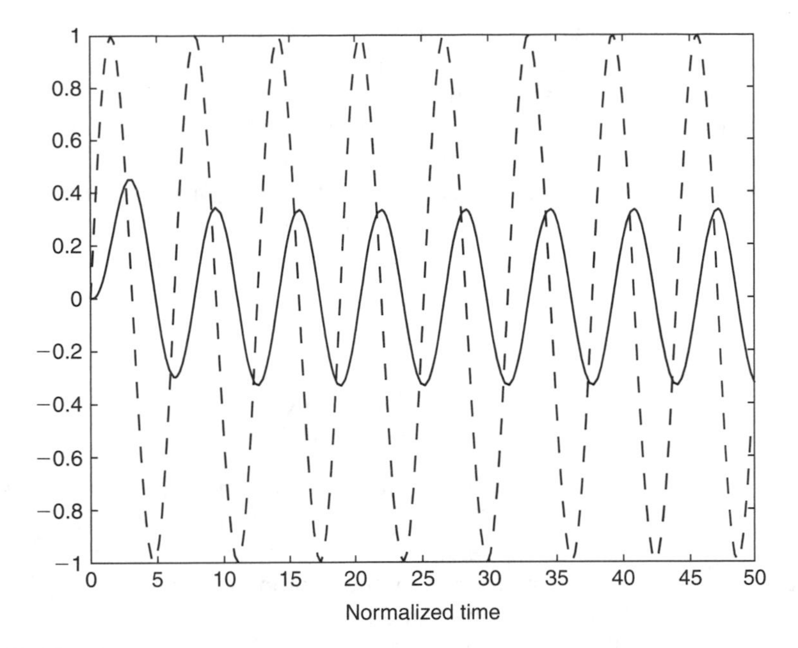

FIGURE 4.6
The potential differences across the source (dashed line) and the capacitor (solid line) in an *RLC* circuit with an ac source [$LC = 1$, $RC = 3$, and $V_s = \sin(t)$].

The results are plotted in Figure 4.6. Note the phase shift between the two potential differences. □

EXAMPLE 4.13

Using the MATLAB ODE solver, solve the problem of relaxation oscillations in lasers.

Solution: Because many readers may not be familiar with the statement of the problem, let us first introduce the physical background to the problem.

A simple gas laser consists of two parallel mirrors sandwiching a tube with a gas, selected for having two excitation levels separated in energy by an amount equal to the energy of the photon quantum that we are attempting to have the laser system produce. In a laser (light amplification by stimulated emission radiation), a pumping mechanism takes the atom to the upper excited level. However, the atom does not stay in this level; it decays to lower levels, including the lower excited level. The dynamics is governed by two physical effects: (1) All excited states of atoms have finite lifetime; and (2) stimulated emission, a quantum mechanical phenomenon, associated with the statistics of the photons (photons are bosons), which predicts that in the presence of an electromagnetic field having a frequency close to that of the frequency of the photon emitted in the transition between the upper excited and lower excited state, the atom emission rate is enhanced and this enhancement is larger the more photons that are present in its vicinity. The rate of change of the number of photons is equal to the rate generated from the decay of the atoms due to stimulated emission, minus the decay due to the finite lifetime of the photon in the resonating cavity. Putting all this together, one is led, in the simplest approximation, to write what are called the rate equations for the number of atoms in the excited state and for the photon numbers in the cavity. These coupled equations, in their simplest forms, are given by

$$\frac{dN}{dt} = P - \frac{N}{\tau_{decay}} - BnN \tag{4.58}$$

$$\frac{dn}{dt} = -\frac{n}{\tau_{cavity}} + BnN \tag{4.59}$$

where N is the normalized number of atoms in the atom's upper excited state, n the normalized number of photons present, P the pumping rate, τ_{decay} the atomic decay time from the upper excited state, due to all effects except that of stimulated emission, τ_{cavity} the lifetime of the photon in the resonant cavity, and B the Einstein coefficient for stimulated emission.

These nonlinear differential equations describe the dynamics of laser operation. Now, come back to relaxation oscillations in lasers, which is the problem at hand. Physically, this is an interplay between N and n. An increase in the photon number causes an increase in stimulated emission, which causes a decrease in the population of the higher excited level. This, in turn, causes a reduction in the photon gain, which tends to decrease the number of photons present, and in turn, decreases stimulated emission. This leads to the build-up of the higher excited state population, which increases the rate of change of photons, with the cycle resuming but such that at each new cycle the amplitude of the oscillations is dampened as compared with the cycle just before it, until finally the system reaches a steady state.

To compute the dynamics of the problem, we proceed into two steps. First, we generate the *function M-file* that contains the rate equations, and then proceed to solve these ODEs by calling the MATLAB ODE solver. We use typical numbers for gas lasers.

Specifically, the *function M-file* representing the laser rate equations is given by

```
function ydot=laser1(t,y)
p=30;                         %pumping rate
gamma=10^(-2);                %inverse natural lifetime
B=3;                          %stimulated emission
                              coefficient
c=30;                         %inverse lifetime of
                              photon in cavity
ydot(1)=p-gamma*y(1)-B*y(1)*y(2);
ydot(2)=-c*y(2)+B*y(1)*y(2);
ydot=[ydot(1);ydot(2)];
```

The *script M-file* to compute the laser dynamics and thus simulate the relaxation oscillations is

```
tspan=[0 3];
yin=[1 1];
[t,y]=ode23(@laser1,tspan,yin);

subplot(3,1,1)
plot(t,y(:,1))
xlabel('Normalized Time')
ylabel('N')

subplot(3,1,2)
plot(t,y(:,2))
xlabel('Normalized Time')
ylabel('n')

subplot(3,1,3)
plot(y(:,1),y(:,2))
xlabel('N')
ylabel('n')
```

As can be observed in Figure 4.7, the oscillations, as predicted, damp-out after a while and the dynamical variables reach a steady state. The phase diagram, shown in the bottom panel, is an alternate method to show how the population of the atomic higher excited state and the photon number density reach the steady state.

Question: Compute analytically from Eq. (4.58) and Eq. (4.59), the steady-state values for the higher excited state population and for the photon number, and compare with the numerically obtained asymptotic values.

In-Class Exercise

Pb. 4.40 By changing the values of the appropriate parameters in the above programs, find separately the effects of increasing or decreasing the value of τ_{cavity}, and the effect of the pumping rate on the magnitude and the persistence of the oscillation.

Example 4.14

Using the rate equations developed in Example 4.13, simulate the Q-switching of a laser.

Solution: First, an explanation of the statement of the problem is provided. In Example 4.13, we showed how, following an initial transient period

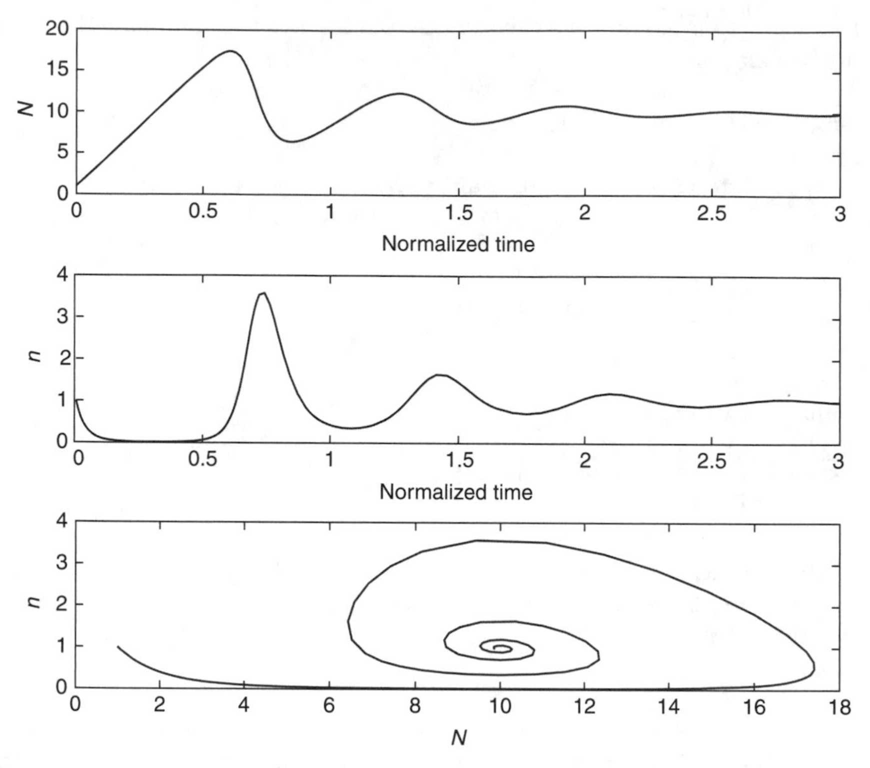

FIGURE 4.7
The dynamics of a laser in the relaxation oscillations regime. Top panel: plot of the higher excited level atoms population as a function of the normalized time; middle panel: plot of the number of photons as a function of the normalized time; bottom panel: phase diagram of the photons number vs. the higher excited level atoms population.

whereby one observes relaxation oscillations, a laser, in the presence of steady pumping, reaches steady-state operation after a while. This is referred to as continuous wave (cw) operation of the laser. In this example, we shall explore the other mode of laser operation, the so-called pulsed mode. In this regime, the experimentalist, through a temporary modification in the absorption properties of the laser resonator, prevents the laser from oscillating, thus leading the higher excited state of the atom to keep building up its population to a very high level before it is allowed to emit any photons. Then, at the desired moment, the laser resonator is allowed back to a state where the optical losses of the resonator are small, thus triggering the excited atoms to dump their stored energy into a short burst of photons. It is this regime that we propose to study in this example.

The laser dynamics are, of course, still described by the rate equations [i.e., Eq. (4.58) and Eq. (4.59)]. What we need to modify from the previous problem are the initial conditions for the system of coupled ODE. At the origin of time [i.e., $t = 0$ or the triggering time], $N(0)$, the initial value of the population of the higher excited state of the atom, is in this instance (because of the induced

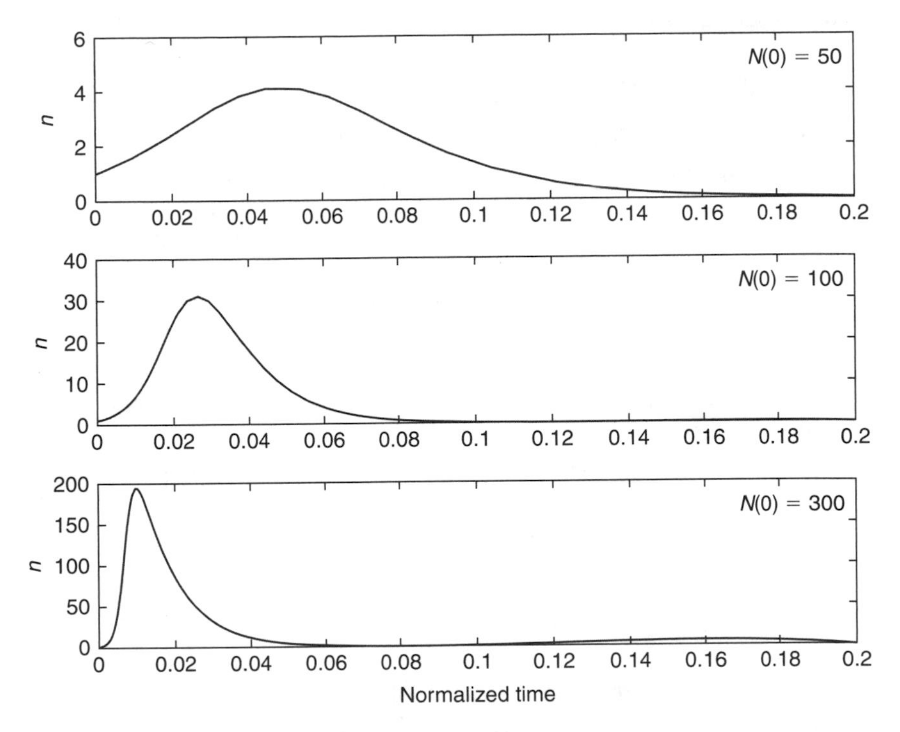

FIGURE 4.8
The temporal profile of the photon burst emitted in a Q-switched laser for different initial values of the excited level atoms population. Top panel: $N(0) = 50$; middle panel: $N(0) = 100$; botton panel: $N(0) = 300$.

build-up) much larger than that of the corresponding photon population $n(0)$. Figure 4.8 shows the ensuing dynamics for the photon population for different values of $N(0)$. We assumed in these simulations the following values for the parameters in the laser1 *function M-file* `(p=0; B=3; c=100; gamma=0.01)`.

In examining Figure 4.8, we observe that as $N(0)$ increases, the pulse's total energy increases — as it should since more energy is stored in the excited atoms. Furthermore, the duration of the produced pulse (i.e., the width of the pulse temporal profile) narrows, and the delay in the position of its peak from the trigger time gets to be smaller as the number of the initial higher excited level atoms increases. □

Homework Problem

Pb. 4.41 Investigate how changes in the values of τ_{cavity} and τ_{decay} modify the duration of the produced pulse. Plot the Q-switched pulse duration as function of each of these variables.

4.8 Integral Equations*

An integral equation is an equation where the unknown function, to be determined, appears under the integral sign. A linear integral equation is one in which nonlinear functions of the unknown function do not appear.

An integral equation for the unknown function $y(x)$ of the form

$$y(x) = F(x) + \lambda \int_a^b K(x,\xi) y(\xi)\, d\xi \tag{4.60}$$

where F and K are given functions, and λ, a, and b are constants, is called a Fredholm equation.

An integral equation for the unknown function $y(x)$ of the form

$$y(x) = F(x) + \lambda \int_a^x K(x,\xi) y(\xi)\, d\xi \tag{4.61}$$

where F and K are given functions, and λ and a are constants, is called a Volterra equation. The difference with the preceding equation is that here the upper limit of the integral is not a constant.

The purpose of this section is to show that any second-order linear ODE can be reduced to an integral equation (actually the result can be generalized to any order ODE). Our ultimate goal is to provide yet another technique for solving ODEs. The advantage of this method is that one can directly incorporate in the form of the equation, the initial conditions or the boundary conditions of the problem. The numerical techniques to solve this class of problems will be illustrated in Chapter 8 as they involve matrix algebra.

Prior to deriving the form of the integral equation corresponding to an arbitrary second-order ODE, we shall derive a useful lemma.

LEMMA For an arbitrary integrable function $f(x)$, the following relation holds:

$$\int_a^x \int_a^{x_n} \int_a^{x_{n-1}} \cdots \int_a^{x_3} \int_a^{x_2} f(x_1)\, dx_1\, dx_2\, dx_3 \cdots dx_{n-1}\, dx_n = \frac{1}{(n-1)!} \int_a^x (x-\xi)^{n-1} f(\xi)\, d\xi \tag{4.62}$$

PROOF First recall Leibnitz's rule for the differentiation of an integral involving a parameter:

$$\frac{d}{dx} \int_{L(x)}^{U(x)} g(x,\xi) d\xi = \int_{L(x)}^{U(x)} \frac{\partial g(x,\xi)}{\partial \xi} d\xi + g(x, U(x)) \frac{dU(x)}{dx} - g(x, L(x)) \frac{dL(x)}{dx} \tag{4.63}$$

*The asterisk indicates more advanced material that may be skipped in a first reading.

Now apply this relation to the differentiation of the function $I_n(x)$ defined through:

$$I_n(x) = \int_a^x (x-\xi)^{n-1} f(\xi)\, d\xi \tag{4.64}$$

We obtain

$$\begin{aligned}\frac{dI_n}{dx} &= (n-1)\int_a^x (x-\xi)^{n-2} f(\xi) d\xi + \left[(x-\xi)^{n-1} f(\xi)\right]_{\xi=x} \\ &= (n-1) I_{n-1}(x) \qquad \text{for } n > 1 \\ &= f(x) \qquad \text{for } n = 1\end{aligned} \tag{4.65}$$

From which we deduce

$$\frac{d^n I_n}{dx^n} = (n-1)!\, f(x) \tag{4.66}$$

$$I_1(x) = \int_a^x f(x_1) dx_1 \tag{4.67}$$

$$I_2(x) = \int_a^x I_1(x_2) dx_2 = \int_a^x \int_a^{x_2} f(x_1) dx_1\, dx_2 \tag{4.68}$$

More generally

$$I_n(x) = (n-1)! \int_a^x \int_a^{x_n} \int_a^{x_{n-1}} \cdots = \int_a^{x_3} \int_a^{x_2} f(x_1)\, dx_1\, dx_2\, dx_3 \cdots dx_{n-1} dx_n \tag{4.69}$$

Note that we made use above of the fact that $I_n(x)$ and its first $(n-1)$ derivatives all vanish when $x = a$.

Next we shall establish the relation between differential and integral equations.

THEOREM The second-order linear ODE given by

$$\frac{d^2 y}{dx^2} + B(x)\frac{dy}{dx} + C(x) y = f(x) \tag{4.70}$$

with the initial conditions $y(a) = y_{init}$ and $y'(a) = y'_{init}$, is equivalent to the integral equation

$$y(x) = F(x) + \int_a^x K(x,\xi)\, y(\xi)\, d\xi \tag{4.71}$$

with

$$K(x,\xi) = (\xi - x)[C(\xi) - B'(\xi)] - B(\xi) \tag{4.72}$$

and

$$F(x) = \int_a^x (x-\xi) f(\xi) d\xi + [B(a)y_{init} + y'_{init}](x-a) + y_{init} \tag{4.73}$$

PROOF Integrating the differential equation, using the initial conditions, we obtain

$$y'(x) - y'_{init} = -\int_a^x B(x_1)\, y'(x_1) dx_1 - \int_a^x C(x_1)\, y(x_1) dx_1 + \int_a^x f(x_1) dx_1 \tag{4.74}$$

Next, integrating the first term on the right by parts, we obtain

$$y'(x) = -B(x)\, y(x) - \int_a^x [C(x_1) - B'(x_1)] y(x_1)\, dx_1 + \int_a^x f(x_1) dx_1 + B(a)y_{init} + y'_{init} \tag{4.75}$$

Performing a second integration, gives

$$\begin{aligned} y(x) = &-\int_0^a B(x_1)\, y(x_1) dx_1 - \int_a^x \int_a^{x_2} [C(x_1) - B'(x_1)] y(x_1)\, dx_1\, dx_2 \\ &+ \int_a^x \int_a^{x_2} f(x_1)\, dx_1\, dx_2 + [B(a)y_{init} + y'_{init}](x-a) + y_{init} \end{aligned} \tag{4.76}$$

Using now the lemma on the terms with double integration, gives the desired result

$$\begin{aligned} y(x) = &-\int_a^x \{B(\xi) + (x-\xi)[C(\xi) - B'(\xi)]\}\, y(\xi)\, d\xi \\ &+ \int_a^x (x-\xi) f(\xi)\, d\xi + (x-a)[B(a)y_{init} + y'_{init}] + y_{init} \end{aligned} \tag{4.77}$$

EXAMPLE 4.15

Find the integral equation that is equivalent to

$$\frac{d^2y}{dx^2} + \omega_0^2\, y = f(x) \tag{4.78}$$

with the initial conditions $y(0) = 1$ and $y'(0) = 0$.

Solution: Here $a = 0$, and the functions $B(x)$ and $C(x)$ are respectively $B = 0$ and $C = \omega_0^2$.

Next, let us compute the K and F functions:

$$K(x, \xi) = (\xi - x)\,\omega_0^2 \tag{4.79}$$

$$F(x) = \int_0^x (x - \xi) f(\xi) d\xi + 1 \tag{4.80}$$

giving for the integral equation

$$y(x) = \omega_0^2 \int_0^x (\xi - x)\, y(\xi)\, d\xi - \int_0^x (\xi - x) f(\xi)\, d\xi + 1 \tag{4.81}$$

□

EXAMPLE 4.16

Find the integral equation that is equivalent to the ODE

$$\frac{d^2y}{dx^2} + \omega_0^2\, y = f(x) \tag{4.82}$$

with the boundary conditions $y(0) = 0$ and $y(l) = 0$.

Solution: Integrate the ODE from 0 to x, and we obtain

$$\frac{dy}{dx} = -\omega_0^2 \int_0^x y(x_1) dx_1 + C \tag{4.83}$$

where C is a constant of integration (actually corresponding to the unknown value of $y'(0)$). Integrating this equation again from 0 to x, we obtain

$$y(x) - y(0) = -\omega_0^2 \int_0^x \int_0^{x_2} y(x_1) dx_1 \, dx_2 + Cx \tag{4.84}$$

Using the above lemma, the first term on the right-hand side can be written as a single integration:

$$y(x) = -\omega_0^2 \int_0^x (x - \xi) y(\xi) \, d\xi + Cx \tag{4.85}$$

We used in the above derivation the boundary condition $y(0) = 0$. We still have to determine the constant C. This can be obtained from the boundary condition $y(l) = 0$.

This gives

$$C = \frac{\omega_0^2}{l} \int_0^l (l - \xi) y(\xi) \, d\xi \tag{4.86}$$

and the integral equation reduces to

$$y(x) = -\omega_0^2 \int_0^x (x - \xi) y(\xi) \, d\xi + \omega_0^2 \frac{x}{l} \int_0^l (l - \xi) y(\xi) \, d\xi \tag{4.87}$$

The second term on the right-hand side can be written as two integrals, the first an integral from 0 to x and the other an integral from x to l. Combining this first integral with the first term on the right-hand side, we obtain

$$y(x) = \omega_0^2 \int_0^l K(x, \xi) y(\xi) \, d\xi \tag{4.88}$$

where

$$K(x, \xi) = \begin{cases} \dfrac{\xi}{l}(l - x) & \text{for } \xi < x \\[2ex] \dfrac{x}{l}(l - \xi) & \text{for } \xi > x \end{cases} \tag{4.89}$$

□

Homework Problems

Pb. 4.42 Show that the integral equation equivalent to the ODE:

$$\frac{d^2y}{dx^2} = g(x)$$

with initial conditions: $y(0) = y_{init}$ and $y'(0) = y'_{init}$, is

$$y(x) = \int_0^x (x - \xi) g(\xi)\, d\xi + y'_{init}\, x + y_{init}$$

Pb. 4.43 Show that the integral equation equivalent to the ODE:

$$\frac{d^2y}{dx^2} + xy = 1$$

with initial conditions: $y(0) = 0$ and $y'(0) = 0$, is

$$y(x) = -\int_0^x \xi(x - \xi) y(\xi)\, d\xi + \frac{1}{2} x^2$$

Pb. 4.44 Show that the integral equation equivalent to the ODE:

$$\frac{d^2y}{dx^2} + B\frac{dy}{dx} + Cy = 0$$

with initial conditions $y(0) = 0$ and $y(1) = 0$, and where B and C are constants, is

$$y(x) = \int_0^1 K(x, \xi)\, y(\xi)\, d\xi$$

with

$$K(x, \xi) = \begin{cases} C\xi(1 - x) + Bx - B & \text{for } \xi < x \\ Cx(1 - \xi) + Bx & \text{for } \xi > x \end{cases}$$

4.9 MATLAB Commands Review

`dblquad`	Double integrate a function of two variables.
`ode23` and **`ode45`**	Ordinary differential equations solvers.
`prod`	Finds the product of all the elements belonging to an array.
`quad` and **`quadl`**	Integrate a function between fixed limits using respectively the Simpson and adaptive Lobatto algorithms.
`trapz`	Finds the integral using the Trapezoid rule.
`triplequad`	Triple integrate a function of three variables.
`unwrap`	Unwraps radian phases by changing absolute jumps greater than pi to their 2*pi complement.

5

Root Solving and Optimization Methods

This chapter is devoted to an exploration of various simple analytical and numerical techniques and MATLAB commands useful for finding the zeros and the extrema of functions of one and two variables.

5.1 Finding the Real Roots of a Function of One Variable

This section explores different techniques for finding the real roots (zeros) of an arbitrary function of one variable. Specifically, we discuss the graphical method, the numerical techniques known as the Direct Iterative and the Newton–Raphson methods, and the built-in **`fzero`** function of MATLAB.

5.1.1 Graphical Method

In the graphical method, we find the zeros of a single variable function by implementing the following steps:

1. Plot the particular function over a suitable domain.
2. Identify the neighborhoods where the curve crosses the x-axis (there may be more than one point); and at each such point, the following steps should be independently implemented.
3. Zoom in on the neighborhood of each intersection point by repeated application of the MATLAB **`axis`** or **`zoom`** commands.
4. Use the crosshair of the **`ginput`** command to read the coordinates of the intersection of the function with the x-axis.

IN-CLASS EXERCISES

Pb. 5.1 Find graphically the solution of each of the following equations:

a. $\sin^2(x) - \frac{1}{2} = 0$

b. $\cos^2(x) - \frac{3}{4} = 0$

in the domain $0 \leq x \leq \pi$.

Pb. 5.2 Verify your graphical answers of Pb. 5.1 with that you would obtain analytically.

5.1.2 Numerical Methods

This section briefly discusses two techniques for finding the zeros of a function in one variable, namely the Direct Iterative and the Newton–Raphson techniques. We do not concern ourselves too much, at this point, with an optimization of the routine execution time, or with the inherent limits of each of the methods, except in the most general way. Furthermore, to avoid the inherent limits of these techniques in some pathological cases, we assume that we plot each function under consideration, verify that it crosses the x-axis, and satisfy ourselves in an empirical way that there does not seem to be any pathology around the intersection point before we embark on the application of the following algorithms. These statements will be made more rigorous to you in future courses in numerical analysis.

5.1.2.1 *The Direct Iterative Method*

This is a particularly useful technique when the equation $f(x) = 0$ can be cast in the form

$$x = F(x) \tag{5.1}$$

$F(x)$ is then called an iteration function, and it can be used for the generation of the sequence

$$x_{k+1} = F(x_k) \tag{5.2}$$

To guarantee that this method gives accurate results in a specific case, the function should be continuous and it should satisfy the contraction condition

$$|F(x_n) - F(x_m)| \leq s|x_n - x_m| \tag{5.3}$$

where $0 \leq s < 1$; that is, the changes in the value of the function are smaller than the changes in the value of the arguments. Under these conditions, we shall prove that the iterative function possesses a fixed point (i.e., that ultimately the difference between two successive iterations can be arbitrarily small).

PROOF Let x_{guess} be the first term in the iteration, then

$$\left|F(x_1) - F(x_{\text{guess}})\right| \leq s\left|x_1 - x_{\text{guess}}\right| \tag{5.4}$$

but since

$$F(x_{\text{guess}}) = x_1 \quad \text{and} \quad F(x_1) = x_2 \tag{5.5}$$

then

$$\left|x_2 - x_1\right| \leq s\left|x_1 - x_{\text{guess}}\right| \tag{5.6}$$

Similarly,

$$\left|F(x_2) - F(x_1)\right| \leq s\left|x_2 - x_1\right| \tag{5.7}$$

translates into

$$\left|x_3 - x_2\right| \leq s\left|x_2 - x_1\right| \leq s^2\left|x_1 - x_{\text{guess}}\right| \tag{5.8}$$

$$\left|x_{m+1} - x_m\right| \leq s^m\left|x_1 - x_{\text{guess}}\right| \tag{5.9}$$

but, because s is a non-negative number smaller than 1, the right-hand side of the inequality in Eq. (5.9) can be made, for large enough value of m, arbitrarily small, and the above iterative procedure does indeed converge to a fixed point.

EXAMPLE 5.1

Find the zero of the function

$$y = x - \sin(x) - 1 \tag{5.10}$$

Solution: At the zero, the iterative form can be written as

$$x(k) = \sin(x(k-1)) + 1 \tag{5.11}$$

The contraction property, required for the application of this method, is valid in this case because the difference between two sines is always smaller than the difference between their arguments. The fixed point can then be obtained by the following MATLAB routine:

```
x(1)=1;                    %value of the initial guess
   for k=2:20
     x(k)=sin(x(k-1))+1;
   end
x
```

which returns

```
Ans
  1.0000 1.8415 1.9636 1.9238 1.9383 1.9332 1.9350
  1.9344 1.9346 1.9345 1.9346 1.9346 1.9346 1.9346
  1.9346 1.9346 1.9346 1.9346 1.9346 1.9346
```

As can be noticed from the above printout, about 11 iterations were required to get the value of the fixed point accurate to one part per 10,000. □

NOTE A more efficient technique to find the answer within a prescribed error tolerance is to write the program with the **while** command, where we can specify the tolerance level desired.

5.1.2.2 The Newton–Raphson Method

This method requires a knowledge of both the function and its derivative. The method makes use of the geometrical interpretation of the derivative being the tangent at a particular point, and that the tangent is the limit of the chord between two close points on the curve. It is based on the fact that if $f(x_a)$ and $f(x_b)$ have opposite signs and the function f is continuous on the interval $[x_a, x_b]$, we know from the Intermediate Value theorem of calculus that there is at least one value x_c between x_a and x_b, such that $f(x_c) = 0$. Sufficient conditions for this method to work are that $f'(x)$ and $f''(x)$ have constant sign on an open interval that contains the solution $f(x) = 0$; in that case, any starting point that is close enough to the solution will give successive Newton's approximations that converge to the solution.

Consider the function $f(x)$ at a point x_1 in the neighborhood of a zero of this function. The equation of the line tangent to this curve at x_1 is given by

$$y - f(x_1) = f'(x_1)(x - x_1) \tag{5.12}$$

This tangent line, as shown in Figure 5.1, crosses the x-axis at a point nearer to the zero of the function $f(x)$ than x_1. If $f'(x_1) \neq 0$, the tangent line crosses the x-axis at the point $P_2(x_2, 0)$. Substituting the coordidnates of the point P_2 in Eq. (5.12), the equation of the tangent line, we obtain

$$x_2 = x_1 - \frac{f(x_1)}{f'(x_1)} \tag{5.13}$$

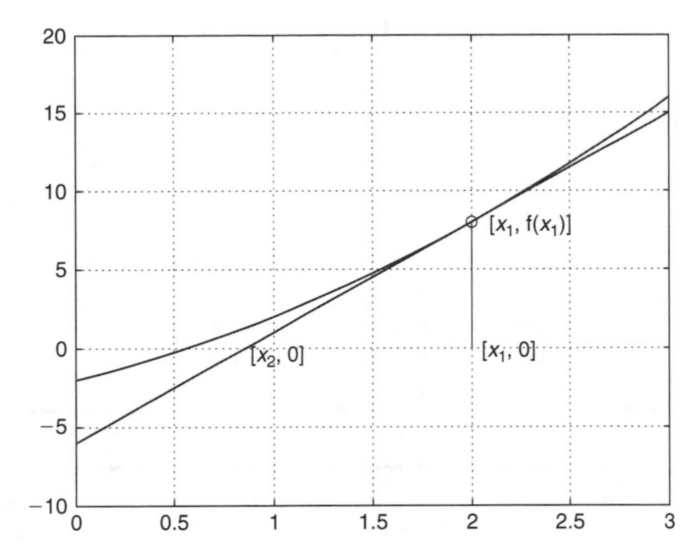

FIGURE 5.1
The first iteration of the Newton–Raphson method.

The process can be repeated to find the points $x_3, x_4, x_5, \ldots$. These points get closer and closer to the zero of the function. The iterative function that generates this sequence is

$$x(k) = x(k-1) - \frac{f(x(k-1))}{f'(x(k-1))} \tag{5.14}$$

This iteration process provides the algorithm for the Newton–Raphson method for obtaining the zero of a function of one variable.

Example 5.2

Find the zero of the function

$$y = \sin(x) - 2\tan(x) + 1 \tag{5.15}$$

in the neighborhood of the point $x = \pi/4$.

Solution: First compute the derivative of the function y:

$$y' = \cos(x) - 2\sec^2(x) \tag{5.16}$$

Second, edit and execute the following program:

```
x(1)=pi/4;
  for k=2:20
       x(k)=x(k-1)-(sin(x(k-1))-2*tan(x(k-1))+1)/...
```

```
        (cos(x(k-1))-2*sec(x(k-1))^2);
    end
  x
```

returns

```
  x =
        0.7854    0.6965    0.6847    0.6845    0.6845
        0.6845    0.6845    0.6845    0.6845    0.6845
        0.6845    0.6845    0.6845    0.6845    0.6845
        0.6845    0.6845    0.6845    0.6845    0.6845
```
□

In-Class Exercises

Pb. 5.3 Write a routine to find the zero of the function $y = x - \sin(x) - 1$ using the Newton–Raphson algorithm.

Pb. 5.4 Compare the answers from the present algorithm with that of the Direct Iterative method, at each iteration step, in the search for the zeros of the function:

$$y = x - \sin(x) - 1$$

Comment on which of the two methods appears to be more effective and converges faster.

Example 5.3

Apply the Newton–Raphson method to find the voltage–current relation in a diode circuit with an ac voltage source.

Solution: The diode is a nonlinear semiconductor electronic device with a voltage current curve that is described, for voltage values larger than the reverse breakdown potential (a negative quantity), by

$$i = I_s(e^{v/kT} - 1) \tag{5.17}$$

where I_s is the reverse saturation current (which is typically in the order of 10^{-6} mA), and $k'T$ is the average thermal energy of an electron divided by its charge at the diode operating temperature (equal to 1/40 V at room temperature). An important application of this device is to use it as a rectifier (a device that passes the current in one direction only).

The problem we want to solve is to find the current through the circuit (shown in Figure 5.2) as a function of time if we are given a sinusoidal time-dependent source potential. The other equation, in addition to Eq. (5.17) that we need to set the problem, is Ohm's law across R. This law, as previously

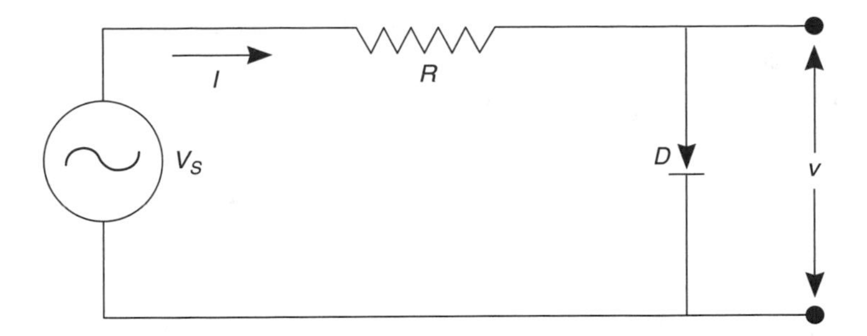

FIGURE 5.2
The diode semirectifier circuit.

noted, states that the current passing through a resistor is equal to the potential difference across the resistor, divided by the value of the resistance:

$$i = \frac{V_s - v}{R} \tag{5.18}$$

Eliminating the current from Eq. (5.17) and Eq. (5.18), we obtain a nonlinear equation in the potential across the diode. Solving this problem is then reduced to finding the roots of the function f defined as

$$f(v) = I_s[\exp(v/k'T) - 1] - \left(\frac{V_s - v}{R}\right) \tag{5.19}$$

where the potential across the diode is the unknown.

In the Newton–Raphson method, we also need for our iteration the derivative of this function

$$f'(v) = \left(\frac{1}{k'T}\right) I_s \exp(v/k'T) + \frac{1}{R} \tag{5.20}$$

For a particular value of V_s, we need to determine v and, from this value of the potential across the diode, we can determine the current in the circuit. However, because we are interested in obtaining the current through the diode for a source potential that is a function of time, we need to repeat the Newton–Raphson iteration for each of the different source voltage values at the different times. The sequence of the computation would proceed as follows:

1. Generate the time array.
2. Generate the source potential for the different elements in the time array.

3. For each time array entry, find the potential across the diode using the Newton–Raphson method.
4. Obtain the current array from the potential array.
5. Plot the source potential and the current arrays as a function of the time array.

Assuming that the source potential is given by

$$V_s = V_0 \sin(2\pi f t) \tag{5.21}$$

and that $f = 60\,\text{Hz}$, $V_0 = 5\text{V}$, $k'T = 0.025\,\text{V}$, $R = 500\,\Omega$, and the saturation current $I_s = 10^{-6}\,\text{mA}$; the following *script M-file* finds the current in this circuit:

```
Is=10^(-9); R = 500; kpT = 1/40; f = 60; V0 = 5;
t=linspace(0,2/f,600);
L=length(t);
K=200;
Vs=(V0*sin(2*pi*t*f))'*ones(1,K);
v=zeros(L,K);
i=zeros(L,K);

 for k=1:K-1
      v(:,k+1) = v(:,k)-(Is*(exp((1/kpT)*v(:,k))-1) ...
                  -(1/R)*(Vs(:,k)-v(:,k)))./...
                  ((1/kpT)*Is*exp((1/kpT)*v(:,k))+1/R);
      i(:,k+1) = (Vs(:,k+1)-v(:,k+1))/R;
 end

plot(t,1000*i(:,K),'b',t,Vs(:,K), 'g')
```

The current (expressed in mA) and the voltage (in V) of the source will appear in your graph window when you execute this program. □

Homework Problem

Pb. 5.5 The apparent simplicity of the Newton–Raphson method is very misleading, suffice it to say that some of the early work on fractals started while considering pathologies in this model.

a. State, to the best of your ability, the conditions that the function, its derivative, and the original guess should satisfy so that this iterate converges to the correct limit.
b. Show that the Newton–Raphson method iterates cannot find the zero of the function below, if the initial guess is other than zero:

$$y = x^{1/3}$$

c. Illustrate, with a simple sketch, the reason that this method fails in part (b).

5.1.3 MATLAB **fzero** Built-in Function

Next, we introduce the use of the MATLAB command **fzero** for finding the zeros of any function in one variable.

The recommended sequence of steps for finding the zeros of a function of one variable are as follows:

1. Edit a *function M-file* for the function under consideration; or if the expression for the function is compact, define an *anonymous function*.
2. Plot the curve of the function over the appropriate domain, and estimate the value of the zeros. Use the **ezplot** command if appropriate.
3. Using each of the estimates for the values of the zeros found in (2) above as an initial "guess" use the command **fzero** to accurately find each of the roots. The syntax is as follows:

```
xroot=fzero(@funname,xguess)
```

EXAMPLE 5.4

Using the **fzero** command find the zero of the function:

$$y = x - \sin(x) - 1 \tag{5.22}$$

Solution: Execute the following commands:

```
Example_54=@(x)x-sin(x)-1;
ezplot(Example_54,[1 3])
```

From Figure 5.3, we approximate $x = 2$, for the initial point in the **fzero** command.

Entering

```
xroot=fzero(Example_54,2)
```

returns

```
xroot =
  1.9346
```

If desired, you can specify the desired accuracy for both the value of the root and the function while using the **fzero** command.

Entering

```
options=optimset ('TolX',1e-10,'TolFun',1e-10);
[xroot, value]=fzero(Example_54,2,options)
```

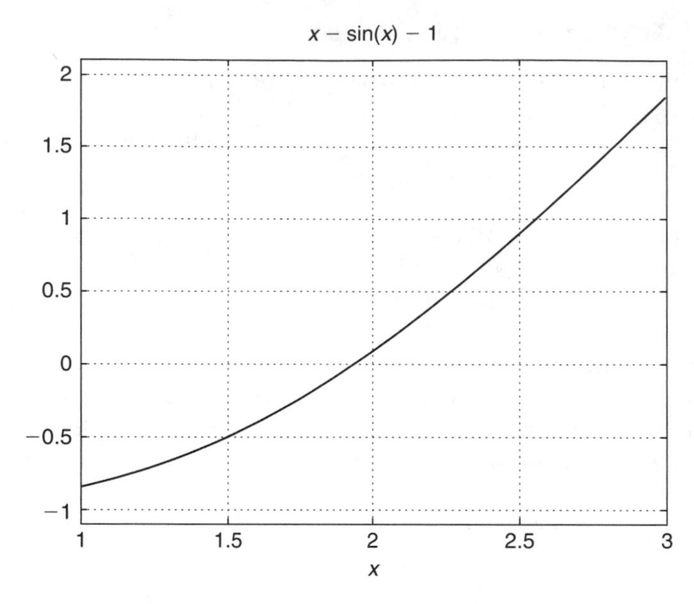

FIGURE 5.3
Graphics for Example 5.4.

returns

```
xroot =
  1.9346
value =
  -7.4748e-011
```
□

In-Class Exercises

In each of the following problems, find the zeros of the following functions over the interval [0, 5], using the **fzero** command:

Pb. 5.6 $f(x) = x^2 - 1$.

Pb. 5.7 $f(x) = \sin^2(x) - 3/4$. Compare your answer with the analytical result.

Pb. 5.8 $f(x) = 2\sin^2(x) - x^2$.

Pb. 5.9 $f(x) = x - \tan(x)$.

5.1.4 Application: Zeros of the Zero-Order Bessel Function

In the following application, we use the command **fzero** to find the zeros of the zero-order Bessel function, and learn in the process some important facts about this often-used special function of applied mathematics.

Bessel functions are solutions to Bessel's differential equations of order n, given by

$$x^2 \frac{d^2y}{dx^2} + x\frac{dy}{dx} + (x - n)y = 0 \tag{5.23}$$

There are special types of Bessel functions referred to as "of the first, second, and third kinds." Bessel functions of integer order appear, *inter alia*, in the expression of the radiation field in cylindrically shaped resonant cavities, and in light diffraction from a circular hole. Bessel functions of half-integer indices (see Pb. 2.26) appear in problems of spherical cavities and scattering of electromagnetic waves. Airy functions, a member of the Bessel functions family, appear in a number of important problems of optics and quantum mechanics.

The recursion formula that relates the Bessel function of any kind of a certain order with those of the same kind of adjacent orders is

$$2nZ_n(x) = xZ_{n-1}(x) + xZ_{n+1}(x) \tag{5.24}$$

where $Z_n(x)$ is the generic designation for all kinds of Bessel functions.

In this application, we concern ourselves only with the Bessel function of the first kind, usually denoted by $J_n(x)$. Its MATLAB call command is **besselj(n,x)**. In the present problem, we are interested in the root structure of the Bessel function of the first kind and of zero order.

In the program that follows, we call the Bessel function from the MATLAB library; however, we could have generated it ourselves using the techniques of Section 4.7 because we know the ODE that it satisfies, and its value and that of its derivative at $x = 0$, namely

$$J_0(x = 0) = 1 \quad \text{and} \quad J_0'(x = 0) = 0$$

The problem that we want to solve here is to find the zeros of $J_0(x)$ and compare to these exact values those obtained from the approximate expression:

$$x_{0,k} = \frac{\pi}{4}(4k-1) + \frac{1}{2\pi(4k-1)} - \frac{31}{6\pi^3(4k-1)^3} + \frac{3779}{15\pi^5(4k-1)^5} + \cdots \tag{5.25}$$

To implement this task, edit and execute the following *script M-file*:

```
for k=1:10
  p(k)=4*k-1;
  x0(k)=fzero('besselj(0,x)',(pi/4)*p(k));
  x0approx(k)=(pi/4)*p(k)+(1/(2*pi))*(p(k)^(-1))-...
               (31/6)*(1/pi^3)*(p(k)^(-3))+...
               (3779/15)*(1/pi^5)*(p(k)^(-5));
end
```

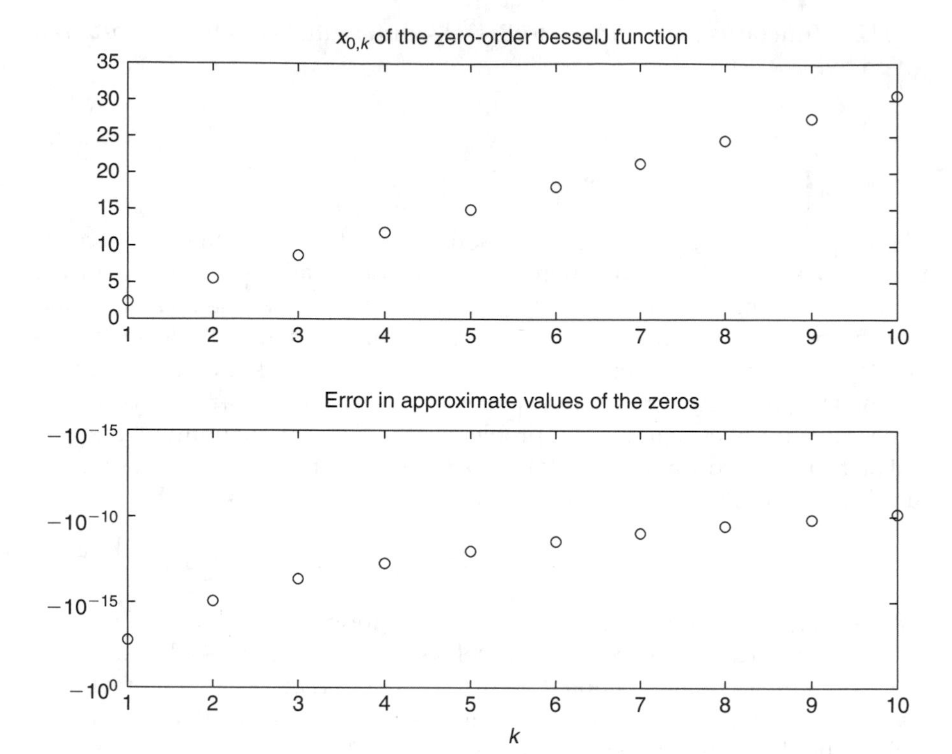

FIGURE 5.4
The first 10 zeros of the Bessel function $J_0(x)$. Top panel: the values of the successive zeros (roots) of $J_0(x)$. Bottom panel: deviation in the values of these zeros between their exact expressions and their approximate values as given in Eq. (5.25).

```
kk=1:10;
subplot(2,1,1);
plot(kk,x0,'o')
title('x_{0,k} of the zero order besselJ function')
subplot(2,1,2);
semilogy(kk,x0-x0approx,'o')
title('Error in Approximate Values of the Zeros')
```

As you can easily observe by examining Figure 5.4, the approximate series is suitable for calculating all (except the smallest) zeros of the function $J_0(x)$ correctly to at least five digits.

5.2 Roots of a Polynomial

While the analytical expressions for the roots of quadratic, cubic, and quartic equations are known, in general, the roots of higher-order polynomials cannot be found analytically. MATLAB has a built-in command that finds all the roots (real and complex) for any polynomial equation. As noted previously, the MATLAB command for finding the polynomial roots is **`r=roots(p)`**.

In interpreting the results from this command, recall the fundamental theorem of algebra, which states the root properties of a polynomial of degree n with real coefficients:

1. The nth polynomial admits n complex roots.
2. Complex roots come in conjugate pairs. [If you are not familiar with complex numbers and with the term complex conjugate (the latter term should pique your curiosity), be a little patient. Help is on the way; Chapter 6 covers the topic of complex numbers].

Inversely, knowing the roots of a polynomial, we can construct the polynomial. The MATLAB command is **`poly(r)`**.

IN-CLASS EXERCISES

Pb. 5.10 Find the roots of the polynomial $p = [1\ 3\ 2\ 1\ 0\ 3]$, and compute their sum and product.

Pb. 5.11 Consider the two polynomials:

$$p_1 = [1\,3\,2\,1\,0\,3] \quad \text{and} \quad p_2 = [3\,2\,1]$$

Find the value(s) of x at which the curves representing these polynomials would intersect.

Pb. 5.12 Find the constants A, B, C, D, and a, b, c, d that permits the following expansion in partial fractions:

$$\frac{1}{x^4 - 25x^2 + 144} = \frac{A}{(x-a)} + \frac{B}{(x-b)} + \frac{C}{(x-c)} + \frac{D}{(x-d)}$$

5.3 Optimization Methods for Functions of One Variable

Many design problems call for the maximization or minimization (optimization) of a particular function belonging to a particular domain. (Recall the resistor circuit [Figure 3.1] in which we wanted to find the maximum power delivered to a load resistor.) In elementary calculus, you learned that you obtain the local extrema of a function by finding the values of the independent variable at which the first derivative of the function is zero. In this section, we explore other techniques for solving this problem when the simple method is not practical. Let us start by reminding ourselves of some terms definitions.

DEFINITIONS

1. The domain is the set of elements to which a function assigns values. The range is the set of values thus obtained.
   If I, the domain of the function $f(x)$, contains the point c, we say that
2. $f(c)$ is the maximum value of the function on I if $f(c) \geq f(x)$ for all $x \in I$.
3. $f(c)$ is the minimum value of the function on I if $f(c) \leq f(x)$ for all $x \in I$.
4. An extremum is the common designation for either the maximum value or the minimum value.

Using the above definitions, we note that the maximum (minimum) may appear at an endpoint of the interval I, or possibly in the interior of the interval. If a maximum (minimum) appears at an endpoint, we describe this extreme point as an endpoint maximum (minimum). If a maximum (minimum) appears in the interior of the interval, we describe this extreme point as a local maximum (minimum). The largest (smallest) value among the maximum (minimum) values (either endpoint or local) is called the global maximum (minimum) and is in a number of instances the object of interest. As pointed out earlier, the following methods are suitable when the direct method is not suitable due to a number of practical complications, most often associated with finding the zeros of transcendental equations. As with finding the zeros of a function, in this instance we will also explore the graphical method, a simple numerical method, and the MATLAB built-in command for finding the extremum.

5.3.1 Graphical Method

In the graphical method, in steps very similar to those described in Section 5.1.1 for finding the zeros of a single variable function, we follow these steps:

1. Plot the particular function over the defined domain.
2. Examine the plot to determine whether the extremum is an endpoint extremum or a local extremum.
3. Zoom in on the neighborhood of the so-identified extremum by repeated application of the MATLAB **`axis`** or **`zoom`** commands.
4. Use the crosshair of the **`ginput`** command to read the coordinates of the extremum. [Be especially careful here. Extra caution is prompted by the fact that the curve is flat (its tangent is parallel to the x-axis) at a local extremum; thus, you may need to re-plot the curve in the neighborhood of this extremum to find, through visual means, accurate results for the coordinates of the extremum. There may be too

few points in the original plot for the zooming technique to provide more than a very rough approximation.]

IN-CLASS EXERCISES

Find, graphically, for each of the following exercises, the coordinates of the global maximum and the global minimum for the following curves in the indicated intervals. Specify the nature of the extremum.

Pb. 5.13 $y = f(x) = \exp(-x^2)\sin^2(x)$ on $-4 < x < 4$.

Pb. 5.14 $y = f(x) = \exp(-x^2)[x^3 + 2x + 3]$ on $-4 < x < 4$.

Pb. 5.15 $y = f(x) = 2\sin(x) - x$ on $0 < x < 2\pi$.

Pb. 5.16 $y = f(x) = \sqrt{1+\sin(x)}$ on $0 < x < 2\pi$.

5.3.2 Numerical Method: The Golden Section Method

We assume that we have plotted the function and have established that a local minimum exists. Our goal at this point is to accurately pinpoint the position and value of this minimum. We detail the derivation of an elementary technique for this search: the *Golden Section Method.* More accurate and efficient techniques for this task have been developed. These are incorporated in the built-in command **`fminbnd`**; this mode of use is discussed in Section 5.3.3.

***The Golden Section Method*:** Assume that, by examining the graph of the function under consideration, we have established that the local minimum $x_{\min} \in [a, b]$. This means that the curve of the function is strictly decreasing in the interval $[a, x_{\min}]$ and is strictly increasing in the interval $[x_{\min}, b]$. Next, choose a number $r < 1/2$, but whose precise value will be determined later, and define the internal points c and d such that

$$c = a + r(b - a) \tag{5.26}$$

$$d = a + (1 - r)(b - a) \tag{5.27}$$

and such that $a < c < d < b$. Next, evaluate the values of the function at c and d. If we find that $f(c) \geq f(d)$, we can assert that $x_{\min} \in [c, b]$; that is, we narrowed the external bounds of the interval. (If the inequality was in the other sense, we could have instead narrowed the outer limit from the right.) If in the second iteration, we fix the new internal points such that the new value of c is the old value of d, then all we have to compute at this step is the new value of d. If we repeat the same iteration k times, until the separation between c and d is smaller than the desired tolerance, then at that point we can assert that

$$x_{\min} = \frac{c(k) + d(k)}{2} \tag{5.28}$$

Now, let us determine the value of r that will allow the above iteration to proceed as described. Translating the above statements into equations, we desire that

$$c(2) = d(1)$$
$$\Rightarrow c(2) = a(2) + r(b(2) - a(2)) = a(1) + (1 - r)(b(1) - a(1)) \tag{5.29}$$

$$a(2) = c(1) = a(1) + r(b(1) - a(1)) \tag{5.30}$$

$$b(2) = b(1) \tag{5.31}$$

Now, replacing the values of $a(2)$ and $b(2)$ from Eq. (5.30) and Eq. (5.31) into Eq. (5.29), we are led to a second-degree equation in r:

$$r^2 - 3r + 1 = 0 \tag{5.32}$$

The desired root is the value of the Golden ratio:

$$r = \frac{3 - \sqrt{5}}{2} \tag{5.33}$$

and, hence, the name of the method.

The following *function M-file* implements the above algorithm:

```
function [xmin,ymin]=goldensection(funname,a,b,...
  tolerance)
r=(3-sqrt(5))/2;
c=a+r*(b-a);
fc=funname(c);
d=a+(1-r)*(b-a);
fd=funname(d);

while d-c>tolerance
    if fc>=fd
      dnew=c+(1-r)*(b-c);
      a=c;
      c=d;
      fc=fd;
      d=dnew;
      fd=funname(dnew);

    else
      cnew=a+r*(d-a);
      b=d;
      d=c;
      fd=fc;
```

```
        c=cnew;
        fc=funname(cnew);
    end
end

xmin=(c+d)/2;
ymin=funname(xmin);
```

For example, if we wanted to find the position of the minimum of the cosine function and its value in the interval $3 < x < 3.5$, accurate to one part per 10,000, we would enter in the command window, the following command:

```
[xmin,ymin]=goldensection(@(x)cos(x),3,3.5,1e-4)
```

5.3.3 MATLAB fminbnd Built-in Function

Following similar steps to those suggested for using the **fzero** command when seeking the zeros of a function, we can use the **fminbnd** command to find the minimum of a function in one variable over a given interval. The recommended sequence of steps is as follows:

1. Edit a *function M-file* for the function under consideration, or create an *anonymous function* if the expression of the function is compact.
2. Plot the curve of the function over the desired domain, to overview the function's shape and make an estimate of the position of the minimum. In particular, from the figure, find a lower bound and an upper bound for the position of the minimum. Denote these bounds by a and b.
3. Use the command **fminbnd** to find accurately the position and value of the minimum to within the desired tolerances. The syntax is as follows:

```
options=optimset('TolX',?,'TolFun',?);
[xmin, value]=fminbnd(@funname,a,b,options)
```

EXAMPLE 5.5

Using the **fminbnd** command find the minimum of the function

$$y = -x^3 \exp(-x^2) \tag{5.34}$$

in the interval $0 \leq x \leq 4$

Solution: Execute the following commands:

```
Example_55=@(x)-x.^3.*exp(-x.^2);
ezplot(Example_55,[0 4])
```

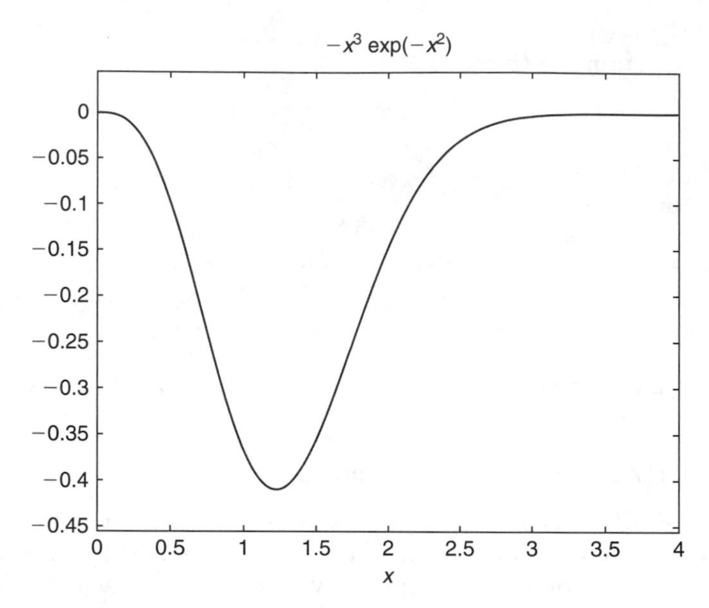

FIGURE 5.5
Graphics for Example 5.5.

From the plot in Figure 5.5, we can estimate that the position of the local minimum is located in the interval [1 1.5]. We enter

```
options=optimset('TolX',1e-10,'TolFun',1e-10);
[xmin, value]=fminbnd(Example_55,1,1.5,options)
```

gives

```
xmin =
  1.2247
value =
  -0.4099
```
□

The same command can be used to find the maximum of a function. The technique is to consider the negative of the function and look for the minimum.

Homework Problems

Pb. 5.17 A certain ellipse is defined by the parametric equations

$$x = 3\cos(t)$$
$$y = 2\sin\left(t + \frac{\pi}{3}\right)$$

(cont'd.)

Homework Problems *(cont'd.)*

Compute for this ellipse the long and short radii, and the angle that the long axis makes with the x-axis.

Pb. 5.18 We have two posts of height 6 and 8 m, and separated by a distance of 21 m. A rope is to run from the top of one post to the ground between the posts and then to the top of the other post (Figure 5.6). Find the configuration that minimizes the length of the rope.

Pb. 5.19 Fermat's principle states that light traveling from point A to point B selects the path which requires the least amount of travel time. Consider the situation in which an engineer in a submarine wants to communicate, using a laser-like pointer, with a detector at the top of the mast of another boat. At what angle θ to the vertical should he point his beam? Assume that the detector is 50 ft above the water surface, the submarine transmitter is 30 ft under the surface, the horizontal distance separating the boat from the submarine is 100 ft, and the velocity of light in water is 3/4 of its velocity in air (Figure 5.7).

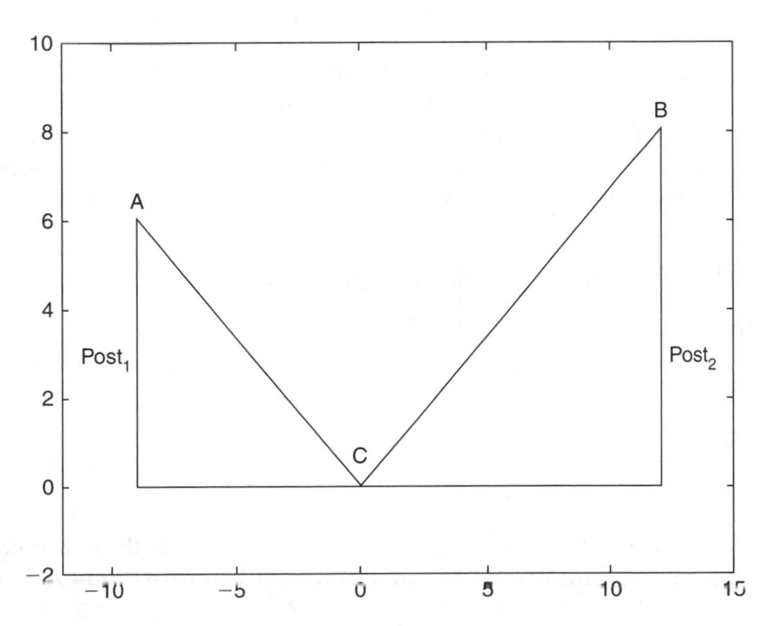

FIGURE 5.6
Schematics for Pb. 5.18. (ACB is the line whose length we want to minimize.)

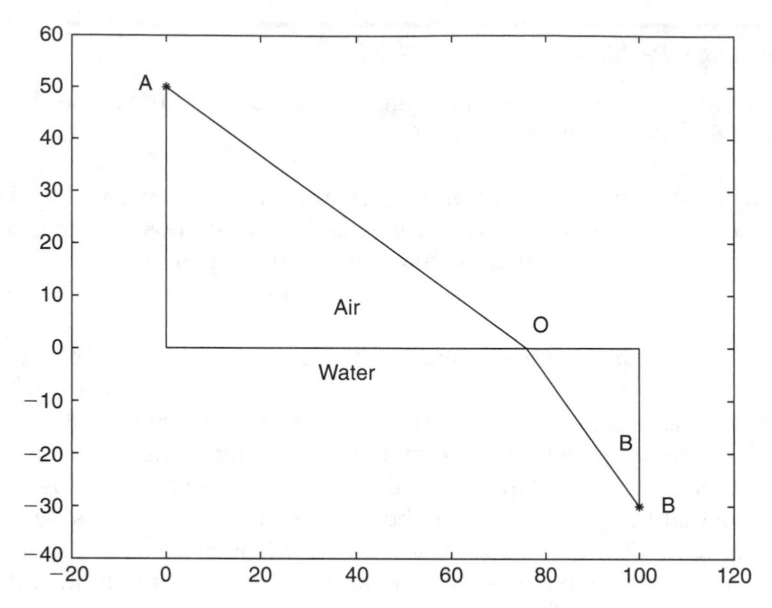

FIGURE 5.7
Schematics for Pb. 5.19. A is the location of the detector at the top of the mast, B is the location of the emitter in the submarine, and BOA is the optical path of the ray of light.

5.4 The Zeros and the Minima of Functions in Two Variables*

In Chapter 1, in the context of applications of the **`contour`** command, we explored graphical techniques for finding the solutions of two arbitrary equations in two unknowns, and for finding the minima of a function in two unknowns. Here, we discuss the MATLAB commands that allow us a more accurate determination of these solutions.

5.4.1 The MATLAB `fsolve` Built-in Command

This command is included in The Optimization Toolbox.

The systematic technique to use this command to solve the system of equations

$$f_1(x, y) = 0 \quad \text{and} \quad f_2(x, y) = 0 \tag{5.35}$$

is as follows:

1. Plot the contours C_1 and C_2 at $z = 0$, which represents the intersection of the two surfaces, $z_1 = f_1(x, y)$ and $z_2 = f_2(x, y)$ with the x–y plane.
2. Estimate the coordinates of the points of intersection of the curves C_1 and C_2.

---

* The asterisk indicates more advanced material that may be skipped in a first reading.

3. Use the approximate values of the coordinates of the point(s) of intersection of the two curves C_1 and C_2, obtained in (2) from the graph, as starting values in the **fsolve** command. The syntax of this command is as follows:

```
x=fsolve(funname,xo,options)
```

It should be noted that here, both x and the function are 2-D arrays.

EXAMPLE 5.6

Using the **fsolve** command, find the solution of the following system of equations:

$$\begin{aligned} & y^2 - 3\sin\left(\frac{\pi x}{4}\right) - 6 = 0 \\ & y - x - 1 = 0 \end{aligned} \tag{5.36}$$

in the domain $0 \le x \le 4$ and $0 \le y \le 4$.

Solution: Execute the following commands:

```
x=0:0.01:4;
y=0:0.01:4;
[X,Y]=meshgrid(x,y);
Z1=Y.^2-3*sin(pi*X/4)-6;
Z2=Y-X-1;
contour(X,Y,Z1,[0 0]);
hold on
contour(X,Y,Z2,[0 0]);
hold off
```

Reading off Figure 5.8, we estimate the point of intersection of the curves C_1 and C_2 to be located at (1.8, 2.9). (Note that we could also have obtained the same graphics as above by using the **ezplot** command to plot an implicitly defined function — see Example C.13.)

Now entering and executing

```
Example_56=@(x) [x(2).^2-3*sin(pi*x(1)/4)-6 ...
            x(2)-x(1)-1];
options=optimset('TolX',1e-10,'TolFun',1e-10);
[xroots, value]=fsolve(Example_56,[1.8 2.9],options)
```

returns

```
xroots =
  2.0000  3.0000
```

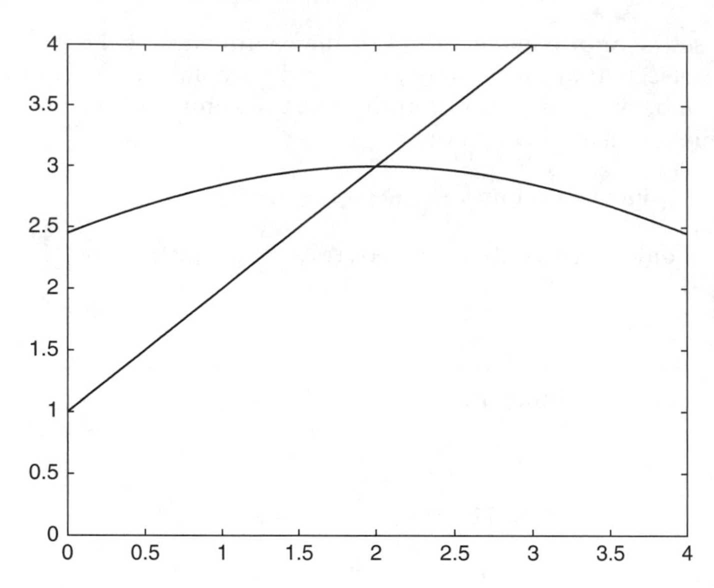

FIGURE 5.8
Graphics for Example 5.6.

```
value =
  1.0e-010 *
    0.4922    0.0000
```

Note that while we used the (x, y) nomenclature in the **contour** command, we should use the vectorized notation for the variables in the functions expressions in the **fsolve** command. □

In-Class Exercises

Pb. 5.20 Find the points of intersection of the circle and ellipse defined by the functions:

$$x^2 + y^2 - 25 = 0$$
$$\frac{(x-8)^2}{36} + \frac{(y-1)^2}{16} - 1 = 0$$

Pb. 5.21 Find the point(s) of intersection of the ellipse and parabola defined by the functions:

$$x^2 + y^2 - 2xy + 2x - y - 5 = 0$$
$$y = x^2 - 2x + 2$$

5.4.2 The MATLAB fminsearch Built-in Command

Analytically, searching for the local extrema of a function of several variables consist of solving the simultaneous equations obtained by setting all partial derivatives of the function equal to zeros. The resulting equations can be solved numerically by using the **fsolve** command discussed in the previous section. Alternately, the command **fminsearch** can be used to find directly the location of a local minimum of a function of several variables, and the value of the function at that point, without having to find analytically the components of the gradient of the function. For simplicity, we shall limit our discussion here to a function of two variables:

$$z = f(x, y) \tag{5.37}$$

If little is known, *a priori*, about the considered function, the systematic steps followed to use effectively the command **fminsearch** are as follows:

1. Plot the contours of the surface $z = f(x, y)$.
2. From the graph of the contours, estimate the location of the local minimum of interest.
3. Use the approximate value of the coordinates of the minimum obtained in (2) as starting values in the **fminsearch** command. The syntax is as follows:

```
[xmin,value]=fminsearch(funname,xo,options)
```

x_0 is the array representing the initial estimate for the location of the minimum, **xmin** the computed position of the minimum, and **value** the value of the function at the computed position of the minimum **xmin**.

EXAMPLE 5.7

Using the **fminsearch** command find the location and the value of the function at that location for

$$z = (x - 2)^2 + (y - 4)^2 + 3 \tag{5.38}$$

in the domain $0 \le x \le 5$ and $0 \le y \le 5$.

Solution:

(i) Define an anonymous function for the surface, and plot its contour lines in the given domain:

```
Example_57a=@(x,y)((x-2).^2+(y-4).^2+3);
ezcontour(Example_57a,[0 5],[0 5])
```

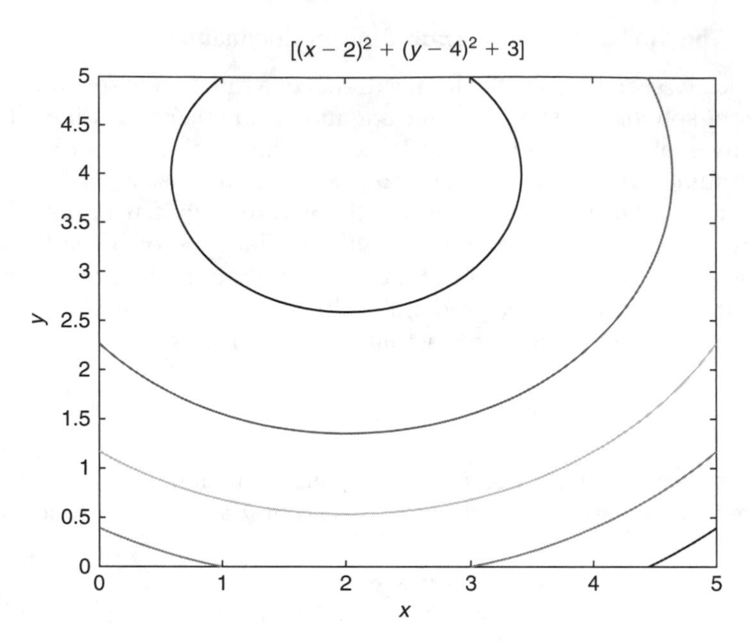

FIGURE 5.9
Graphics for Example 5.7.

(ii) Examine the contours graph in Figure 5.9 and estimate the location of the local minimum to be at

$$x_0 = (1.7, 4.2)$$

(iii) Execute the following commands:

```
Example_57b=@(x)((x(1)-2).^2+(x(2)-4).^2+3);
options=optimset('TolX',1e-8,'TolFun',1e-8);
[xmin, value]=fminsearch(Example_57b,[1.7 ...
               4.2],options)
```

gives

```
xmin =
  2.0000    4.0000
value =
  3
```

Note that while we use the (x, y) nomenclature in the contour command, we should use the vectorized notation for the variables of the function in the **fminsearch** command. □

IN-CLASS EXERCISE

Pb. 5.22 Find the minimum distance between the point $P(1, 1, 0)$ and the cone defined by the equation

$$z = 2(x^2 + y^2)^{1/2}$$

EXAMPLE 5.8

Assume that we have experimentally obtained a set of data points and we wish to fit them into some line:

$$y = ax + b \tag{5.39}$$

The "best fit" is by convention defined to be the line that minimizes the sum of the squares of the vertical distances of the data points from this line, i.e., we choose a and b, such that

$$\begin{aligned} f(a, b) = {} & (ax_1 + b - y_1)^2 + (ax_{21} + b - y_2)^2 + (ax_3 + b - y_3)^2 \\ & + \cdots + (ax_N + b - y_N)^2 \end{aligned} \tag{5.40}$$

is a minimum.

Consider the data points (0.25, 2.27), (0.5, 2.49), (1, 3.05), (2, 3.95), (2.5, 4.51), and (3, 4.97).

We want to find a and b in this case.

Solution: Editing and executing the following program:

```
%let a=z(1) and b=z(2)
x=[0.25 0.5 1 2 2.5 3];
y=[2.27 2.49 3.05 3.95 4.51 4.97];
squaredistance=@(z)(sum((z(1)*x+z(2)-y).^2));
[z,value]=fminsearch(squaredistance,[1 1])
```

returns

```
z =
   0.9842    2.0227
value =
   0.0049
```

i.e., the best fit is for a =0.9842 and b =2.0227 □

Homework Problems

Pb. 5.23 Using analytical technique (i.e., partial derivatives are zero at the local minimum), show that the best fit of the data is obtained when

$$a = \frac{\left(\sum_{1}^{N} x_i\right)\left(\sum_{1}^{N} y_i\right) - N\left(\sum_{1}^{N} x_i y_i\right)}{\left(\sum_{1}^{N} x_i\right)^2 - N\left(\sum_{1}^{N} x_i^2\right)}$$

$$b = \frac{\left(\sum_{1}^{N} x_i\right)\left(\sum_{1}^{N} x_i y_i\right) - \left(\sum_{1}^{N} x_i^2\right)\left(\sum_{1}^{N} y_i\right)}{\left(\sum_{1}^{N} x_i\right)^2 - N\left(\sum_{1}^{N} x_i^2\right)}$$

Pb. 5.24 In this problem we propose to apply the developed optimization techniques to the important problem of the optical narrow band transmission filter. This filter, in very wide use in optics, consists of two parallel semireflective surfaces (i.e., mirrors) with reflection coatings R_1 and R_2, and separated by a distance L. Assuming that the material between the mirrors has an index of refraction n and that the incoming beam of light has frequency ω and is making an angle θ_i with the normal to the semireflective surfaces, then the ratio of the transmitted light intensity to the incident intensity is

$$T = \frac{I_{\text{transm.}}}{I_{\text{incid.}}} = \frac{(1-R_1)(1-R_2)}{\left(1-\sqrt{R_1R_2}\right)^2 + 4\sqrt{R_1R_2}\,\sin^2\left(\pi\frac{\omega}{\omega_0}\right)}$$

where $\omega_0 = \pi c/nL\cos(\theta_t)$, $\sin(\theta_i) = n\sin(\theta_t)$, and θ_t is the angle that the transmitted light makes with the normal to the mirror surfaces.

In the following activities, we want to understand how the above transmission filter responds as a function of the specified parameters. Choose the following parameters:

$$R_1 = R_2 = 0.8$$
$$0 \le \omega \le 4\omega_0$$

a. Plot T vs. ω/ω_0 for the above frequency range.

b. At what frequencies does the transmission reach a maximum, and a minimum?

(cont'd.)

Homework Problems *(cont'd.)*

c. Devise two methods by which you can tune the filter so that the maximum of the filter transmission is centered around a particular physical frequency.

d. How sharp is the filter? By sharp, we mean: what is the width of the transmission band that allows through at least 50% of the incident light? Define the width relative to ω_0.

e. Answer question (**d**) with the values of the reflection coatings given now by

$$R_1 = R_2 = 0.9$$
$$0 \leq \omega \leq 4\omega_0$$

Does the sharpness of the filter increase or decrease with an increase of the reflection coefficients of the coating surfaces for the two mirrors?

f. Choosing $\omega = \omega_0$, plot a 3-D mesh of T as a function of the reflection coefficients R_1 and R_2. Show, both graphically and numerically, that the best performance occurs when the reflection coatings are the same.

g. Plot the contrast function defined as $C = \frac{T_{max}}{T_{min}}$ as a function of the reflection coefficients R_1 and R_2. How should you choose your mirrors for maximum contrast?

5.5 Finding the Minima of Functions with Constraints Present*

In the previous section, we discussed the techniques for finding the minimum of a function of multiple variables with no constraints present. Most of the optimization problems encountered in engineering for functions of multiple variables are, in contrast, problems with constraints present. These constraints can be of technical/physical nature or often in design problems can be of economic, environmental, or other statutory or regulatory nature. In this section, we shall look for the minima of functions of two variables with one or more constraints using analytical and MATLAB techniques. The more general problems of functions with more than two variables can be solved through the same steps, but will not be examined in detail here.

We shall first discuss the Lagrange multiplier theorem, and then explore the use of the MATLAB command **`fmincon`**, from the Optimization ToolBox, for solving this class of problems.

*The asterisk indicates more advanced material that may be skipped in a first reading.

It should not escape the reader's attention that for a simple constraint function where y can be written as a simple function of x, this problem can be reduced to the optimization of a function in one variable.

5.5.1 Lagrange Multipliers

Mathematically, the problem is often posed as follows: Find the point $P(x, y)$ which makes the value of $f(x, y)$ an extremum subject to the constraint $c(x, y) = d$, where d is a constant.

THEOREM If the functions $f(x, y)$ and $c(x, y)$ are differentiable at $P(x, y)$ and $\vec{\nabla} c \neq 0$, then the extrema of $f(x, y)$ subject to the condition $c(x, y) = d$ can occur only at points where

$$\vec{\nabla} f = \lambda \vec{\nabla} c \tag{5.41}$$

(λ is called the Lagrange multiplier.)

PROOF If the directional derivative of the function $f(x, y)$ in the direction of the tangent to the curve $c(x, y) = d$ is not zero at a point on the curve, then the values of $f(x, y)$ are either increasing or decreasing as the point $P(x, y)$ moves along the curve. This means that the extrema of $f(x, y)$ on the curve $c(x, y) = d$ can only occur where the directional derivative of $f(x, y)$ in the direction of the curve is zero, therefore $\vec{\nabla} f$ must be orthogonal to the curve at those points. Since, in the present case, $\vec{\nabla} f$ and $\vec{\nabla} c$ are coplanar and are both orthogonal to the curve $c(x, y) = d$ at the extrema, then they should be parallel to each other. The proportionality constant relating their magnitude is the Lagrange multiplier.

The problem of finding the extrema of $f(x, y)$ subject to the constraint $c(x, y) = d$ reduces then to solving the following system of equations:

$$\begin{aligned} \frac{\partial f(x,y)}{\partial x} &= \lambda \frac{\partial c(x,y)}{\partial x} \\ \frac{\partial f(x,y)}{\partial y} &= \lambda \frac{\partial c(x,y)}{\partial y} \\ c(x,y) &= d \end{aligned} \tag{5.42}$$

EXAMPLE 5.9

Find the local extrema of the function $f = cx + dy$, subject to the constraint $\frac{x^2}{a^2} + \frac{y^2}{b^2} = 1.$

Solution: The system of equations to solve are

$$\begin{aligned} c &= \lambda \frac{2x}{a^2} \\ d &= \lambda \frac{2y}{b^2} \\ \frac{x^2}{a^2} + \frac{y^2}{b^2} &= 1 \end{aligned} \tag{5.43}$$

Eliminating λ from the first two equations, we find that at the extremum:

$$\frac{x}{a^2 c} = \frac{y}{b^2 d} \tag{5.44}$$

Substituting in the third equation, we obtain the value of x at the extremum:

$$x^2 = a^2 \left(\frac{a^2 c^2}{a^2 c^2 + b^2 d^2} \right) \tag{5.45}$$

giving for y at the extremum the value:

$$y^2 = b^2 \left(\frac{b^2 d^2}{a^2 c^2 + b^2 d^2} \right) \tag{5.46}$$

We have here two extrema, a maximum and a minimum, corresponding to the different signs of the square root of the above expressions. □

5.5.2 MATLAB `fmincon` Built-in Function

This command incorporates within its inputs, the possibility of having the constraints as equalities or inequalities, allow for both linear and nonlinear constraints, and can look for solutions in a bounded region of space. The syntax for this command is

```
[x,value]=fmincon(fun,x0,A,B,Aeq,Beq,lb,ub,nonlcon)
```

`fun` is the function that we are looking to minimize, **`x0`** is the initial guess for the solution, the linear constraints are written in matrix forms, **`A*x<=B`**, **`Aeq*x=Beq`**, the search domain is defined through **`x>lb`** (lower bound), **`x<ub`** (upper bound), and the function **`nonlcon`** is a user defined function that accepts **`x`** and returns the vectors **`c`** and **`ceq`** where **`c(x)<=0`** and `ceq(x)=0`. (The rule is to set to `[]` any of the equalities, inequalities, or bounds that do not apply in the particular problem considered.)

Example 5.10

Find the local minima for each of the given functions subject to the given constraints:

$$\text{(a) } f = \left((x-1)^2 + (y-1)^2\right)^{1/2} + \left((x-3)^2 + (y-1)^2\right)^{1/2} \tag{5.47}$$

subject to the constraints $x \geq 0$, $y \geq 0$, and $y + x = 4$.

$$\text{(b) } f = 4x + 3y \tag{5.48}$$

subject to the constraint

$$\frac{x^2}{25} + \frac{y^2}{25} = 1$$

Solution:

(a) Entering

```
Example_510a=@(x)sqrt((x(1)-1).^2+(x(2)-1).^2)-...
                  sqrt((x(1)-3).^2+(x(2)-1).^2);
Aeq=[1 1]; Beq=[4];
[x value]=fmincon(Example_510a,[1.5; 1.5],...
              [],[],[Aeq],[Beq],[0;0],[],[])
```

gives

```
x =
     0
     4
value=
     -1.0804
```

(b) Define in a *function M-file* the function:

```
function [c,ceq]=Example_510b(x)
c=[];
ceq=x(1).^2+x(2).^2-25;
```

Entering in the command window

```
[x,value]=fmincon(@(x)4*x(1)+3*x(2),[-5;2],...
          [],[],[],[],[],[],@(x)Example_510b(x))
```

gives

```
x=
     -4.0000
     -3.0000
value=
     -25.000
```
□

Homework Problems

Pb. 5.25 A particle is moving on the ellipse defined by the equation

$$\frac{x^2}{4} + \frac{y^2}{9} = 1$$

Which point $P(x, y)$ on this ellipse has the sum of its distances from the points $A(-2, 3)$, $B(2, 1)$, and $C(3, 5)$ a minimum?

Pb. 5.26 A computer manufacturer owns two assembly plants. Each plant can produce each of three models of computers, Model-A, Model-B, and Model-C. The daily production capacity for the two plants are:

- Plant I: 100 Model-A, 300 Model-B, and 500 Model-C
- Plant II: 200 Model-A, 200 Model-B, and 200 Model-C

The marketing department estimates that over the next quarter, the company needs to produce at least 8,000 units of Model-A, 16,000 units of Model-B, and 20,000 units of Model-C.

a. Using the `fmincon` command, help the operation manager schedule the number of production-days each of the plants is to be put on line, if his goal is to minimize the production costs. Assume that it costs an additional $30,000 per day per plant when a plant is on line.

b. Look-up in the MATLAB Help the `linprog` command, and use it to solve this problem.

5.6 MATLAB Commands Review

`besselj`	The built-in BesselJ function.
`fminbnd`	Finds the minimum value of a single variable function or a restricted domain.

fminsearch	Finds the local minimum of a multivariable function.
fmincon	Finds the local minimum of a multivariable function with constraints present.
fsolve	Finds a root to a system of nonlinear equations assuming an initial guess.
fzero	Finds the zero of a single variable function assuming an initial guess.
roots	Finds the roots of a polynomial if the polynomial coefficients are given.
poly	Assembles a polynomial from its roots.

6

Complex Numbers

In this chapter, we start by reviewing the basic algebraic and geometrical properties of complex numbers; we then proceed to show the power of complex numbers as tools for solving a number of basic electrical engineering problems.

6.1 Introduction

Since $x^2 > 0$ for all real numbers x, the equation $x^2 = -1$ admits no real number as a solution. To deal with this problem, mathematicians in the eighteenth century introduced the imaginary number $i = \sqrt{-1} = j$. (So as not to confuse the usual symbol for a current with this quantity, electrical engineers prefer the use of the j symbol. MATLAB accepts either symbol, but always gives the answer with the symbol i.)

Expressions of the form

$$z = a + jb \tag{6.1}$$

where a and b are real numbers, are called complex numbers. As illustrated in Section 6.2, this representation has properties similar to that of an ordered pair (a, b), which is represented by a point in the 2-D plane.

The real number a is called the real part of z, and the real number b is called the imaginary part of z. These numbers are referred to by the symbols $a = \text{Re}(z)$ and $b = \text{Im}(z)$.

When complex numbers are represented geometrically in the x–y coordinate system, the x-axis is called the real axis, the y-axis is called the imaginary axis, and the plane is called the complex plane.

6.2 The Basics

In this section, you will learn how, using MATLAB, you can represent a complex number in the complex plane. It also shows how the addition (or subtraction) of two complex numbers, or the multiplication of a complex number by a real number or by j, can be interpreted geometrically.

Example 6.1

Plot in the complex plane, the three points (P_1, P_2, P_3) representing the complex numbers $z_1 = 1$, $z_2 = j$, $z_3 = -1$.

Solution: Enter and execute the following commands in the command window:

```
z1=1;
z2=j;
z3=-1;
plot(real(z1),imag(z1),'*')
axis([-2 2 -2 2])
axis('square')
hold on
plot(real(z2),imag(z2),'o')
plot(real(z3),imag(z3),'x')
hold off
```

Alternately, a complex number as an argument in the **plot** command is interpreted by MATLAB to mean: take the real part of the complex number to be the x-coordinate and the imaginary part of the complex number to be the y-coordinate. This alternative syntax works when the imaginary part of the complex number is not zero. □

6.2.1 Addition

Next, we define addition for complex numbers. The rule can be directly deduced from analogy of addition of two vectors in a plane: the x-component of the sum of two vectors is the sum of the x-components of each of the vectors, and similarly for the y-component. Therefore

if
$$z_1 = a_1 + jb_1 \tag{6.2}$$

and
$$z_2 = a_2 + jb_2 \tag{6.3}$$

then
$$z_1 + z_2 = (a_1 + a_2) + j(b_1 + b_2) \tag{6.4}$$

The addition or subtraction rules for complex numbers are geometrically translated through the parallelogram rules for the addition and subtraction of vectors.

Example 6.2

Find the sum and difference of the complex numbers:

$$z_1 = 1 + 2j \quad \text{and} \quad z_2 = 2 + j$$

Solution: Grouping the real and imaginary parts separately, we obtain

$$z_1 + z_2 = 3 + 3j$$

and

$$z_1 - z_2 = -1 + j \quad \square$$

PREPARATORY EXERCISE

Pb. 6.1 Given the complex numbers z_1, z_2, and z_3 corresponding to the vertices P_1, P_2, and P_3 of a parallelogram, find z_4 corresponding to the fourth vertex P_4. (Assume that P_4 and P_2 are opposite vertices of the parallelogram.) Verify your answer graphically for the case:

$$z_1 = 2 + j, \quad z_2 = 1 + 2j, \quad z_3 = 4 + 3j$$

6.2.2 Multiplication by a Real or Imaginary Number

If we multiply the complex number $z = a + jb$ by a real number k, the resultant complex number is given by

$$k \times z = k \times (a + jb) = ka + jkb \tag{6.5}$$

What happens when we multiply by j?

Let us, for a moment, return to Example 6.1. We note the following properties for the three points P_1, P_2, and P_3:

1. The three points are equidistant from the origin of the axis.
2. The point P_2 is obtained from the point P_1 by a $\pi/2$ counterclockwise rotation.
3. The point P_3 is obtained from the point P_2 through another $\pi/2$ counterclockwise rotation.

We also note, by examining the algebraic forms of z_1, z_2, and z_3 that

$$z_2 = jz_1 \quad \text{and} \quad z_3 = jz_2 = j^2 z_1 = -z_1$$

That is, multiplying by j is geometrically equivalent to a counterclockwise rotation by an angle of $\pi/2$.

6.2.3 Multiplication of Two Complex Numbers

The multiplication of two complex numbers follows the same rules of algebra for real numbers, but considers $j^2 = -1$. This yields if

$$z_1 = a_1 + jb_1 \quad \text{and} \quad z_2 = a_2 + jb_2$$

$$\Rightarrow z_1 z_2 = (a_1 a_2 - b_1 b_2) + j(a_1 b_2 + b_1 a_2) \qquad (6.6)$$

Preparatory Exercises

Solve the following problems analytically:

Pb. 6.2 Find $z_1 z_2$, z_1^2, z_2^2 for the following pairs:

a. $z_1 = 3j; \quad z_2 = 1 - j$

b. $z_1 = 4 + 6j; \quad z_2 = 2 - 3j$

c. $z_1 = \frac{1}{3}(2 + 4j); \quad z_2 = \frac{1}{2}(1 - 5j)$

d. $z_1 = \frac{1}{3}(2 - 4j); \quad z_2 = \frac{1}{2}(1 + 5j)$

Pb. 6.3 Find the real quantities m and n in each of the following equations:

a. $mj + n(1 + j) = 3 - 2j$

b. $m(2 + 3j) + n(1 - 4j) = 7 + 5j$

(*Hint*: Two complex numbers are equal if separately the real and imaginary parts are equal.)

Pb. 6.4 Write the answers in standard form: (i.e., $a + jb$)

a. $(3 - 2j)^2 - (3 + 2j)^2$

b. $(7 + 14j)^7$

c. $\left[(2 + j)\left(\frac{1}{2} + 2j\right)\right]^2$

d. $j(1 + 7j) - 3j(4 + 2j)$

Pb. 6.5 Show that for all complex numbers z_1, z_2, and z_3, we have the following properties:

$$z_1 z_2 = z_2 z_1 \text{ (commutativity property)}$$

$$z_1(z_2 + z_3) = z_1 z_2 + z_1 z_3 \text{ (distributivity property)}$$

Pb. 6.6 Consider the triangle $\Delta(ABC)$ shown in Figure 6.1, in which D is the midpoint of the BC segment, and let the point G be defined such that

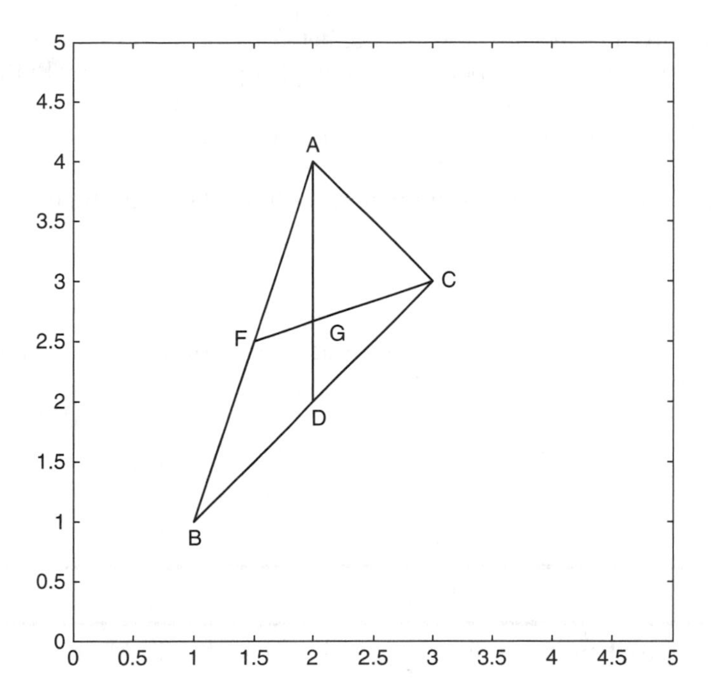

FIGURE 6.1
The center of mass of a triangle.

$(GD) = \frac{1}{3}(AD)$. Assuming that z_A, z_B, z_C are the complex numbers representing the points (A, B, C).

a. Find the complex number z_G that represents the point G.
b. Show that $(CG) = \frac{2}{3}(CF)$ if F is the midpoint of the segment AB.

6.3 Complex Conjugation and Division

DEFINITION The complex conjugate of a complex number z, which is denoted by $\bar{z}$, is given by

$$\bar{z} = a - jb \quad \text{if } z = a + jb \tag{6.7}$$

That is, $\bar{z}$ is obtained from z by reversing the sign of Im(z). Geometrically, z and $\bar{z}$ form a pair of symmetric points with respect to the real axis (x-axis) in the complex plane.

In MATLAB, complex conjugation is written as **`conj(z)`**.

DEFINITION The modulus of a complex number $z = a + jb$, denoted by $|z|$, is given by

$$|z| = \sqrt{a^2 + b^2} \tag{6.8}$$

Geometrically, it represents the distance between the origin and the point representing the complex number z in the complex plane, which by Pythagorean theorem is given by the same quantity.

In MATLAB, the modulus of z is denoted by **abs(z)**.

THEOREM For any complex number z, we have the result that:

$$|z|^2 = \bar{z}z \tag{6.9}$$

PROOF Using the above two definitions for the complex conjugate and the norm, we can write

$$\bar{z}z = (a - jb)(a + jb) = a^2 + b^2 = |z|^2$$

IN-CLASS EXERCISE

Solve the problem analytically, and then use MATLAB to verify your answers.

Pb. 6.7 Let $z = 3 + 4j$. Find $|z|$, $\bar{z}$, and $z\bar{z}$. Verify the above theorem.

6.3.1 Application to Division

Using the above definitions and theorem, we now want to compute the inverse of a complex number with respect to the multiplication operation. We write the results in standard form.

$$z^{-1} = \frac{1}{z} = \frac{1}{(a + jb)}\left(\frac{a - jb}{a - jb}\right) = \frac{a - jb}{a^2 + b^2} = \frac{\bar{z}}{|z|^2} \tag{6.10}$$

from which we deduce that

$$\text{Re}\left(\frac{1}{z}\right) = \frac{\text{Re}(z)}{[\text{Re}(z)]^2 + [\text{Im}(z)]^2} \tag{6.11}$$

and

$$\text{Im}\left(\frac{1}{z}\right) = \frac{-\text{Im}(z)}{[\text{Re}(z)]^2 + [\text{Im}(z)]^2} \tag{6.12}$$

To summarize the above results, and to help you build your syntax for the quantities defined in this section, edit the following *script M-file* and execute it:

```
z=3+4*j
z_bar=conj(z)
modul_z=abs(z)
modul2_z=sqrt(z*conj(z))  %alternate way to compute
                          % module
inv_z=1/z
re_inv_z=real(1/z)
im_inv_z=imag(1/z)
```

IN-CLASS EXERCISES

Pb. 6.8 Analytically and numerically, obtain in the standard form an expression for each of the following quantities:

$$\text{a. } \frac{3+4j}{2+5j} \qquad \text{b. } \frac{\sqrt{3}+j}{(1-j)(3+j)} \qquad \text{c. } \left[\frac{1-2j}{2+3j}-\frac{3+j}{2j}\right]$$

Pb. 6.9 For any pair of complex numbers z_1 and z_2, show that

$$\begin{aligned} \overline{z_1+z_2} &= \overline{z}_1+\overline{z}_2 \\ \overline{z_1-z_2} &= \overline{z}_1-\overline{z}_2 \\ \overline{z_1 z_2} &= \overline{z}_1\overline{z}_2 \\ \overline{(z_1/z_2)} &= \overline{z}_1/\overline{z}_2 \\ \overline{\overline{z}} &= z \end{aligned}$$

6.4 Polar Form of Complex Numbers

If we use polar coordinates, we can write the real and imaginary parts of a complex number $z = a + jb$ in terms of the modulus of z and the polar angle θ:

$$a = r\cos(\theta) = |z|\cos(\theta) \tag{6.13}$$

$$b = r\sin(\theta) = |z|\sin(\theta) \tag{6.14}$$

and the complex number z can then be written in polar form as

$$z = |z|\cos(\theta) + j|z|\sin(\theta) = |z|(\cos(\theta) + j\sin(\theta)) \tag{6.15}$$

The angle θ is called the argument of z and is usually evaluated in the interval $-\pi \leq \theta \leq \pi$. However, we still have the same complex number if we added to the value of θ an integer multiple of 2π.

$$\begin{aligned} \theta &= \arg(z) \\ \tan(\theta) &= \frac{b}{a} \end{aligned} \tag{6.16}$$

From the above results, it is obvious that the argument of the complex conjugate of a complex number is equal to minus the argument of this complex number.

In MATLAB, the command for arg(z) is **`angle(z)`**.

IN-CLASS EXERCISE

Pb. 6.10 Find the modulus and argument for each of the following complex numbers:

$$z_1 = 1 + 2j; \quad z_2 = 2 + j; \quad z_3 = 1 - 2j; \quad z_4 = -1 + 2j; \quad z_5 = -1 - 2j$$

Plot these points. Can you detect any geometrical pattern? Generalize.

The main advantage of writing complex numbers in polar form is that it makes the multiplication and division operations more transparent, and provides a simple geometric interpretation to these operations, as shown below.

6.4.1 New Insights into Multiplication and Division of Complex Numbers

Consider the two complex numbers z_1 and z_2 written in polar form:

$$z_1 = |z_1|(\cos(\theta_1) + j\sin(\theta_1)) \tag{6.17}$$

$$z_2 = |z_2|(\cos(\theta_2) + j\sin(\theta_2)) \tag{6.18}$$

Their product $z_1 z_2$ is given by

$$z_1 z_2 = |z_1||z_2| \begin{bmatrix} (\cos(\theta_1)\cos(\theta_2) - \sin(\theta_1)\sin(\theta_2)) \\ +\ j(\sin(\theta_1)\cos(\theta_2) + \cos(\theta_1)\sin(\theta_2)) \end{bmatrix} \tag{6.19}$$

But using the trigonometric identities for the sine and cosine of the sum of two angles:

$$\cos(\theta_1 + \theta_2) = \cos(\theta_1)\cos(\theta_2) - \sin(\theta_1)\sin(\theta_2) \tag{6.20}$$

$$\sin(\theta_1 + \theta_2) = \sin(\theta_1)\cos(\theta_2) + \cos(\theta_1)\sin(\theta_2) \tag{6.21}$$

the product of two complex numbers can then be written in the simpler form:

$$z_1 z_2 = |z_1|\,|z_2|\,[\cos(\theta_1 + \theta_2) + j\sin(\theta_1 + \theta_2)] \tag{6.22}$$

That is, when multiplying two complex numbers, the modulus of the product is the product of the moduli, while the argument is the sum of arguments:

$$|z_1 z_2| = |z_1|\,|z_2| \tag{6.23}$$

$$\arg(z_1 z_2) = \arg(z_1) + \arg(z_2) \tag{6.24}$$

The above result can be generalized to the product of n complex numbers and the result is

$$|z_1 z_2 \ldots z_n| = |z_1|\,|z_2|\,\ldots\,|z_n| \tag{6.25}$$

$$\arg(z_1 z_2 \ldots z_n) = \arg(z_1) + \arg(z_2) + \cdots + (z_n) \tag{6.26}$$

A particular form of this expression is the De Moivre theorem, which states that

$$(\cos(\theta) + j\sin(\theta))^n = \cos(n\theta) + j\sin(n\theta) \tag{6.27}$$

The above results suggest that the polar form of a complex number may be written as a function of an exponential function because of the additivity of the arguments upon multiplication. We revisit this issue later.

In-Class Exercises

Pb. 6.11 Show that $\dfrac{z_1}{z_2} = \dfrac{|z_1|}{|z_2|}[\cos(\theta_1 - \theta_2) + j\sin(\theta_1 - \theta_2)]$.

Pb. 6.12 Explain, using Eqs. (6.23) and (6.24) why multiplication of any complex number by j is equivalent to a rotation of the point representing this number in the complex plane by $\pi/2$.

Pb. 6.13 By what angle must we rotate point $P(3, 4)$ to transform it to point $P'(4, 3)$?

Pb. 6.14 The points $z_1 = 1 + 2j$ and $z_2 = 2 + j$ are adjacent vertices of a regular hexagon.

a. Find the vertex z_3 that is also a vertex of the same hexagon and that is adjacent to z_2 ($z_3 \neq z_1$).

b. Write a difference equation which generates the other vertices of this hexagon.
c. Develop a program to plot this hexagon.

Pb. 6.15 Show that the points A, B, and C representing the complex numbers z_A, z_B, and z_C in the complex plane lie on the same straight line if and only if

$$\frac{z_A - z_C}{z_B - z_C} \quad \text{is real.}$$

Pb. 6.16 Determine the coordinates of the P' point obtained from the point $P(2, 4)$ through a reflection around the line $y = \frac{x}{2} + 2$.

Pb. 6.17 Consider two points A and B representing, in the complex plane, the complex numbers z_1 and $1/\bar{z}_1$. Let P be any point on the circle of radius 1 and centered at the origin (the unit circle). Show that the ratio of the length of the line segments PA and PB is the same, regardless of the position of point P on the unit circle.

Pb. 6.18 Find the polar form of each of the following quantities:

$$\frac{(1+j)^{15}}{(1-j)^9} \qquad \sqrt{(-1+j)(j+1)} \qquad (1+j+j^2+j^3)^{99}$$

6.4.2 Roots of Complex Numbers

Given the value of the complex number z, we are interested here in finding the solutions of the equation

$$v^n = z \tag{6.28}$$

Let us write both the solutions and z in polar forms;

$$v = \rho(\cos(\alpha) + j\sin(\alpha)) \tag{6.29}$$

$$z = r(\cos(\theta) + j\sin(\theta)) \tag{6.30}$$

From the De Moivre theorem, the expression for $v^n = z$ can be written as

$$\rho^n(\cos(n\alpha) + j\sin(n\alpha)) = r(\cos(\theta) + j\sin(\theta)) \tag{6.31}$$

Comparing the moduli of both sides, we deduce by inspection that

$$\rho = \sqrt[n]{r} \tag{6.32}$$

The treatment of the argument should be done with great care. Recalling that two angles have the same cosine and sine if they are equal or differ from each other by an integer multiple of 2π, we can then deduce that

$$n\alpha = \theta + 2k\pi \quad k = 0, \pm 1, \pm 2, \pm 3, \ldots \tag{6.33}$$

Therefore, the general expression for the roots is:

$$z^{1/n} = r^{1/n}\left(\cos\left(\frac{\theta}{n} + \frac{2k\pi}{n}\right) + j\sin\left(\frac{\theta}{n} + \frac{2k\pi}{n}\right)\right) \\ \text{with } k = 0, 1, 2, \ldots, (n-1) \tag{6.34}$$

Note that the roots reproduce themselves outside the range $k = 0, 1, 2, \ldots, (n - 1)$.

IN-CLASS EXERCISE

Pb. 6.19 Calculate the roots of the equation $z^5 - 32 = 0$, and plot them in the complex plane.

- **a.** What geometric shape does the polygon with the solutions as vertices form?
- **b.** What is the sum of these roots? (Derive your answer both algebraically and geometrically.)

6.4.3 The Function $y = e^{j\theta}$

As alluded to previously, the expression $\cos(\theta) + j\sin(\theta)$ behaves very much as if it was an exponential; because of the additivity of the arguments of each term in the argument of the product, we denote this quantity by

$$e^{j\theta} = \cos(\theta) + j\sin(\theta) \tag{6.35}$$

PROOF Compute the Taylor expansion for both sides of the above equation. The series expansion for $e^{j\theta}$ is obtained by evaluating Taylor's formula at $x = j\theta$, giving (see Appendix D)

$$e^{j\theta} = \sum_{n=0}^{\infty} \frac{1}{n!}(j\theta)^n \tag{6.36}$$

When this series expansion for $e^{j\theta}$ is written in terms of its even part and odd part, we have the result

$$e^{j\theta} = \sum_{m=0}^{\infty} \frac{1}{(2m)!}(j\theta)^{2m} + \sum_{m=0}^{\infty} \frac{1}{(2m+1)!}(j\theta)^{2m+1} \tag{6.37}$$

However, since $j^2 = -1$, this last equation can also be written as

$$e^{j\theta} = \sum_{m=0}^{\infty} \frac{(-1)^m}{(2m)!}(\theta)^{2m} + j\sum_{m=0}^{\infty} \frac{(-1)^m}{(2m+1)!}(\theta)^{2m+1} \tag{6.38}$$

which, by inspection, can be verified to be the sum of the Taylor expansions for the cosine and sine functions.

In this notation, the product of two complex numbers z_1 and z_2 is: $r_1 r_2 e^{j(\theta_1+\theta_2)}$ It is then a simple matter to show that

if
$$z = r\exp(j\theta) \tag{6.39}$$

then
$$\bar{z} = r\exp(-j\theta) \tag{6.40}$$

and
$$z^{-1} = \frac{1}{r}\exp(-j\theta) \tag{6.41}$$

from which we can deduce Euler's equations:

$$\cos(\theta) = \frac{\exp(j\theta) + \exp(-j\theta)}{2} \tag{6.42}$$

and

$$\sin(\theta) = \frac{\exp(j\theta) - \exp(-j\theta)}{2j} \tag{6.43}$$

Example 6.3

Use MATLAB to generate the graph of the unit circle in the complex plane.

Solution: Because all points on the unit circle are equidistant from the origin and their distance to the origin (their modulus) is equal to 1, we can generate the circle by plotting the N roots of unity, taking a very large value for N. This can be implemented by executing the following *script M-file.*

```
N=720;
z=exp(j*2*pi*[1:N]./N);
plot(z)
axis square
```
□

In-Class Exercises

Pb. 6.20 Using the exponential form of the n-roots of unity, and the expression for the sum of a geometric series (given in Appendix D), show that the sum of these roots is zero.

Pb. 6.21 Compute the following sums:

a. $1 + \cos(x) + \cos(2x) + \cdots + \cos(nx)$
b. $\sin(x) + \sin(2x) + \cdots + \sin(nx)$
c. $\cos(a) + \cos(a + b) + \cdots + \cos(a + nb)$
d. $\sin(a) + \sin(a + b) + \cdots + \sin(a + nb)$

Pb. 6.22 Verify numerically that for $z = x + jy$,

$$\lim_{n \to \infty} \left(1 + \frac{z}{n}\right)^n = \exp(x)(\cos(y) + j\sin(y))$$

For what values of y is this quantity pure imaginary?

Homework Problems

Pb. 6.23 Plot the curves determined by the following parametric representations:

a. $z = 1 - jt \quad 0 \le t \le 2$
b. $z = t + jt^2 \quad -\infty < t < \infty$

c. $z = 2(\cos(t) + j\sin(t)) \quad \frac{\pi}{2} < t < \frac{3\pi}{2}$

d. $z = 3(t + j - j\exp(-jt)) \quad 0 < t < \infty$

Pb. 6.24 Find the expression $f(x,y)=0$ and plot the families of curves defined by each of the corresponding equations:

a. $\mathrm{Re}\left(\frac{1}{z}\right) = 2$ **b.** $\mathrm{Im}\left(\frac{1}{z}\right) = 2$

c. $\mathrm{Re}(z^2) = 4$ **d.** $\mathrm{Im}(z^2) = 4$

e. $\left|\frac{z-3}{z+3}\right| = 1$ **f.** $\arg\left(\frac{z-3}{z+3}\right) = \frac{\pi}{4}$

g. $|z - 1| = 3$ **h.** $|z| = \mathrm{Im}(z) + 4$

(cont'd.)

Homework Problems (*cont'd.*)

Pb. 6.25 Find the image of the line $\text{Re}(z) = 1$ upon the transformation $z' = z^2 + z$. (First obtain the result analytically, and then verify it graphically.)

Pb. 6.26 Consider the following bilinear transformation: $z' = \dfrac{az + b}{cz + d}$

Show how with proper choices of the constants a, b, c, and d, we can generate all transformations of planar geometry (i.e., scaling, rotation, translation, and inversion).

Pb. 6.27 Plot the curves C' generated by the points P' that are the images of points on the circle centered at (3, 4) and of radius 5 under the transformation of the preceding problem, with the following parameters:

Case 1: $a = \exp(j\pi/4)$, $b = 0$, $c = 0$, $d = 1$
Case 2: $a = 1$, $b = 3$, $c = 0$, $d = 1$
Case 3: $a = 0$, $b = 1$, $c = 1$, $d = 0$

6.5 Analytical Solutions of Constant Coefficients ODE

Finding the solutions of an ODE with constant coefficients is conceptually very similar to solving the linear difference equation with constant coefficients. We repeat the exercise here for its pedagogical benefits and to bring out some of the finer technical details peculiar to the ODEs of particular interest for later discussions.

The linear differential equation of interest is given by

$$a_n \frac{d^n y}{dt^n} + a_{n-1} \frac{d^{n-1} y}{dt^{n-1}} + \cdots + a_1 \frac{dy}{dt} + a_0 y = u(t) \tag{6.44}$$

In this section, we find the solutions of this ODE for the cases that $u(t) = 0$ and $u(t) \neq 0$.

The solutions for the first case are referred to as the homogeneous solutions. By substitution, it is a trivial matter to verify that if $y_1(t)$ and $y_2(t)$ are solutions, then $c_1 y_1(t) + c_2 y_2(t)$, where c_1 and c_2 are constants, is also a solution. This is, as previously mentioned, referred to as the superposition principle for linear systems.

If $u(t) \neq 0$, the general solution of the ODE will be the sum of the corresponding homogeneous solution and the particular solution peculiar to the specific details of $u(t)$. Furthermore, by inspection, it is clear that if the source can be decomposed into many components, then the particular solution can be written as the sum of the particular solutions for the different components and with the same weights as in the source. This property characterizes a linear system.

DEFINITION A system L is considered linear if

$$L(c_1u_1(t) + c_2u_2(t) + \cdots + c_nu_n(t)) = c_1L(u_1(t)) + c_2L(u_2(t)) + \cdots + c_nL(u_n(t)) \tag{6.45}$$

where the c's are constants and the u's are time-dependent source signals.

6.5.1 Transient Solutions

To obtain the homogeneous solutions, we set $u(t) = 0$. We guess that the solution to this homogeneous differential equation is $y = \exp(st)$. You may wonder why we made this guess; the secret is in the property of the exponential function, whose derivative is proportional to the function itself. That is

$$\frac{d(\exp(st))}{dt} = s\exp(st) \tag{6.46}$$

Through this substitution, the above ODE reduces to an algebraic equation, and the solution of this algebraic equation then reduces to finding the roots of the polynomial:

$$a_ns^n + a_{n-1}s^{n-1} + \cdots + a_1s + a_0 = 0 \tag{6.47}$$

We learned in Chapter 5 the MATLAB command for finding these roots, when needed. Now, using the superposition principle, and assuming all the roots are distinct, the general solution of the homogeneous differential equation is given by

$$y_{\text{homog.}} = c_1\exp(s_1t) + c_2\exp(s_2t) + \cdots + c_n\exp(s_nt) \tag{6.48}$$

where $s_1, s_2, \ldots, s_n$ are the above roots and $c_1, c_2, \ldots, c_n$ are constant determined from the initial conditions of the solution and all its derivatives to order $n - 1$.

NOTE In the case that two or more of the roots are equal, it is easy to verify that the solution of the homogeneous ODE includes, instead of a constant multiplied by the exponential term corresponding to that root, a polynomial multiplying the exponential function. The degree of this polynomial is $(m - 1)$ if m is the degeneracy of the root in question.

EXAMPLE 6.4

Find the transient solutions to the second-order differential equation

$$a\frac{d^2y}{dt^2} + b\frac{dy}{dt} + cy = 0 \tag{6.49}$$

Solution: The characteristic polynomial associated with this ODE is the second-degree equation given by

$$as^2 + bs + c = 0 \tag{6.50}$$

The roots of this equation are $s_\pm = \dfrac{-b \pm \sqrt{b^2 - 4ac}}{2a}$

The nature of the solutions is very dependent on the sign of the discriminant $(b^2 - 4ac)$:

- If $b^2 - 4ac > 0$, the two roots are distinct and real. Call these roots α_1 and α_2; the solution is then

  $$y_{\text{homog.}} = c_1 \exp(\alpha_1 t) + c_2 \exp(\alpha_2 t) \tag{6.51}$$

  In many physical problems of interest, we desire solutions that are zero at infinity, that is, decay over a finite time. This requires that both α_1 and α_2 be negative; or if only one of them is negative, that the c coefficient of the exponentially increasing solution be zero. This class of solutions is called the over-damped class.
- If $b^2 - 4ac = 0$, the two roots are equal, and we call this root $\alpha_{\text{degen.}}$. In this case, we can verify, by inspection, that the function $t \exp(\alpha_{\text{degen.}})$ is also a solution of the second-order ODE, and the homogeneous solution to the differential equation is of the form:

  $$y_{\text{homog.}} = (c_1 + c_2 t) \exp(\alpha_{\text{degen.}} t) \tag{6.52}$$

  As pointed earlier, the polynomial, multiplying the exponential function, is of degree 1 here because the degeneracy of the root is of degree 2. This class of solutions is referred to as the critically damped class.
- If $b^2 - 4ac < 0$, the two roots are complex conjugates of each other, and their real part is negative for physically interesting cases. If we denote these roots by $s_\pm = -\alpha \pm j\beta$, the solutions to the homogeneous differential equations take the form:

  $$y_{\text{homog.}} = \exp(-\alpha t)(c_1 \cos(\beta t) + c_2 \sin(\beta t)) \tag{6.53}$$

  This class of solutions is referred to as the under-damped class. □

In-Class Exercises

Find and plot the transient solutions to the following homogeneous equations, using the indicated initial conditions:

Pb. 6.28 $a = 1, b = 3, c = 2 \quad y(t = 0) = 1 \quad y'(t = 0) = -3/2$

Pb. 6.29 $a = 1, b = 2, c = 1 \quad y(t = 0) = 1 \quad y'(t = 0) = 2$

Pb. 6.30 $a = 1, b = 2, c = 2 \quad y(t = 0) = 4 \quad y'(t = 0) = -4$

6.5.2 Solution in the Presence of a Source: Green Function Technique

We discussed above the general technique for finding the homogeneous solutions for constant coefficients ODE. We illustrated the technique by applying it to a second-order differential equation. In this section, we shall discuss a general technique for finding the solution of the inhomogeneous equation for the case of a linear oscillator subject to a time-dependent source term. This technique is called the Green function method. Essentially, what we shall prove is that once the Green function for the particular ODE is found, finding the inhomogeneous solution to this equation reduces to performing an integration. In Appendix C, we discuss another technique for finding the solution of inhomogeneous solutions of constant coefficients ODE, using Laplace transforms.

The second-order differential equation describing the linear oscillator is

$$a\frac{d^2y}{dt^2} + b\frac{dy}{dt} + cy = f(t) \tag{6.54}$$

To obtain the solution of this equation, we shall sequentially find the solutions of this equation when the source term is a step function, an impulse function, a spike function, and finally an arbitrary function.

(a) *Response to a step function:* The step function at the arbitrary point t_0 is the product of a scalar by the unit step function defined in Section 3.9. Specifically

$$H(t_0) = AU(t_0) = \begin{cases} 0 & \text{for } t < t_0 \\ A & \text{for } t > t_0 \end{cases} \tag{6.55}$$

We shall solve this problem with the initial conditions

$$y(t_0) = 0 \text{ and } y'(t_0) = 0. \tag{6.56}$$

Furthermore, we shall assume that its homogeneous solution can be written in the form of Eq. (6.53). The particular solution to this ODE is simply

$$y_{\text{partic.}}(t) = \frac{A}{c} \tag{6.57}$$

and the general solution is the sum of the homogeneous and the particular solutions. Applying the above initial conditions yields

$$c_1 = -\frac{A}{c}, \quad c_2 = -\frac{\alpha A}{\beta c} \tag{6.58}$$

The general solution for the step function is for $t > t_0$,

$$y(t) = \frac{A}{c}\left[1 - e^{-\alpha(t-t_0)}\cos(\beta(t-t_0)) - \frac{\alpha}{\beta}e^{-\alpha(t-t_0)}\sin(\beta(t-t_0))\right] \quad (6.59)$$

(b) *Response to an impulse function:* If we consider the impulse (rectangle) function as the difference between the step functions $H(t_0)$ and $H(t_0 + \tau)$, where τ is the pulse width, we can, using the superposition principle, immediately deduce the solution of the ODE for the impulse function for $t > t_0 + \tau$:

$$y(t) = \frac{A}{c}e^{-\alpha(t-t_0)}\begin{bmatrix} e^{\alpha\tau}\cos(\beta(t-t_0-\tau)) - \cos(\beta(t-t_0)) + \\ \frac{\alpha}{\beta}e^{\alpha\tau}\sin(\beta(t-t_0-\tau)) - \frac{\alpha}{\beta}\sin(\beta(t-t_0)) \end{bmatrix} \quad (6.60)$$

(c) *Response to a spike (Dirac delta) function:* The spike function is an impulse function with zero width, but such that the product of its height by its width is equal to 1. Therefore, to find the solution in this case, we take the solution for the impulse and expand it in powers of τ, keeping only the constant and linear terms in τ. Noting that to first order in τ, the above factors can be written as

$$\begin{aligned} e^{\alpha\tau} &\approx 1 + \alpha\tau \\ \cos(\beta(t-t_0-\tau)) &= \cos(\beta(t-t_0))\cos(\beta\tau) + \sin(\beta(t-t_0))\sin(\beta\tau) \\ &\approx \cos(\beta(t-t_0)) + (\beta\tau)\sin(\beta(t-t_0)) \\ \sin(\beta(t-t_0-\tau)) &= \sin(\beta(t-t_0))\cos(\beta\tau) - \cos(\beta(t-t_0))\sin(\beta\tau) \\ &\approx \sin(\beta(t-t_0)) - (\beta\tau)\cos(\beta(t-t_0)) \end{aligned} \quad (6.61)$$

Substituting these in the solution of the ODE for the impulse function, we obtain the solution for the spike's excitation:

$$y(t) = \frac{(A\tau)}{c}e^{-\alpha(t-t_0)}\left[\left(\beta + \frac{\alpha^2}{\beta}\right)\sin(\beta(t-t_0))\right] \quad (6.62)$$

Recalling that

$$A\tau = 1,\ \alpha = -\frac{b}{2a}, \quad \text{and} \quad \beta^2 = \frac{4ac - b^2}{4a^2} \quad (6.63)$$

the solution for the spike centered at t_0 is

$$y(t) = G(t, t_0) = \begin{cases} \dfrac{1}{\beta a} e^{-\alpha(t-t_0)} \sin(\beta(t-t_0)) & \text{for } t > t_0 \\ 0 & \text{for } t < t_0 \end{cases} \tag{6.64}$$

This solution is what is referred to as the Green function of the ODE.

(d) *Response to an arbitrary function:* An arbitrary function can be approximated by a series of impulses of heights $f_n(t_n)$ centered at the points t_n's and such that each of these impulses has a width τ.
Using the superposition principle, the solution for all including the N impulses for $t_N < t < t_{N+1}$ is given by

$$y(t) = \sum_{n=-\infty}^{N} \frac{f_n(t_n)\tau}{\beta a} e^{-\alpha(t-t_n)} \sin(\beta(t-t_n)) \tag{6.65}$$

This expression can be written in the limit that $\tau \to 0$ as an integral, and the solution reduces to

$$y(t) = \int_{-\infty}^{t} \frac{f(t')}{\beta a} e^{-\alpha(t-t')} \sin(\beta(t-t'))\, dt' \tag{6.66}$$

which is in the form promised earlier.

Homework Problem

Pb. 6.31 Find the response of the linear oscillator to an excitation of the form

$$f(t) = \begin{cases} F_0 e^{-\gamma t} & \text{for } t \geq 0 \\ 0 & \text{for } t < 0 \end{cases}$$

Discuss and identify the different limits of the solution.
Plot the solutions for the following cases:

a. $\alpha = 0.1\sqrt{c/a}$, $\gamma = 0.3\sqrt{c/a}$.
b. $\alpha = 0.2\sqrt{c/a}$, $\gamma = 0.2\sqrt{c/a}$.
c. $\alpha = 0.3\sqrt{c/a}$, $\gamma = 0.1\sqrt{c/a}$.

6.5.3 Steady-State Solutions

In this subsection, we find the particular solutions of the ODEs when the driving force is a single-term sinusoidal of an infinite extent. A different technique than that of Section 6.5.2 and which is more suitable for the steady-state solution is presented here.

As pointed out previously, because of the superposition principle, it is also possible to write the steady-state solution for any combination of such inputs. This, combined with the Fourier series techniques (briefly discussed in Chapter 7), will also allow you to write the solution for any periodic function.

We will discuss in detail the particular solution for the first- and the second-order differential equations because these represent, as previously shown in Section 4.7, important cases in circuit analysis.

EXAMPLE 6.5

Find the particular solution to the first-order differential equation:

$$a\frac{dy}{dt} + by = A\cos(\omega t) \qquad (6.67)$$

Solution: We guess that the particular solution of this ODE is a sinusoidal of the form:

$$\begin{aligned} y_{\text{partic.}}(t) = B\cos(\omega t - \phi) &= B[\cos(\phi)\cos(\omega t) + \sin(\phi)\sin(\omega t)] \\ &= B_c\cos(\omega t) + B_s\sin(\omega t) \end{aligned} \qquad (6.68)$$

Our task now is to find B_c and B_s that would force Eq. (6.68) to be the solution of Eq. (6.67). Therefore, we substitute this trial solution in the differential equation and require that, separately, the coefficients of $\sin(\omega t)$ and $\cos(\omega t)$ terms match on both sides of the resulting equation. These requirements are necessary for the trial solution to be valid at all times. The resulting conditions are:

$$B_s = \frac{a\omega}{b}B_c \qquad B_c = \frac{Ab}{a^2\omega^2 + b^2} \qquad (6.69)$$

from which we can also deduce the polar form of the solution, giving

$$B^2 = \frac{A^2}{a^2\omega^2 + b^2} \qquad \tan(\phi) = \frac{a\omega}{b} \qquad (6.70)$$

□

EXAMPLE 6.6

Find the particular solution to the second-order differential equation:

$$a\frac{d^2y}{dt^2} + b\frac{dy}{dt} + cy = A\cos(\omega t) \tag{6.71}$$

Solution: Again, take the trial particular solution to be of the form

$$\begin{aligned} y_{\text{partic.}}(t) = B\cos(\omega t - \phi) &= B[\cos(\phi)\cos(\omega t) + \sin(\phi)\sin(\omega t)] \\ &= B_c \cos(\omega t) + B_s \sin(\omega t) \end{aligned} \tag{6.72}$$

Repeating the same steps as in Example 6.5, we find

$$B_s = \frac{b\omega}{(c - a\omega^2)^2 + \omega^2 b^2} A \qquad B_c = \frac{(c - a\omega^2)}{(c - a\omega^2)^2 + \omega^2 b^2} \tag{6.73}$$

$$B^2 = \frac{A^2}{(c - a\omega^2)^2 + \omega^2 b^2} \qquad \tan(\phi) = \frac{b\omega}{c - a\omega^2} \tag{6.74}$$

□

6.5.4 Applications to Circuit Analysis

An important application of the sinusoidal forms for the particular solutions is in circuit analysis with inductors, resistors, and capacitors as elements in the presence of ac source potentials. We describe later a more efficient analytical method (phasor representation) for solving this kind of problem; however, we believe that it is important that you also become familiar with the present technique.

6.5.4.1 RC Circuit

Referring to the *RC* circuit shown in Figure 4.4, we derived the differential equation that the potential difference across the capacitor must satisfy; namely

$$RC\frac{dV_C}{dt} + V_C = V_0 \cos(\omega t) \tag{6.75}$$

This is a first-order differential equation, the particular solution of which is given in Example 6.5 if we were to identify the coefficients in the ODE as follows: $a = RC$, $b = 1$, $A = V_0$.

6.5.4.2 RLC Circuit

Referring to the circuit, shown in Figure 4.5, the voltage across the capacitor satisfies the following ODE:

$$LC\frac{d^2V_c}{dt^2} + RC\frac{dV_C}{dt} + V_C = V_0\cos(\omega t) \qquad (6.76)$$

This equation can be identified with that given in Example 6.6 if the ODE coefficients are specified as follows: $a = LC$, $b = RC$, $c = 1$, $A = V_0$.

IN-CLASS EXERCISES

Pb. 6.32 This problem pertains to the *RC* circuit:

a. Write the output signal V_C in the amplitude-phase representation.
b. Plot the gain response as a function of a normalized frequency that you will have to select. (The gain of a circuit is defined as the ratio of the amplitude of the output signal over the amplitude of the input signal.)
c. Determine the phase response of the system (i.e., the relative phase of the output signal to that of the input signal as function of the frequency) also as function of the normalized frequency.
d. Can this circuit be used as a filter (i.e., a device that lets through only a specified frequency band)? Specify the parameters of this band.

Pb. 6.33 This problem pertains to the *RLC* circuit:

a. Write the output signal V_C in the amplitude-phase representation.
b. Defining the resonance frequency of this circuit as $\omega_0 = 1/\sqrt{L/C}$ plot the gain curve and the phase curve for the following cases: $\omega_0 L/R = 0.1, 1, 10$.
c. Find at which frequency is the gain maximum, and find the width of the gain curve, for $\frac{\omega_0 L}{R} \gg 1$.
d. Can you think of a possible application for this circuit?

Pb. 6.34 Assume that the source potential in the *RLC* circuit has five frequency components at ω, 2ω, … , 5ω of equal amplitude. Plot the input and output potentials as a function of time over the interval $0 < \omega t < 8\pi$. Assume that

$$\omega = \omega_0 = \frac{1}{\sqrt{LC}} \quad \text{and} \quad \frac{\omega_0 L}{R} = 10.$$

6.6 Phasors

A technique in widespread use to compute the steady-state solutions of systems with sinusoidal input is the method of phasors. In this and the following two sections, we define phasors, learn how to use them to add two or more cw signals having the same frequency, and how to find the particular solution of an ODE with a sinusoidal driving function.

There are two key ideas behind the phasor representation of a signal:

1. A real, sinusoidal time-varying signal may be represented by a complex time-varying signal.
2. This complex signal can be represented as the product of a complex number that is independent of time and a complex signal that is dependent on time.

EXAMPLE 6.7

Decompose the signal $V = A\cos(\omega t + \phi)$ according to the above prescription.

Solution: This signal can, using the polar representation of complex numbers, also be written as

$$V = A\cos(\omega t + \phi) = \text{Re}[A\exp(j(\omega t + \phi))] = \text{Re}[Ae^{j\phi}e^{j\omega t}] \tag{6.77}$$

where the phasor, denoted with a tilde on top of its corresponding signal symbol, is given by

$$\tilde{V} = Ae^{j\phi} \tag{6.78}$$

(*Warning*: Do not mix up the tilde symbol that we use here, to indicate a phasor, with the overbar that denotes complex conjugation.)

Having achieved the above goal of separating the time-independent part of the complex number from its time-dependent part, we now learn how to manipulate these objects. A lot of insight can be immediately gained if we note that this form of the phasor is exactly in the polar form of a complex number, with clear geometric interpretation for its magnitude and phase. □

6.6.1 Phasor of Two Added Signals

The sum of two signals with common frequencies but different amplitudes and phases is

$$V_{tot.} = A_{tot.}\cos(\omega t + \phi_{tot.}) = A_1\cos(\omega t + \phi_1) + A_2\cos(\omega t + \phi_2) \tag{6.79}$$

To write the above result in phasor notation, note that the above sum can also be written as follows:

$$\begin{aligned} V_{tot.} &= \text{Re}[A_1 \exp(j(\omega t + \phi_1)) + A_2 \exp(j(\omega t + \phi_2))] \\ &= \text{Re}[(A_1 e^{j\phi_1} + A_2 e^{j\phi_2}) e^{j\omega t}] \end{aligned} \tag{6.80}$$

and where

$$\tilde{V}_{tot.} = A_{tot.} e^{j\phi_{tot.}} = \tilde{V}_1 + \tilde{V}_2 \tag{6.81}$$

PREPARATORY EXERCISE

Pb. 6.35 Write the analytical expression for $A_{tot.}$ and $\phi_{tot.}$ in Eq. (6.81) as functions of the amplitudes and phases of signals 1 and 2.

(*Hint:* Use the trigonometric relations valid for an arbitrary triangle.)

6.6.2 Total Phasor of Many Signals

The above result can, of course, be generalized to the sum of many cw signals; specifically

$$\begin{aligned} V_{tot.} &= A_{tot.} \cos(\omega t + \phi_{tot.}) = \sum_{n=1}^{N} A_n \cos(\omega t + \phi_n) \\ &= \text{Re}\left[\sum_{n=1}^{N} A_n \exp(j\omega t + j\phi_n)\right] = \text{Re}\left[e^{j\omega t} \sum_{n=1}^{N} A_n e^{j\phi_n}\right] \end{aligned} \tag{6.82}$$

and

$$\tilde{V}_{tot.} = \sum_{n=1}^{N} \tilde{V}_n \tag{6.83}$$

$$\Rightarrow A_{tot.} = \left|\tilde{V}_{tot.}\right| \tag{6.84}$$

$$\phi_{tot.} = \arg(\tilde{V}_{tot.}) \tag{6.85}$$

That is, the resultant field can be obtained through the simple operation of adding all the complex numbers (phasors) that represent each of the individual signals.

Example 6.8

Given 10 signals, the phasor of each of the form $A_n e^{j\phi_n}$, where the amplitude and phase for each have the functional forms $A_n = 1/n$ and $\phi_n = n^2$ write a MATLAB program to compute the resultant sum phasor.

Solution: Edit and execute the following *script M-file*:

```
N=10;
n=1:N;
amplitude_n=1./n;
phase_n=n.^2;
phasor_n=amplitude_n.*exp(j.*phase_n);
phasor_tot=sum(phasor_n);
amplitude_tot=abs(phasor_tot)
phase_tot=angle(phasor_tot)
```
□

In-Class Exercises

Pb. 6.36 For the same functional forms for the amplitude and phase as given in Example 6.8, plot the total amplitude and the total phase as function of N.

Pb. 6.37 Show that if you add N signals with the same magnitude and frequency but with phases equally distributed over the $[0, 2\pi]$ interval, the resultant phasor will be zero. (*Hint:* Remember the result for the sum of the roots of unity.)

Pb. 6.38 Show that the resultant signal from adding N signals having the same frequency has the largest amplitude when all the individual signals are in phase (this situation is referred to as maximal constructive interference).

Pb. 6.39 In this problem, we consider what happens if the frequency and amplitude of N different signals are still equal, but the different phases of the signals are randomly distributed over the $[0, 2\pi]$ interval. Find the amplitude of the resultant signal if $N = 1000$, and compare it with the maximal constructive interference result. (*Hint*: Recall that the **rand(1,N)** command generates a 1-D array of N random numbers uniformly distributed in the interval $[0, 1]$.)

Pb. 6.40 The service provided to your home by the electric utility company is a two-phase service. This means that two 110-V/60-Hz hot lines plus a neutral (ground) line terminate in your panel. The hot lines are π out of phase.

a. Which signal would you use to drive your clock radio or your toaster?

b. What configuration will you use to drive your oven or your dryer?

Pb. 6.41 In most industrial environments, electric power is delivered in what is called a three-phase service. This consists of three 110-V/60-Hz lines with phases given by $(0, 2\pi/3, 4\pi/3)$. What is the maximum voltage that you can obtain from any combination of two of these signals?

Pb. 6.42 Two- and three-phase power can be extended to N-phase power. In such a scheme, the N-110-V/60-Hz signals are given by

$$V_n = 110\cos\left(120\pi t + \frac{2\pi n}{N}\right) \quad \text{and} \quad n = 0, 1, \ldots, N-1$$

While the sum of the voltage of all the lines is zero, the average power is not. Find the total power, assuming that the power from each line is proportional to the square of its time-dependent voltage. (*Hint:* Use the double angle formula for the cosine function.)

$$p_n(t) = A_n^2 \cos^2\left(\omega t + \frac{2\pi n}{N}\right) \quad \text{and} \quad P = \sum_{n=0}^{N-1} p_n$$

6.7 Interference and Diffraction of Electromagnetic Waves

6.7.1 The Electromagnetic Wave

Electromagnetic waves (em waves) are manifest as radio and TV broadcast signals, microwave communication signals, light of any color, X-rays, γ-rays, etc. While these waves have different sources and methods of generation and require different kinds of detectors, they do share some general characteristics. They differ from each other only in the value of their frequencies. Indeed, it was one of the greatest intellectual achievements of the nineteenth century when Maxwell developed the system of equations, now named in his honor, to describe these waves' commonality. The most important of these properties is that they all travel in vacuum with, what is called, the speed of light c ($c = 3 \times 10^8$ m/s). The detailed study of these waves is the subject of many electrophysics subspecialties.

Electromagnetic waves are traveling waves. To understand their mathematical nature, consider a typical expression for the electric field associated with such waves:

$$E(z, t) = E_0 \cos[kz - \omega t] \tag{6.86}$$

Here, E_0 is the amplitude of the wave, z the spatial coordinate parallel to the direction of propagation of the wave, and k the wave number.

Note that if we plot the field for a fixed time, for example, at $t = 0$, the field takes the shape of a sinusoidal function in space:

$$E(z, t = 0) = E_0 \cos[kz] \tag{6.87}$$

From the above equation, one deduces that the wave number $k = 2\pi/\lambda$, where λ is the wavelength of the wave (i.e., the length after which the wave shape reproduces itself).

Now let us look at the field when an observer, located at $z = 0$, would measure it as a function of time. Then

$$E(z = 0, t) = E_0 \cos[\omega t] \tag{6.88}$$

The temporal period, that is, the time after which the wave shape reproduces itself, is $T = 2\pi/\omega$ where ω is the angular frequency of the wave.

Next, we want to relate the wave number to the angular frequency. To do that, consider an observer located at $z = 0$. The observer measures the field at $t = 0$ to be E_0. At time Δt later, he should measure the same field, whether he uses Eq. (6.87) or Eq. (6.88) if he takes $\Delta z = c\Delta t$, the distance that the wave crest has moved, and where c is the speed of propagation of the wave. From this, one deduces that the wave number and the angular frequency are related by $kc = \omega$. This relation holds true for all electromagnetic waves; that is, as the frequency increases, the wavelength decreases.

If two traveling waves have the same amplitude and frequency, but one is traveling to the right while the other is traveling to the left, the result is a standing wave. The following program permits visualization of this standing wave:

```
x=0:0.01:5;
a=1;
k=2*pi;
w=2*pi;
t=0:0.05:2;
  for m=1:41;
      z1=cos(k*x-w*t(m));
      z2=cos(k*x+w*t(m));
      z=z1+z2;
      plot(x,z,'r');
      axis([0 5 -3 3]);
      M(:,m)=getframe;
  end
movie(M,20)
```

Compare the spatio-temporal profile of the resultant to that for a single wave (i.e., set `z2` = 0).

6.7.2 Addition of Two Electromagnetic Waves

In many practical instances, we are faced with the problem that two em waves originating from the same source, but following different spatial paths, meet again at a certain position. We want to find the total field at this position resulting from adding the two waves. We first note that, in the simplest case where the amplitude of the two fields are kept equal, the effect of the different paths is only to dephase one of the waves from the other by an amount $\Delta\phi = k\Delta l$, where Δl is the path difference. In effect, the total field is given by

$$E_{tot.}(t) = E_0 \cos[\omega t + \phi_1] + E_0 \cos[\omega t + \phi_2] \tag{6.89}$$

where $\Delta\phi = \phi_1 - \phi_2$. This form is similar to those studied in the addition of two phasors and we will hence describe the problem in this language.

The resultant phasor is

$$\tilde{E}_{tot.} = \tilde{E}_1 + \tilde{E}_2 \tag{6.90}$$

PREPARATORY EXERCISE

Pb. 6.43 Graph a polar plot of the modulus of the resultant phasor given in Eq. (6.90) as a function of E_0 and $\Delta\phi$, for two signals of equal modulus but different phases.

6.7.3 Generalization to N-waves

The addition of two electromagnetic waves can be generalized to the case of N-waves.

EXAMPLE 6.9

Find the resultant field of equal-amplitude N-waves, each phase-shifted from the preceding by the same $\Delta\phi$.

Solution: The problem consists of computing an expression of the following kind:

$$\tilde{E}_{tot.} = \tilde{E}_1 + \tilde{E}_2 + \cdots + \tilde{E}_n = E_0(1 + e^{j\Delta\phi} + e^{j2\Delta\phi} + \cdots + e^{j(N-1)\Delta\phi}) \tag{6.91}$$

We have encountered such an expression previously. This sum is that corresponding to the sum of a geometric series. Computing this sum, the modulus square of the resultant phasor is

$$\begin{aligned} \left|\tilde{E}_{tot.}\right|^2 &= E_0^2 \frac{(1 - e^{jN\Delta\phi})}{(1 - e^{j\Delta\phi})} \frac{(1 - e^{-jN\Delta\phi})}{(1 - e^{-j\Delta\phi})} \\ &= E_0^2 \left(\frac{1 - \cos(N\Delta\phi)}{1 - \cos(\Delta\phi)}\right) = E_0^2 \left(\frac{\sin^2(N\Delta\phi/2)}{\sin^2(\Delta\phi/2)}\right) \end{aligned} \tag{6.92}$$

Because the source is the same for each of the components, the modulus of each phasor is related to the source amplitude by $E_0 = E_{source}/N$. The results are expressed usually as function of the source field. □

IN-CLASS EXERCISES

Pb. 6.44 Plot the normalized square modulus of the resultant of N-waves as a function of $\Delta\phi$ for different values of N (5, 50, and 500) over the interval $\frac{-\pi}{2} < \Delta\phi < \frac{\pi}{2}$.

Pb. 6.45 Find the dependence of the height of the central peak of Eq. (6.92) on N.

Pb. 6.46 Find the phase shift that corresponds to the position of the first minimum of Eq. (6.92).

Pb. 6.47 Compare the plots of the modulus square of the resultant phasor from N equal amplitude waves, each shifted from the precedent by the same shift, and that of exactly the same configuration but this time with the odd-numbered sources switched off.

Pb. 6.48 In an antenna array with the field representing N aligned, equally spaced individual antennae excited by the same source is given by Eq. (6.91). If the line connecting the point of observation to the center of the array is making an angle θ with the antenna array, the phase shift is $\Delta\phi = (2\pi/\lambda)\, d \cos(\theta)$, where λ is the wavelength of radiation and d the spacing between two consecutive antennae. Draw the polar plot of the total intensity as function of the angle θ for a spacing $d = \lambda/2$ for different values of N (2, 4, 6, and 10).

Pb. 6.49 Do the results of Pb. 6.48 suggest to you a strategy for designing a multiantenna system with sharp directivity? Can you think of a method, short of moving the antennae around, that permits this array to sweep a range of angles with maximum directivity?

Pb. 6.50 The following program simulates a 25-element array-swept radar beam:

```
th=0:0.01:pi;
t=-0.5*sqrt(3):0.025*sqrt(3):0.5*sqrt(3);
N=25;
  for m=1:41;
      I=(1/N^2)*(sin(N*((pi/2)*cos(th)+...
        (pi/2)*t(m))).^2)./((sin((pi/2)*...
        cos(th)+(pi/2)*t(m))).^2);
      polar(th,I);
      M(:,m)=getframe;
  end
movie(M,10)
```

a. Determine the range of the sweeping angle.
b. Can you think of an electronic method for implementing this task?

6.8 Solving ac Circuits with Phasors: The Impedance Method

In Section 6.5, we examined the conventional technique for solving some simple ac circuits problems. We suggested that using phasors may speed up the determination of the solution. This is the subject of this chapter section.

We will treat, using this technique, the simple *RLC* circuit already solved through other means in order to give you a measure of the simplifications that can be achieved in circuit analysis through this technique. We then proceed to use the phasor technique to investigate another circuit configuration: the infinite *LC* ladder. The power of the phasor technique will also be put to use when we solve topologically much more difficult circuit problems than the one-loop category encountered thus far. Essentially, a straightforward algebraic technique can give the voltages and currents for any circuit. We illustrate the latter case in Chapter 8.

Recalling that the voltage drops across resistors, inductors, and capacitors can all be expressed as function of the current, its derivative, and its integral, our goal is to find a technique to replace these operators by simple algebraic operations. The key to achieving this goal is to realize that:

if

$$I = I_0 \cos(\omega t + \phi) = \mathrm{Re}[e^{j\omega t}(I_0 e^{j\phi})] \tag{6.93}$$

then

$$\frac{dI}{dt} = -I_0 \omega \sin(\omega t + \phi) = \mathrm{Re}[e^{j\omega t}(I_0(j\omega)e^{j\phi})] \tag{6.94}$$

and

$$\int I dt = \frac{I_0}{\omega} \sin(\omega t + \phi) = \mathrm{Re}\left[e^{j\omega t}\left(I_0\left(\frac{1}{j\omega}\right)e^{j\phi}\right)\right] \tag{6.95}$$

From Eq. (4.25) to Eq. (4.27) and Eq. (6.93) to Eq. (6.95), we can deduce that the phasors representing the voltages across resistors, inductors, and capacitors can be written as follows:

$$\tilde{V}_R = \tilde{I}R = \tilde{I}Z_R \tag{6.96}$$

$$\tilde{V}_L = \tilde{I}(j\omega L) = \tilde{I}Z_L \tag{6.97}$$

$$\tilde{V}_C = \frac{\tilde{I}}{(j\omega C)} = \tilde{I}Z_C \tag{6.98}$$

The terms multiplying the current phasor on the RHS of each of the above equations are called the resistor, the inductor, and the capacitor impedances, respectively.

6.8.1 *RLC* Circuit Phasor Analysis

Let us revisit this problem first discussed in Section 4.7. Using Kirchoff's voltage law and Eq. (6.96) to Eq. (6.98), we can write the following relation between the phasor of the current and that of the source potential:

$$\tilde{V}_s = \tilde{I}R + \tilde{I}(j\omega L) + \frac{\tilde{I}}{(j\omega C)} = \tilde{I}\left[R + j\omega L + \frac{1}{j\omega C}\right] \qquad (6.99)$$

That is, we can immediately compute the modulus and the argument of the phasor of the current if we know the values of the circuit components, the source voltage phasor, and the frequency of the source.

IN-CLASS EXERCISES

Using the expression for the circuit resonance frequency ω_0 previously introduced in Pb. 6.33, for the *RLC* circuit:

Pb. 6.51 Show that the system's total impedance can be written as

$$Z = R + j\omega_0 L\left(\nu - \frac{1}{\nu}\right), \qquad \text{where } \nu = \frac{\omega}{\omega_0} = \omega\sqrt{LC}$$

Pb. 6.52 Show that $Z(\nu) = \bar{Z}(1/\nu)$; and from this result, deduce the value of ν at which the impedance is entirely real.

Pb. 6.53 Find the magnitude and the phase of the total impedance.

Pb. 6.54 Selecting for the values of the circuit elements $LC = 1$, $RC = 3$, and $\omega = 1$, compare the results that you obtain through the phasor analytical method with the numerical results for the voltage across the capacitor in an *RLC* circuit that you found while solving Eq. (4.36).

6.8.1.1 The Transfer Function

As you would have discovered solving Pb. 6.54, the ratio of the phasor of the potential difference across the capacitor with that of the ac source can be directly calculated once the value of the current phasor is known. This ratio is called the transfer function for this circuit if the voltage across the capacitor is taken as the output of this circuit. It is obtained by combining Eq. (6.98) and Eq. (6.99) and is given by

$$\frac{\tilde{V}_c}{\tilde{V}_s} = \frac{1}{(j\omega RC - \omega^2 LC + 1)} = H(\omega) \tag{6.100}$$

The transfer function concept can be generalized to any ac circuit. It refers to the ratio of the output voltage phasor to the input voltage phasor. It incorporates all the relevant information on the details of the circuit. It is the standard form for representing the response of a circuit to a single sinusoidal function input.

> ***Homework Problem***
>
> **Pb. 6.55** Plot the magnitude and the phase of the transfer function given in Eq. (6.100) as a function of ω, for $LC = 1$, $RC = 0.03$.

6.8.2 The Infinite *LC* Ladder

The *LC* ladder consists of an infinite repetition of the basic elements shown in Figure 6.2.

Using the definition of impedances, the phasors of the n and $(n + 1)$ voltages and currents are related through

$$\tilde{V}_n - \tilde{V}_{n+1} = Z_1 \tilde{I}_n \tag{6.101}$$

$$\tilde{V}_{n+1} = (\tilde{I}_n - \tilde{I}_{n+1}) Z_2 \tag{6.102}$$

From Eq. (6.101), we deduce the following expressions for $\tilde{I}_n$ and $\tilde{I}_{n+1}$:

$$\tilde{I}_n = \frac{\tilde{V}_n - \tilde{V}_{n+1}}{Z_1} \tag{6.103}$$

$$\tilde{I}_{n+1} = \frac{\tilde{V}_{n+1} - \tilde{V}_{n+2}}{Z_1} \tag{6.104}$$

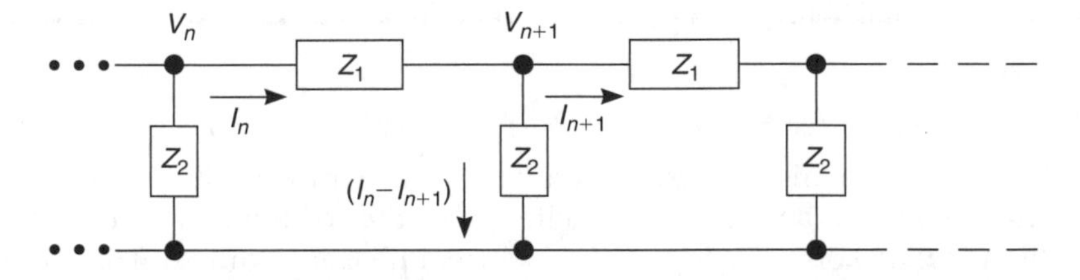

FIGURE 6.2
The circuit of an infinite *LC* ladder.

Substituting these values for the currents in Eq. (6.102), we deduce a second-order difference equation for the voltage phasor:

$$\tilde{V}_{n+2} - \left(\frac{Z_1}{Z_2} + 2\right)\tilde{V}_{n+1} + \tilde{V}_n = 0 \tag{6.105}$$

The solution of this difference equation can be directly obtained by the techniques discussed in Chapter 2 for obtaining solutions of homogeneous difference equations. The physically meaningful solution is given by

$$\lambda = 1 + \frac{1}{Z_2}\left\{\frac{Z_1}{2} - \sqrt{\frac{Z_1^2}{4} + Z_2 Z_1}\right\} \tag{6.106}$$

and the voltage phasor at node n is then given by

$$\tilde{V}_n = \tilde{V}_s \lambda^n \tag{6.107}$$

We consider the model where $Z_1 = j\omega L$ and $Z_2 = 1/(j\omega C)$, respectively, for an inductor and a capacitor. The expression for λ then takes the following form:

$$\lambda = \left(1 - \frac{\nu^2}{2}\right) - j\left(\nu^2 - \frac{\nu^4}{4}\right)^{1/2} \tag{6.108}$$

where the normalized frequency is defined by $\nu = \omega/\omega_0 = \omega\sqrt{LC}$. We plot in Figure 6.3 the magnitude and the phase of the root λ as function of the normalized frequency.

As can be directly observed from an examination of Figure 6.3, the magnitude of λ is equal to 1 (i.e., the magnitude of $\tilde{V}_n$ is also 1) for $\nu < \nu_{\text{cutoff}} = 2$, while it drops precipitously after that, with the dropoff in the potential much steeper with increasing node number. Physically, this represents extremely short penetration through the ladder for signals with frequencies larger than the cutoff frequency. Furthermore, note that for $\nu < \nu_{\text{cutoff}} = 2$, the phase of $\bar{V}_n$ increases linearly with the index n; and because it is negative, it corresponds to a delay in the signal as it propagates down the ladder, which corresponds to a finite velocity of propagation for the signal.

Before we leave this ladder circuit, it is worth addressing a practical concern. While it is impossible to realize an infinite-dimensional ladder, the above conclusions do not change by much if we replace the infinite ladder by a finite ladder and we terminate it after a while by a resistor with resistance equal to $\sqrt{L/C}$.

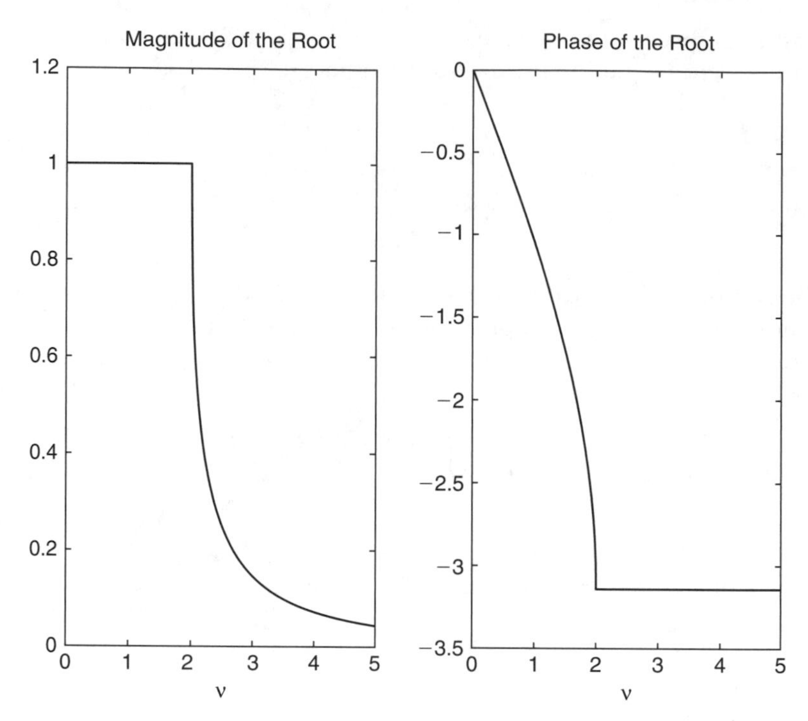

FIGURE 6.3
The magnitude (left panel) and the phase (right panel) of the characteristic root of the infinite *LC* ladder.

IN-CLASS EXERCISE

Pb. 6.56 Repeat the analysis given above for the *LC* ladder circuit, if instead we were to:

a. Interchange the positions of the inductors and the capacitors in the ladder circuit. Based on this result and the above *LC* result, can you design a bandpass filter with a flat response?

b. Interchange the inductor elements by resistors. In particular, compute the input impedance of this circuit.

6.9 Transfer Function for a Difference Equation with Constant Coefficients*

In Section 6.8.1, we found the transfer function for what essentially was a simple ODE. In this section, we generalize the technique to find the transfer

* The asterisk indicates more advanced material that may be skipped in a first reading.

function of a difference equation with constant coefficients. The form of the difference equation is given by

$$\begin{aligned} y(k) &= b_0 u(k) + b_1 u(k-1) + \cdots + b_m u(k-m) \\ &\quad - a_1 y(k-1) - a_2 y(k-2) - \cdots - a_n y(k-n) \end{aligned} \tag{6.109}$$

Along the same route that we followed in the phasor treatment of ODE, assume that both the input and output are of the form:

$$u(k) = Ue^{j\Omega k} \quad \text{and} \quad y(k) = Ye^{j\Omega k} \tag{6.110}$$

where Ω is a normalized frequency; typically, in electrical engineering applications, the real frequency multiplied by the sampling time. Replacing these expressions in the difference equation, we obtain:

$$\frac{Y}{U} = \frac{\sum_{l=0}^{m} b_l e^{-j\Omega l}}{1 + \sum_{l=1}^{n} a_l e^{-j\Omega l}} = \frac{\sum_{l=0}^{m} b_l z^{-l}}{1 + \sum_{l=1}^{n} a_l z^{-l}} \equiv H(z) \tag{6.111}$$

where, by convention, $z = e^{j\Omega}$.

EXAMPLE 6.10

Find the transfer function of the following difference equation:

$$y(k) = u(k) + \frac{2}{3} y(k-1) - \frac{1}{3} y(k-2) \tag{6.112}$$

Solution: By direct substitution into Eq. (6.111), we find

$$H(z) = \frac{1}{1 - \frac{2}{3} z^{-1} + \frac{1}{3} z^{-2}} = \frac{z^2}{z^2 - \frac{2}{3} z + \frac{1}{3}} \tag{6.113}$$

It is to be noted that the transfer function is a ratio of two polynomials. The zeros of the numerator are called the zeros of the transfer function, while the zeros of the denominator are called its poles. If the coefficients of the difference equations are real, then by the fundamental theorem of algebra, the zeros and the poles are either real or are pairs of complex conjugate numbers.

The transfer function fully describes any linear system. As will be shown in linear systems courses, the z-transform of the transfer function gives the weights for the solution of the difference equation, while the values of the poles of the transfer function determine what are called the system modes of the solution. These are the modes intrinsic to the circuit, and they do not depend on the specific form of the input function.

Furthermore, it is worth noting that the study of recursive filters, the backbone of digital signal processing, can be simply reduced to a study of the transfer function under different configurations. In Applications 2 and 3 that follow, we briefly illustrate two particular digital filters in wide use. □

Application 1

Using the transfer function formalism, we want to estimate the accuracy of the three integrating schemes discussed in Chapter 4. We want to compare the transfer function of each of those algorithms to that of the exact result, obtained upon integrating exactly the function $e^{j\omega t}$.

The exact result for integrating the function $e^{j\omega t}$ is, of course, $e^{j\omega t}/j\omega$, thus giving for the exact transfer function for integration the expression:

$$H_{exact} = \frac{1}{j\omega} \tag{6.114}$$

Before proceeding with the computation of the transfer function for the different numerical schemes, let us pause for a moment and consider what we are actually doing when we numerically integrate a function. We go through the following steps:

1. We discretize the time interval over which we integrate; that is, we define the sampling time Δt, such that the discrete points abscissa are given by $k(\Delta t)$, where k is an integer.
2. We write a difference equation for the integral relating its values at the discrete points with its values and that of the integrand at discrete points with equal or smaller indices.
3. We obtain the value of the integral by iterating the defining difference equation.

The test function used for the estimation of the integration methods accuracy is written at the discrete points as

$$y(k) = e^{jk\omega(\Delta t)} \tag{6.115}$$

The difference equations associated with each of the numerical integration schemes are

$$I_T(k+1) = I_T(k) + \frac{\Delta t}{2}(y(k+1) + y(k)) \tag{6.116}$$

$$I_{MP}(k+1) = I_{MP}(k) + \Delta t\, y(k+1/2) \tag{6.117}$$

$$I_S(k+1) = I_S(k-1) + \frac{\Delta t}{3}(y(k+1) + 4y(k) + y(k-1)) \tag{6.118}$$

leading to the following expressions for the respective transfer functions:

$$H_T = \frac{\Delta t}{2} \frac{e^{j\omega(\Delta t)} + 1}{e^{j\omega(\Delta t)} - 1} \tag{6.119}$$

$$H_{MP} = \Delta t \frac{e^{j\omega(\Delta t)/2}}{e^{j\omega(\Delta t)} - 1} \tag{6.120}$$

$$H_S = \frac{\Delta t}{3} \frac{(e^{j\omega(\Delta t)} + 4 + e^{-j\omega(\Delta t)})}{e^{j\omega(\Delta t)} - e^{-j\omega(\Delta t)}} \tag{6.121}$$

The measures of accuracy of the integration scheme are the ratios of these transfer functions to that of the exact expression. These are given, respectively, by

$$R_T = \frac{(\omega\Delta t/2)}{\sin(\omega\Delta t/2)} \cos(\omega\Delta t/2) \tag{6.122}$$

$$R_{MP} = \frac{(\omega\Delta t/2)}{\sin(\omega\Delta t/2)} \tag{6.123}$$

$$R_S = \left(\frac{\omega\Delta t}{3}\right) \frac{\cos(\omega\Delta t) + 2}{\sin(\omega\Delta t)} \tag{6.124}$$

Table 6.1 gives the value of this ratio as a function of the number of sampling points, per oscillation period, selected in implementing the different integration subroutines.

As can be noted, the error is less than 1% for any of the discussed methods as long as the number of points in one oscillation period is larger than 20, although the degree of accuracy is best, as we expected based on geometrical arguments, for Simpson's rule.

In a particular application, where a finite number of frequencies are simultaneously present, the choice of (Δt) for achieving a specified level of accuracy in the integration subroutine should ideally be determined using the shortest of the periods present in the integrand.

TABLE 6.1

Accuracy of the Different Elementary Numerical Integrating Methods

Number of Sampling Points in a Period	R_T	R_{MP}	R_S
100	0.9997	1.0002	1.0000
50	0.9986	1.0007	1.0000
40	0.9978	1.0011	1.0000
30	0.9961	1.0020	1.0000
20	0.9909	1.0046	1.0001
10	0.9591	1.0206	1.0014
5	0.7854	1.1107	1.0472

Application 2

As mentioned earlier, the transfer function technique is the prime tool for the analysis and design of digital filters. In this and the following application, we illustrate its use in the design of a low-pass digital filter and a digital prototype bandpass filter.

The low-pass filter, as its name indicates, filters out the high-frequency components from a signal.

Its defining difference equation is given by

$$y(k) = (1 - a)y(k - 1) + au(k) \tag{6.125}$$

giving for its transfer function the expression:

$$H(z) = \frac{a}{1-(1-a)z^{-1}} \tag{6.126}$$

Written as a function of the normalized frequency, it is given by

$$H(e^{j\Omega}) = \frac{ae^{j\Omega}}{e^{j\Omega} - (1-a)} \tag{6.127}$$

We plot, in Figure 6.4, the magnitude and the phase of the transfer function as a function of the normalized frequency for the value of $a = 0.1$. Note that the gain is equal to 1 for $\Omega = 0$, and decreases monotonically thereafter.

To appreciate the operation of this filter, consider a sinusoidal signal that has been contaminated by the addition of noise. We can simulate the noise by adding to the original signal an array consisting of random numbers with maximum amplitude equal to 20% of the original signal. The top panel of Figure 6.5 represents the contaminated signal. If we pass this signal through a low-pass filter, the lower panel of Figure 6.5 shows the outputted filtered signal.

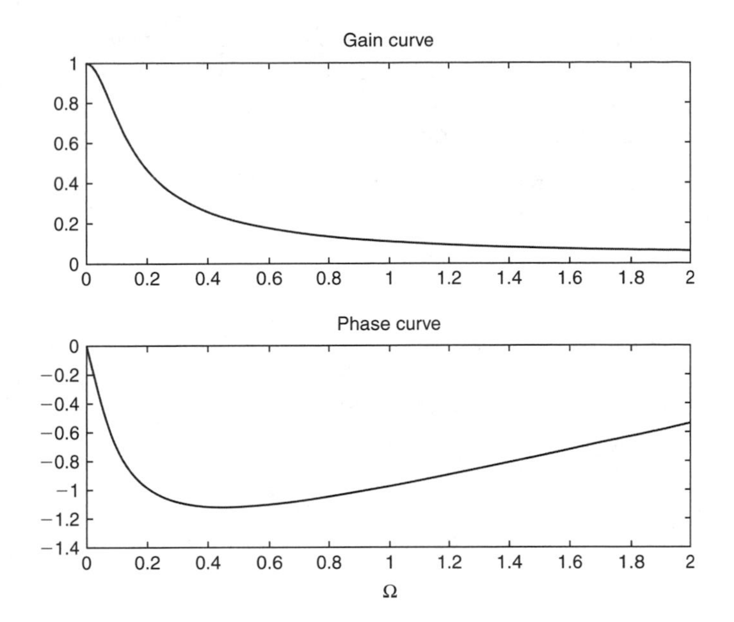

FIGURE 6.4
The gain (top panel) and phase (bottom panel) responses of a low-pass filter as a function of the frequency.

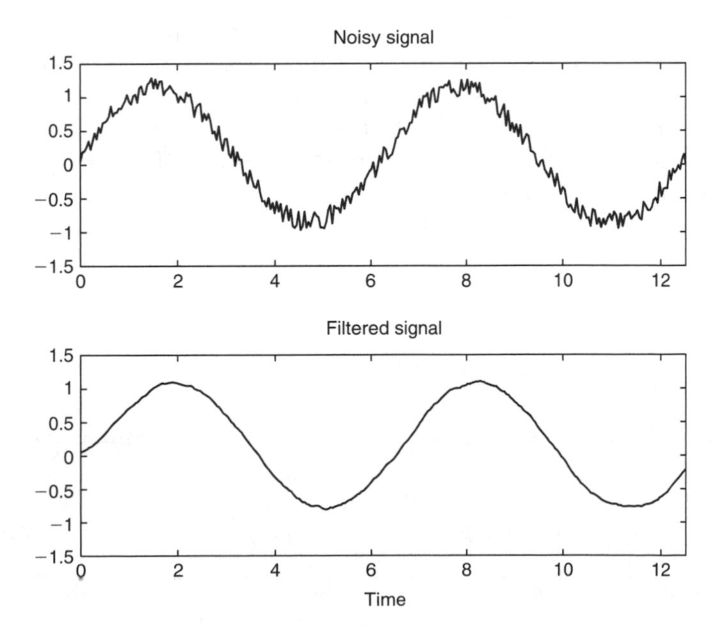

FIGURE 6.5
The action of a low-pass filter. Top panel: profile of the signal contaminated by noise. Bottom panel: profile of the filtered signal.

As can be observed, the noise, which is a high-frequency signal, has been filtered out and the signal shape has been almost restored to its original shape before that noise was added.

The following *script M-file* simulates the above operations:

```
t=linspace(0,4*pi,300);
N=length(t);
s=sin(t);
n=0.3*rand(1,N);
u=s+n;
y(1)=u(1);
  for k=2:N
    y(k)=+0.9*y(k-1)+0.1*u(k);
  end
subplot(2,1,1)
plot(t,u)
axis([0 4*pi -1.5 1.5]);
title('Noisy Signal')
subplot(2,1,2)
plot(t,y)
title('Filtered Signal')
axis([0 4*pi -1.5 1.5]);
```

Application 3

The digital prototype bandpass filter ideally filters out from a signal all frequencies lower than a given frequency and higher than another frequency. In practice, the cutoffs are not so sharp and the lower and higher cutoff frequencies of the bandpass are defined as those at which the gain curve (i.e., the magnitude of the transfer function as function of the frequency) is at $(1/\sqrt{2})$ its maximum value.

The difference equation that describes this prototype filter is

$$y(k) = \left\{(1-r)\sqrt{1-2r\cos(2\Omega_0)+r^2}\right\}u(k) + 2r\cos(\Omega_0)y(k-1) - r^2y(k-2) \tag{6.128}$$

where Ω_0 is the normalized frequency with maximum gain and r a number close to 1.

The purpose of the following analysis is, given the lower and higher cutoff normalized frequencies, to find the quantities Ω_0 and r in the above difference equation.

The transfer function for the above difference equation is given by

$$H(z) = \frac{g_0 z^2}{z^2 - 2r\cos(\Omega_0)z + r^2} \tag{6.129}$$

where

$$g_0 = (1-r)\sqrt{1-2r\cos(2\Omega_0)+r^2} \tag{6.130}$$

and

$$z = e^{j\Omega}$$

The gain of this filter, or equivalently the magnitude of the transfer function, is

$$\left|H(e^{j\Omega})\right| = \frac{(1-r)\sqrt{1-2r\cos(2\Omega_0)+r^2}}{(1+Ar+Br^2+Ar^3+r^4)} \tag{6.131}$$

where

$$A = -4\cos(\Omega)\cos(\Omega_0) \tag{6.132}$$

$$B = 4\cos^2(\Omega) + 4\cos^2(\Omega_0) - 2 \tag{6.133}$$

The lower and upper cutoff frequencies are defined, as previously noted, by the condition

$$\left|H(e^{j\Omega_{(1,2)}})\right| = \frac{1}{\sqrt{2}} \tag{6.134}$$

Substituting condition (6.134) in the gain expression (6.131) leads to the conclusion that the cutoff frequencies are obtained from the solutions of the following quadratic equation:

$$\cos^2(\Omega) - \left[\frac{(1+r^2)\cos(\Omega_0)}{r}\right]\cos(\Omega) + \frac{(1-r)^2}{4r^2}[4r\cos(2\Omega_0)-(1-r)^2] + \cos^2(\Omega_0) = 0 \tag{6.135}$$

Adding and subtracting the roots of this equation, we deduce after some straightforward algebra, the following determining equations for Ω_0 and r:

1. r is the root in the interval [0, 1] of the following eighth-degree polynomial.

$$r^8 + (a-b)r^6 - 8ar^5 + (14a - 2b - 2)r^4 - 8ar^3 + (a-b)r^2 + 1 = 0 \tag{6.136}$$

where

$$a = (\cos(\Omega_1) + \cos(\Omega_2))^2 \tag{6.137}$$

$$b = (\cos(\Omega_1) - \cos(\Omega_2))^2 \tag{6.138}$$

2. Ω_0 is given by

$$\Omega_0 = \cos^{-1}\left[\frac{ra^{1/2}}{1+r^2}\right] \tag{6.139}$$

EXAMPLE 6.11

Write a program to determine the parameters r and Ω_0 of a prototype bandpass filter if the cutoff frequencies and the sampling time are given.

Solution: The following *script M-file* implements the above target:

```
f1=? ;                       %enter the lower cutoff
f2=? ;                       %enter the upper cutoff
tau=? ;                      %enter the sampling time
w1=2*pi*f1*tau;
w2=2*pi*f2*tau;
a=(cos(w1)+cos(w2))^2;
b=(cos(w1)-cos(w2))^2;
p=[1 0 a-b -8*a 14*a-2*b-2 -8*a a-b 0 1];
rr=roots(p);
r=rr(find(rr>0 & rr<1 & imag(rr)==0))
w0=acos((r*a^(1/2))/(1+r^2));
f0=(1/(2*pi*tau))*w0
```

In Figure 6.6, we show the gain and phase response for this filter, for the case that the cutoff frequencies are chosen to be 1000 and 1200 Hz, and the sampling rate is 10 μs.

To test the action of this filter, we input into it a signal that consists of a mixture of a sinusoid having a frequency at the frequency of the maximum gain of this filter and a number of its harmonics; for example,

$$u(t) = \sin(2\pi f_0 t) + 0.5\sin(4\pi f_0 t) + 0.6\sin(6\pi f_0 t) \tag{6.140}$$

We show in Figure 6.7 the input and the filtered signals. As expected from an analysis of the gain curve, only the fundamental frequency signal has survived. The amplitude of the filtered signal settles to that of the fundamental frequency signal following a short transient period.

NOTE Before leaving this topic, it is worth noting that the above prototype bandpass filter can have sharper cutoff features (i.e., decreasing the value of

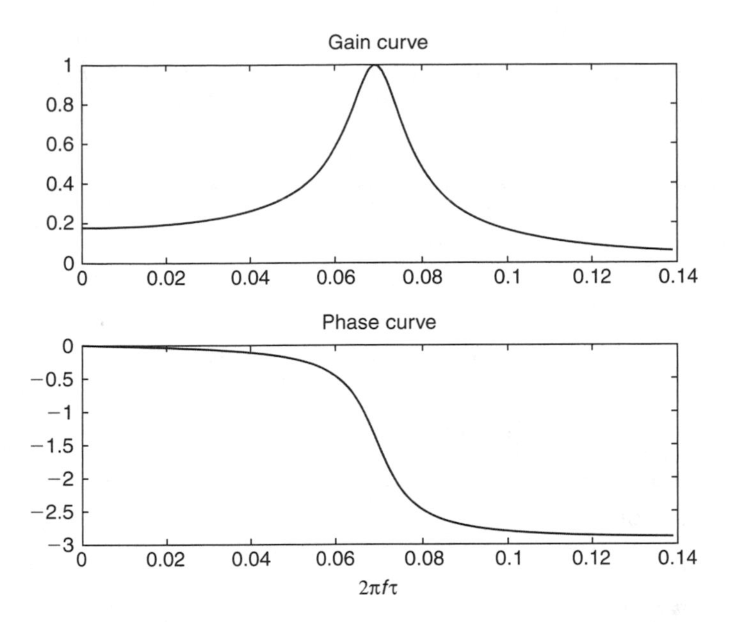

FIGURE 6.6
The transfer function of a prototype bandpass filter. Top panel: plot of the gain curve as function of the normalized frequency. Bottom panel: plot of the phase curve as function of the normalized frequency.

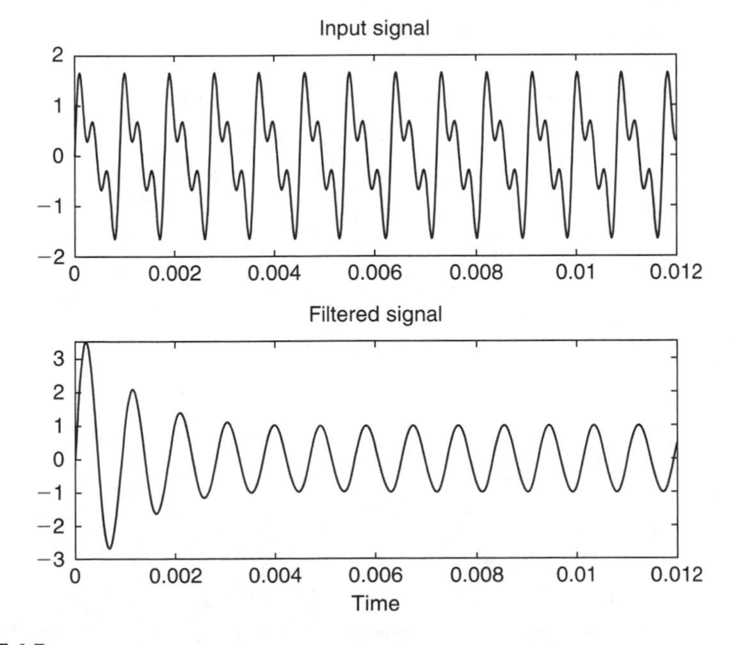

FIGURE 6.7
The filtering action of a prototype bandpass filter. Top panel: input signal consists of a combination of a fundamental frequency signal (equal to the frequency corresponding to the filter maximum gain) and two of its harmonics. Bottom panel: filtered signal.

the gain curve for frequencies below the lower cutoff and higher than the upper cutoff) through having many of these prototype filters in cascade. This will be a topic of study in future linear systems or filter design courses. □

IN-CLASS EXERCISES

Pb. 6.57 Work out the missing algebraic steps in the derivation leading to Eq. (6.136) through Eq. (6.139).

Pb. 6.58 Given the following values for the lower and upper cutoff frequencies and the sampling time:

$$f_1 = 200\,\text{Hz};\ f_2 = 400\,\text{Hz};\ \tau = 10^{-5}\,\text{s}$$

find f_0 and plot the gain curve as function of the normalized frequency for the bandpass prototype filter.

6.10 MATLAB Commands Review

`abs`	Computes the modulus of a complex number.
`angle`	Computes the argument of a complex number.
`conj`	Computes the complex conjugate of a complex number.
`imag`	Computes the imaginary part of a complex number.
`real`	Computes the real part of a complex number.

7

Vectors

In this chapter, we consider both finite- and infinite-dimensional vectors. Using the Dirac notation, we show that the two cases can be conceptually treated in similar fashions. Very basic results from vector calculus in 2-D are used to solve the problem of planetary motion. In the last section, the Fourier series and the expansion in Legendre polynomials are used to illustrate the expansion in particular basis functions in infinite-dimensional spaces.

7.1 Vectors in Two Dimensions

A vector in 2-D is defined by its length and the angle it makes with a reference axis (usually the x-axis). This vector is represented graphically by an arrow. The tail of the arrow is called the initial point of the vector and the tip of the arrow is the terminal point. Two vectors are equal when both their length and angle with a reference axis are equal.

7.1.1 Addition

The sum of two vectors $\vec{u} + \vec{v} = \vec{w}$ is a vector constructed graphically as follows. At the tip of the first vector, draw a vector equal to the second vector, such that its tail coincides with the tip of the first vector. The resultant vector has as its tail that of the first vector, and as its tip, the tip of the just-drawn second vector (the Parallelogram rule) (see Figure 7.1).

The negative of a vector is that vector whose tip and tail have been exchanged from those of the vector. This leads to the conclusion that the difference of two vectors is the other diagonal in the parallelogram (Figure 7.2).

7.1.2 Multiplication of a Vector by a Real Number

If we multiply a vector $\vec{v}$ by a real number k, the result is a vector whose length is k times the length of $\vec{v}$, and whose direction is that of $\vec{v}$ if k is positive, and opposite if k is negative.

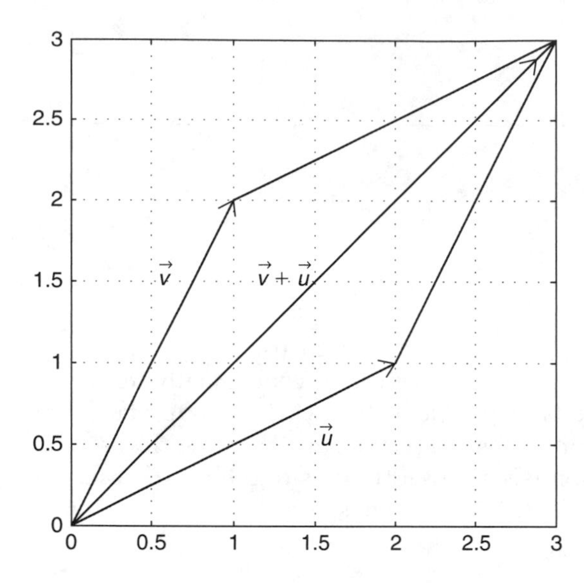

FIGURE 7.1
Sum of two vectors.

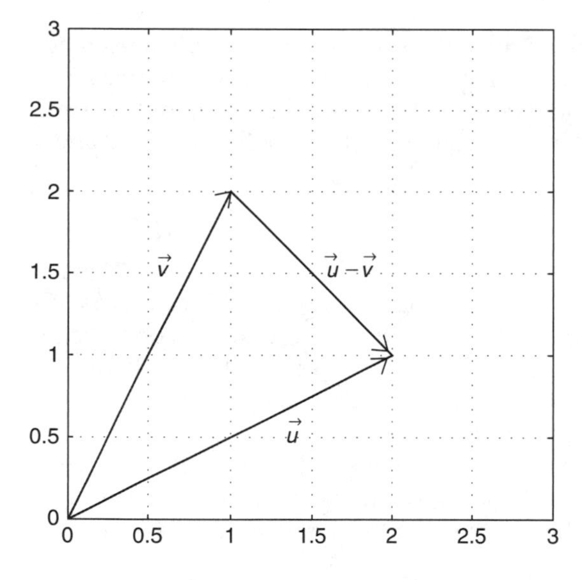

FIGURE 7.2
Difference of two vectors.

7.1.3 Cartesian Representation

It is most convenient for a vector to be described by its projections on the x-axis and on the y-axis, respectively; these are denoted by (v_1, v_2) or (v_x, v_y). In this representation

$$\vec{u} = (u_1, u_2) = (u_1)\hat{e}_1 + (u_2)\hat{e}_2 \tag{7.1}$$

where $\hat{e}_1$ and $\hat{e}_2$ are the unit vectors (length is 1) parallel to the x- and y-axis, respectively. In terms of this representation, we can write the zero vector, the sum of two vectors, and the multiplication of a vector by a real number as follows:

$$\vec{0} = (0, 0) = 0\hat{e}_1 + 0\hat{e}_2 \tag{7.2}$$

$$\vec{u} + \vec{v} = \vec{w} = (u_1 + v_1, u_2 + v_2) = (u_1 + v_1)\hat{e}_1 + (u_2 + v_2)\hat{e}_2 \tag{7.3}$$

$$k\vec{u} = (ku_1, ku_2) = (ku_1)\hat{e}_1 + (ku_2)\hat{e}_2 \tag{7.4}$$

PREPARATORY EXERCISE

Pb. 7.1 Using the above definitions and properties, prove the following identities:

$$\begin{aligned}
&\vec{u} + \vec{v} = \vec{v} + \vec{u} \\
&(\vec{u} + \vec{v}) + \vec{w} = \vec{u} + (\vec{v} + \vec{w}) \\
&\vec{u} + \vec{0} = \vec{0} + \vec{u} = \vec{u} \\
&\vec{u} + (-\vec{u}) = \vec{0} \\
&k(l\vec{u}) = (kl)\vec{u} \\
&k(\vec{u} + \vec{v}) = k\vec{u} + k\vec{v} \\
&(k + l)\vec{u} = k\vec{u} + l\vec{u}
\end{aligned}$$

The norm of a vector is the length of this vector. Using the Pythagorean theorem, its square is

$$\|\vec{u}\|^2 = u_1^2 + u_2^2 \tag{7.5}$$

and therefore the unit vector in the $\vec{u}$ direction, denoted by $\hat{e}_u$, is given by

$$\hat{e}_u = \frac{1}{\sqrt{u_1^2 + u_2^2}}(u_1, u_2) \tag{7.6}$$

All of the above can be generalized to 3-D, or for that matter to n dimensions. For example,

$$\hat{e}_u = \frac{1}{\sqrt{u_1^2 + u_2^2 + \cdots + u_n^2}}(u_1, u_2, \ldots, u_n) \tag{7.7}$$

7.1.4 MATLAB Representation of Vectors

MATLAB distinguishes between two kinds of vectors: the column vector and the row vector. As long as the components of the vectors are all real, the difference between the two is in the structure of the array. In the column vector case, the array representation is vertical and in the row vector case, it is horizontal. This distinction is made for the purpose of including in a consistent structure the formulation of the dot product and the definition of matrix multiplication.

Example 7.1

Type and execute the following commands:

```
>>V=[1 3 5 7]
>>W=[1;3;5;7]
>>V'
>>U=3*V
>>Z=U+V
>>Y=V+W            %you cannot add a row vector and
                   %a column vector
```

You would have observed that:

1. The difference in the representation of the column and row vectors is in the manner they are separated inside the square brackets.
2. The single quotation mark following a vector with real components changes that vector from being a column vector to a row vector, and vice versa.
3. Multiplying a vector by a scalar simply multiplies each component of this vector by this scalar.
4. You can add two vectors of the same kind and the components would be adding by pairs.
5. You cannot add two vectors of different kinds; the computer will give you an error message alerting you that you are adding two quantities of different structures. □

The MATLAB command for obtaining the norm of a vector is **norm**. Using this notation, it is easy to define the unit vector in the same direction as a given vector.

Example 7.2

Find the length of the vector u = [1 5 3 2] and the unit vector parallel to it.

```
u=[1 5 3 2]
length_u=norm(u)              %length of vector u
length2_u=sqrt(sum(u.*u))     %length from definition
unit_u=u/(norm(u))            %unit vector parallel to u
length_unit_u=norm(unit_u)    %verify length of unit
                               vector □
```

7.2 Dot (or Scalar) Product

If the angle between the vectors $\vec{u}$ and $\vec{v}$ is θ, then the dot product of the two vectors is

$$\vec{u} \cdot \vec{v} = \|\vec{u}\| \|\vec{v}\| \cos(\theta) \tag{7.8}$$

The dot product can also be expressed as a function of the vector components. Referring to Figure 7.3, we know from trigonometry the relation relating the length of one side of a triangle with the length of the other two sides and the cosine of the angle between the other two sides. This relation is the generalized Pythagorean theorem. Referring to Figure 7.3, this gives

$$\|PQ\|^2 = \|\vec{u}\|^2 + \|\vec{v}\|^2 - 2\|\vec{u}\| \|\vec{v}\| \cos(\theta) \tag{7.9}$$

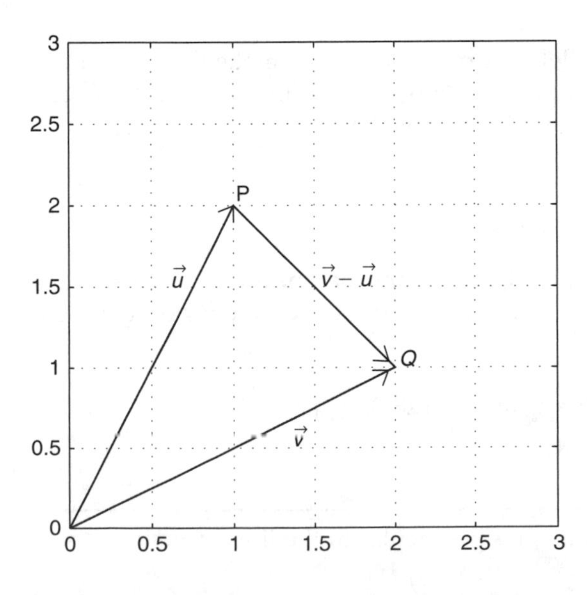

FIGURE 7.3
The geometry of the generalized Pythagorean theorem.

but since

$$\vec{PQ} = \vec{v} - \vec{u} \tag{7.10}$$

$$\Rightarrow \|\vec{u}\|\|\vec{v}\|\cos(\theta) = \frac{1}{2}(\|\vec{u}\|^2 + \|\vec{v}\|^2 - \|\vec{v} - \vec{u}\|^2) \tag{7.11}$$

and the dot product can be written as

$$\vec{u} \cdot \vec{v} = \frac{1}{2}(u_1^2 + u_2^2 + v_1^2 + v_2^2 - (v_1 - u_1)^2 - (v_2 - u_2)^2) = u_1v_1 + u_2v_2 \tag{7.12}$$

In an n-dimensional space, the above expression is generalized to

$$\vec{u} \cdot \vec{v} = u_1v_1 + u_2v_2 + \cdots + u_nv_n \tag{7.13}$$

and the norm square of the vector can be written as the dot product of the vector with itself; that is,

$$\|\vec{u}\|^2 = \vec{u} \cdot \vec{u} = u_1^2 + u_2^2 + \cdots + u_n^2 \tag{7.14}$$

Example 7.3

Parallelism and orthogonality of two vectors in a plane. Let the vectors $\vec{u}$ and $\vec{v}$ be given by: $\vec{u} = 3\hat{e}_1 + 4\hat{e}_2$ and $\vec{v} = a\hat{e}_1 + 7\hat{e}_2$. What is the value of a if the vectors are parallel, and if the vectors are orthogonal?

Solution:
Case 1: If the vectors are parallel, this means that they make the same angle with the x-axis. The tangent of this angle is equal to the ratio of the vector y-component to its x-component. This means that

$$\frac{a}{7} = \frac{3}{4} \Rightarrow a = 21/4$$

Case 2: If the vectors are orthogonal, this means that the angle between them is 90°, and their dot product will be zero because the cosine for that angle is zero. This implies that

$$3a + 28 = 0 \Rightarrow a = -28/3 \qquad \square$$

Example 7.4

Find the unit vector in 2-D that is perpendicular to the line $ax + by + c = 0$.

Solution: Choose two arbitrary points on this line. Denote their coordinates by (x_1, y_1) and (x_2, y_2); being on the line, they satisfy the equation of the line:

$$ax_1 + by_1 + c = 0$$
$$ax_2 + by_2 + c = 0$$

Subtracting the first equation from the second equation, we obtain

$$a(x_2 - x_1) + b(y_2 - y_1) = 0$$

which means that $(a, b) \perp (x_2 - x_1, y_2 - y_1)$, and the unit vector perpendicular to the line is

$$\hat{e}_\perp = \left(\frac{a}{\sqrt{a^2 + b^2}}, \frac{b}{\sqrt{a^2 + b^2}}\right) \quad \square$$

EXAMPLE 7.5

Find the angle that the lines $3x + 2y + 2 = 0$ and $2x - y + 1 = 0$ make with each other.

Solution: The angle between two lines is equal to the angle between their normal unit vectors. The unit vectors normal to each of the lines are, respectively,

$$\hat{n}_1 = \left(\frac{3}{\sqrt{13}}, \frac{2}{\sqrt{13}}\right) \quad \text{and} \quad \hat{n}_2 = \left(\frac{2}{\sqrt{5}}, \frac{-1}{\sqrt{5}}\right)$$

Having the two orthogonal unit vectors to the planes, it is simple to compute the angle between the planes:

$$\cos(\theta) = \hat{n}_1 \cdot \hat{n}_2 = \frac{4}{\sqrt{65}} \Rightarrow \theta = 1.0517 \text{ radians} \quad \square$$

7.2.1 MATLAB Representation of the Dot Product

In MATLAB, the dot product of two vectors can be performed using the result given in Eq. (7.13) or directly by either the command `dot`, or through the product of a row vector by a column vector of the same length, which is a special case of a matrix multiplication that we will cover with greater details in the next chapter.

Example 7.6

Find the dot product of the vectors:

$u = [1\ 5\ 3\ 7]$ and $v = [2\ 4\ 6\ 8]$

Solution: Type and execute each of the following commands:

```
>>u=[1 5 3 7]
>>v=[2 4 6 8]
>>a=sum(u.*v)          %result 7.13
>>b=dot(u,v)
>>c=u*v'
>>d=v*u'
>>e=u*v                %you cannot multiply two rows
>>m=u'*v               %the result is a matrix
```

Analyzing the results returned by MATLAB for the above entries, we note that the dot product of two vectors can be obtained by the multiplication of a row vector on the left and a column vector, of the same length, on the right. Either of the two vectors can be the row vector while the other will be the column vector. If the order of a row and column are exchanged, we obtain a 2-D array structure (i.e., a matrix, the subject of Chapter 8). On the other hand, if we multiply two row vectors, MATLAB gives an error message about the nonmatching of dimensions (more on this in the next chapter). □

In-Class Exercises

Pb. 7.2 Generalize the analytical technique, as previously used in Example 7.4 for finding the normal to a line in 2-D, to find the unit vector in 3-D that is perpendicular to the plane:

$$ax + by + cz + d = 0$$

(*Hint*: A vector is perpendicular to a plane if it is perpendicular to two non-collinear vectors in that plane.)

Pb. 7.3 Find, in 2-D, the distance of the point $P(x_0, y_0)$ from the line $ax + by + c = 0$. (*Hint*: Remember the geometric definition of the dot product.)

Pb. 7.4 Prove the following identities:

$$\vec{u} \cdot \vec{v} = \vec{v} \cdot \vec{u},\ \vec{u} \cdot (\vec{v} + \vec{w}) = \vec{u} \cdot \vec{v} + \vec{u} \cdot \vec{w},\ k \cdot (\vec{u} \cdot \vec{v}) = (k\vec{u}) \cdot \vec{v}$$

7.3 Components, Direction Cosines, and Projections

7.3.1 Components

The components of a vector are the values of each element in the defining n-tuplet representation. For example, consider the vector $\vec{u} = [1\,5\,3\,7]$ in real 4-D. We say that its first, second, third, and fourth components are 1, 5, 3, and 7, respectively. (We are maintaining, in this section, the arrow notation for the vectors, irrespective of the dimension of the space.)

The simplest basis of a n-dimensional vector space is the collection of n unit vectors, each having only one of their components that is nonzero and such that the location of this nonzero element is different for each of these basis vectors. This basis is not unique.

For example, in 4-D space, the canonical four-unit orthonormal basis vectors are given, respectively, by

$$\hat{e}_1 = [1 \quad 0 \quad 0 \quad 0] \tag{7.15}$$

$$\hat{e}_2 = [0 \quad 1 \quad 0 \quad 0] \tag{7.16}$$

$$\hat{e}_3 = [0 \quad 0 \quad 1 \quad 0] \tag{7.17}$$

$$\hat{e}_4 = [0 \quad 0 \quad 0 \quad 1] \tag{7.18}$$

and the vector $\vec{u}$ can be written as a linear combination of the basis vectors:

$$\vec{u} = u_1\hat{e}_1 + u_2\hat{e}_2 + u_3\hat{e}_3 + u_4\hat{e}_4 \tag{7.19}$$

The basis vectors are chosen to be orthonormal, which means that in addition to requiring each one of them to have unit length, they are also orthogonal two by two to each other. These properties of the basis vectors leads us to the following important result: the mth component of a vector is obtained by taking the dot product of the vector with the corresponding unit vector, that is,

$$u_m = \hat{e}_m \cdot \vec{u} \tag{7.20}$$

7.3.2 Direction Cosines

The direction cosines are defined by

$$\cos(\gamma_m) = \frac{u_m}{\|\vec{u}\|} = \frac{\hat{e}_m \cdot \vec{u}}{\|\vec{u}\|} \tag{7.21}$$

In 2-D or 3-D, these quantities have the geometrical interpretation of being the cosine of the angles that the vector $\vec{u}$ makes with the x-, y-, and z-axes.

7.3.3 Projections

The projection of a vector $\vec{u}$ over a vector $\vec{a}$ is a vector whose magnitude is the dot product of the vector $\vec{u}$ with the unit vector in the direction of $\vec{a}$, denoted by $\hat{e}_a$, and whose orientation is in the direction of $\hat{e}_a$:

$$proj_{\vec{a}}(\vec{u}) = (\vec{u} \cdot \hat{e}_a)\hat{e}_a = \frac{\vec{u} \cdot \vec{a}}{\|\vec{a}\|} \frac{\vec{a}}{\|\vec{a}\|} = \frac{\vec{u} \cdot \vec{a}}{\|\vec{a}\|^2} \vec{a} \tag{7.22}$$

The component of $\vec{u}$ that is perpendicular to $\vec{a}$ is obtained by subtracting from $\vec{u}$ the projection vector of $\vec{u}$ over $\vec{a}$.

MATLAB Implementation

Assume that we have the vector $\vec{u} = \hat{e}_1 + 5\hat{e}_2 + 3\hat{e}_3 + 7\hat{e}_4$ and the vector $\vec{a} = 2\hat{e}_1 + 3\hat{e}_2 + \hat{e}_3 + 4\hat{e}_4$. We desire to obtain the components of each vector, the projection of $\vec{u}$ over $\vec{a}$, and the component of $\vec{u}$ orthogonal to $\vec{a}$.

Type, execute, and interpret at each step, each of the following commands using the above definitions:

```
u=[1 5 3 7]
a=[2 3 1 4]
prj_u_over_a=((u*a')/(norm(a)^2))*a
comp_u_orth_to_a=u-prj_u_over_a
c=prj_u_over_a*comp_u_orth_to_a'
```

The last quantity is zero, up to machine round-up errors because the projection of $\vec{u}$ over $\vec{a}$ and the component of $\vec{u}$ orthogonal to $\vec{a}$ are perpendicular.

7.4 The Dirac Notation and Some General Theorems*

Thus far, we have established some key practical results in real finite-dimensional vector spaces; namely:

1. A vector can be decomposed into a linear combination of the basis vectors.
2. The dot product of two vectors can be written as the multiplication of a row vector by a column vector, each of whose elements are the components of the respective vectors.

* The asterisk indicates more advanced material that may be skipped in a first reading.

3. The norm of a vector, a nonnegative quantity, is the square root of the dot product of the vector with itself.
4. The unit vector parallel to a specific vector is that vector divided by its norm.
5. The projection of a vector on another can be deduced from the dot product of the two vectors.

To facilitate the statement of these results in a notation that will be suitable for infinite-dimensional vector spaces (which is very briefly introduced in Section 7.7), Dirac in his elegant formulation of quantum mechanics introduced a simple notation which we now present.

The Dirac notation represents the row vector by what he called the "bra-vector" and the column vector by what he called the "ket-vector," such that when a dot product is obtained by joining the two vectors, the result will be the scalar "bra-ket" quantity. Specifically

$$\text{Column vector } \vec{u} \Rightarrow |u\rangle \tag{7.23}$$

$$\text{Row vector } \vec{v} \Rightarrow \langle v| \tag{7.24}$$

$$\text{Dot product } \vec{v} \cdot \vec{u} \Rightarrow \langle v|u\rangle \tag{7.25}$$

The orthonormality of the basis vectors is written as

$$\langle m|n\rangle = \delta_{m,n} \tag{7.26}$$

where the basis vectors are referred to by their indices, and where $\delta_{m,n}$ is the Kroenecker delta, equal to 1 when its indices are equal, and zero otherwise.

The norm of a vector, a nonnegative quantity, is given by

$$(norm\ of\ |u\rangle)^2 = \|u\|^2 = \langle u|u\rangle \tag{7.27}$$

The Decomposition rule is written as

$$|u\rangle = \sum_n c_n\ |n\rangle \tag{7.28}$$

where the components are obtained by multiplying Eq. (7.28) on the LHS by $\langle m|$. Using Eq. (7.26), we deduce:

$$\langle m|u\rangle = \sum_n c_n \langle m|n\rangle = \sum_n c_n \delta_{m,n} = c_m \tag{7.29}$$

Next, using the Dirac notation, we present the proofs of two key theorems of vector algebra: the Cauchy–Schwartz inequality and the triangle inequality.

7.4.1 Cauchy–Schwartz Inequality

Let $|u\rangle$ and $|v\rangle$ be any nonzero vectors; then

$$\left|\langle u|v\rangle\right|^2 \leq \langle u|u\rangle\langle v|v\rangle \tag{7.30}$$

PROOF Let $\varepsilon = \pm 1$, ($\varepsilon^2 = 1$); then

$$\langle u|v\rangle = \varepsilon\left|\langle u|v\rangle\right| \quad \text{such that} \begin{cases} \varepsilon = 1 & \text{if } \langle u|v\rangle \geq 0 \\ \varepsilon = -1 & \text{if } \langle u|v\rangle \leq 0 \end{cases} \tag{7.31}$$

Now, consider the ket $|\varepsilon u + tv\rangle$; its norm is always nonnegative. Computing this norm square, we obtain:

$$\begin{aligned} \langle \varepsilon u + tv|\varepsilon u + tv\rangle &= \varepsilon^2\langle u|u\rangle + \varepsilon t\langle u|v\rangle + t\varepsilon\langle v|u\rangle + t^2\langle v|v\rangle \\ &= \langle u|u\rangle + 2\varepsilon t\langle u|v\rangle + t^2\langle v|v\rangle \\ &= \langle u|u\rangle + 2t\left|\langle u|v\rangle\right| + t^2\langle v|v\rangle \end{aligned} \tag{7.32}$$

The RHS of this quantity is a positive quadratic polynomial in t, and can be written in the standard form:

$$at^2 + bt + c \geq 0 \tag{7.33}$$

The non-negativity of this quadratic polynomial means that it can have at most one real root. This means that the discriminant must satisfy the inequality

$$b^2 - 4ac \leq 0 \tag{7.34}$$

Replacing a, b, and c by their values from Eq. (7.32), we obtain

$$4\left|\langle u|v\rangle\right|^2 - 4\langle u|u\rangle\langle v|v\rangle \leq 0 \tag{7.35}$$

$$\Rightarrow \left|\langle u|v\rangle\right|^2 \leq \langle u|u\rangle\langle v|v\rangle \tag{7.36}$$

which is the desired result. Note that the equality holds if and only if the two vectors are linearly dependent (i.e., one vector is equal to a scalar multiplied by the other vector).

EXAMPLE 7.7

Show that for any three nonzero numbers, u_1, u_2, and u_3, the following inequality always holds:

$$9 \leq (u_1 + u_2 + u_3)\left(\frac{1}{u_1} + \frac{1}{u_2} + \frac{1}{u_3}\right) \tag{7.37}$$

PROOF Choose the vectors $|v\rangle$ and $|w\rangle$ such that

$$|v\rangle = \left|u_1^{1/2}, u_2^{1/2}, u_3^{1/2}\right\rangle \tag{7.38}$$

$$|w\rangle = \left|\left(\frac{1}{u_1}\right)^{1/2}, \left(\frac{1}{u_2}\right)^{1/2}, \left(\frac{1}{u_3}\right)^{1/2}\right\rangle \tag{7.39}$$

then

$$\langle v|w\rangle = 3 \tag{7.40}$$

$$\langle v|v\rangle = (u_1 + u_2 + u_3) \tag{7.41}$$

$$\langle w|w\rangle = \left(\frac{1}{u_1} + \frac{1}{u_2} + \frac{1}{u_3}\right) \tag{7.42}$$

Applying the Cauchy–Schwartz inequality in Eq. (7.36) establishes the desired result. The above inequality can be trivially generalized to n-elements, which leads to the following important result for the equivalent resistance for resistors all in series or all in parallel. □

Application

The equivalent resistance of n-resistors all in series and the equivalent resistance of the same n-resistors all in parallel obey the relation

$$n^2 \leq \frac{R_{\text{series}}}{R_{\text{parallel}}} \tag{7.43}$$

PROOF The proof is straightforward. Using Eq. (7.37) and recalling Ohm's law for n resistors $\{R_1, R_2, \ldots, R_n\}$, the equivalent resistances for this combination, when all resistors are in series or are all in parallel, are given respectively by

$$R_{\text{series}} = R_1 + R_2 + \cdots + R_n \tag{7.44}$$

and

$$\frac{1}{R_{\text{parallel}}} = \frac{1}{R_1} + \frac{1}{R_2} + \cdots + \frac{1}{R_n} \tag{7.45}$$

Question: Can you derive a similar theorem for capacitors all in series and all in parallel? (Remember that the equivalent capacitance law is different for capacitors than for resistors.)

7.4.2 Triangle Inequality

This is, as the name implies, a generalization of a theorem from Euclidean geometry in 2-D that states that the length of one side of a triangle is smaller or equal to the sum of the the other two sides. Its generalization is

$$\|\vec{u} + \vec{v}\| \leq \|\vec{u}\| + \|\vec{v}\| \tag{7.46}$$

PROOF Using the relation between the norm and the dot product, we have

$$\begin{aligned} \|\vec{u} + \vec{v}\|^2 &= \langle u + v \,|\, u + v \rangle = \langle u \,|\, v \rangle + 2\langle u \,|\, v \rangle + \langle v \,|\, v \rangle \\ &= \|\vec{u}\|^2 + 2\langle u \,|\, v \rangle + \|\vec{v}\|^2 \leq \|\vec{u}\|^2 + 2|\langle u \,|\, v \rangle| + \|\vec{v}\|^2 \end{aligned} \tag{7.47}$$

Using the Cauchy–Schwartz inequality for the dot product appearing in the previous inequality, we deduce that

$$\|\vec{u} + \vec{v}\|^2 \leq \|\vec{u}\|^2 + 2\|\vec{u}\|\|\vec{v}\| + \|\vec{v}\|^2 = (\|\vec{u}\| + \|\vec{v}\|)^2 \tag{7.48}$$

which establishes the theorem.

Homework Problems

Pb. 7.5 Using the Dirac notation, generalize to n-dimensions the 2-D geometry Parallelogram theorem, which states that the sum of the squares of the diagonals of a parallelogram is equal to twice the sum of the squares of the side; or that

$$\|\vec{u} + \vec{v}\|^2 + \|\vec{u} - \vec{v}\|^2 = 2\|\vec{u}\|^2 + 2\|\vec{v}\|^2$$

Pb. 7.6 Referring to the inequality of Eq. (7.43), which relates the equivalent resistances of n-resistors in series and in parallel, under what conditions does the equality hold?

7.5 Cross Product and Scalar Triple Product

In this section and in Section 7.6 and Section 7.7, we restrict our discussions to vectors in a 3-D space, and use the more familiar conventional vector notation.

7.5.1 Cross Product

DEFINITION If two vectors are given by $\vec{u} = (u_1, u_2, u_3)$ and $\vec{v} = (v_1, v_2, v_3)$ then their cross product, denoted by $\vec{u} \times \vec{v}$, is a vector given by

$$\vec{u} \times \vec{v} = (u_2v_3 - u_3v_2, u_3v_1 - u_1v_3, u_1v_2 - u_2v_1) \tag{7.49}$$

By simple substitution, we can infer the following properties for the cross product as summarized in the preparatory exercises below.

PREPARATORY EXERCISES

Pb. 7.7 Show, using the above definition for the cross product, that

a. $\vec{u} \cdot (\vec{u} \times \vec{v}) = \vec{v} \cdot (\vec{u} \times \vec{v}) = 0 \Rightarrow \vec{u} \times \vec{v}$ is orthogonal to both $\vec{u}$ and $\vec{v}$

b. $\|\vec{u} + \vec{v}\|^2 + \|\vec{u} - \vec{v}\|^2 = 2\|\vec{u}\|^2 +$ called the Lagrange Identity

c. $\vec{u} \times \vec{v} = -(\vec{v} \times \vec{u})$ Noncommutativity

d. $\vec{u} \times (\vec{v} + \vec{w}) = \vec{u} \times \vec{v} + \vec{u} \times \vec{w}$ Distributive property

e. $k(\vec{u} \times \vec{v}) = (k\vec{u}) \times \vec{v} = \vec{u} \times (k\vec{v})$

f. $\vec{u} \times \vec{0} = \vec{0}$

g. $\vec{u} \times \vec{u} = \vec{0}$

Pb. 7.8 Verify the following relations for the basis unit vectors:

$$\hat{e}_1 \times \hat{e}_2 = \hat{e}_3; \quad \hat{e}_2 \times \hat{e}_3 = \hat{e}_1; \quad \hat{e}_3 \times \hat{e}_1 = \hat{e}_2$$

Pb. 7.9 The Right Hand three fingers rule is used to determine the direction of a vector equal to the cross product of two other vectors. This rule states that if the index finger is aligned in the direction of the first vector, and the middle finger is aligned with the second vector, then the thumb gives the direction of the cross product of the first vector by the second vector. Use this rule to confirm the results for the different cross products of the different combinations of basis vectors in 3-D obtained in the previous problem.

7.5.2 Geometric Interpretation of the Cross Product

As noted in Pb. 7.7a, the cross product is a vector that is perpendicular to its two constituents. This determines the resultant vector's direction. To determine its magnitude, consider the Lagrange identity. If the angle between $\vec{u}$ and $\vec{v}$ is θ, then

$$\|\vec{u}\times\vec{v}\|^2 = \|\vec{u}\|^2\,\|\vec{v}\|^2 - \|\vec{u}\|^2\,\|\vec{v}\|^2\cos^2(\theta) \tag{7.50}$$

and

$$\|\vec{u}\times\vec{v}\| = \|\vec{u}\|\,\|\vec{v}\|\sin(\theta) \tag{7.51}$$

that is, the magnitude of the cross product of two vectors is the area of the parallelogram formed by these vectors.

7.5.3 Scalar Triple Product

DEFINITION If $\vec{u}$, $\vec{v}$, and $\vec{w}$ are vectors in 3-D, then $\vec{u}\cdot(\vec{v}\times\vec{w})$ is called the scalar triple product of $\vec{u}$, $\vec{v}$, and $\vec{w}$.

PROPERTY

$$\vec{u}\cdot(\vec{v}\times\vec{w}) = \vec{v}\cdot(\vec{w}\times\vec{u}) = \vec{w}\cdot(\vec{u}\times\vec{v}) \tag{7.52}$$

This property can be trivially proven by writing out the components expansions of the three quantities.

7.5.3.1 Geometric Interpretation of the Scalar Triple Product

If the vectors' $\vec{u}$, $\vec{v}$, and $\vec{w}$ original points are brought to the same origin, these three vectors define a parallelepiped. The absolute value of the scalar triple product can then be interpreted as the volume of this parallelepiped. We have shown earlier that $\vec{v}\times\vec{w}$ is a vector that is perpendicular to both $\vec{v}$ and $\vec{w}$, and whose magnitude is the area of the base parallelogram. From the definition of the scalar product, dotting this vector with $\vec{u}$ will give a scalar that is the product of the area of the parallelepiped base multiplied by the parallelepiped height, whose magnitude is exactly the volume of the parallelepiped.

The circular permutation property of Eq. (7.52) then has a very simple geometric interpretation: in computing the volume of a parallelepiped, it does not matter which surface we call base.

MATLAB Representation

The cross product of the vectors $\vec{u} = (u_1, u_2, u_3)$ and $\vec{v} = (v_1, v_2, v_3)$ is found using the `cross(u,v)` command.

The triple scalar product of the vectors $\vec{u}$, $\vec{v}$, and $\vec{w}$ is found through a command of the form `dot(u,cross(v,w))`. Make sure that the vectors defined as arguments of these functions are defined as 3-D vectors, so that the commands work and the results make sense.

Example 7.8

Given the vectors $\vec{u} = (2, 1, 0)$, $\vec{v} = (0, 3, 0)$, $\vec{w} = (1, 2, 3)$, find the cross product of the separate pairs of these vectors, and the volume of the parallelepiped formed by the three vectors.

Solution: Type, execute, and interpret at each step, each of the following commands, using the above definitions:

```
u=[2 1 0]
v=[0 3 0]
w=[1 2 3]
u_cross_v=cross(u,v)
u_cross_w=cross(u,w)
v_cross_w=cross(v,w)
parallelepiped_vol=abs(cross(u,v)*w')
```

□

In-Class Exercises

Pb. 7.10 Compute the shortest distance from New York to London. [*Hints*: (1) The path along a great circle is the shortest path between two points on a sphere; (2) the angle between the radial unit vectors passing through each of the cities can be obtained from their respective latitude and longitude.]

Pb. 7.11 Find two unit vectors that are orthogonal to both vectors given by

$$\vec{a} = (2, -1, 2) \quad \text{and} \quad \vec{b} = (1, 2, -3)$$

Pb. 7.12 Find the area of the triangle with vertices at the points:

$$A(0, -1, 1), \quad B(3, 1, 0), \quad \text{and} \quad C(-2, 0, 2)$$

Pb. 7.13 Find the volume of the parallelepiped formed by the three vectors

$$\vec{u} = (1, 2, 0), \quad \vec{v} = (0, 3, 0), \quad \vec{w} = (1, 2, 3)$$

Pb. 7.14 Determine the equation of a plane that passes through the point (1, 1, 1) and is normal to the vector (2, 1, 2).

Pb. 7.15 Find the angle of intersection of the planes:

$$x + y - z = 0 \quad \text{and} \quad x - 3y + z - 1 = 0$$

Pb. 7.16 Find the distance between the point (3, 1, −2) and the plane $z = 2x - 3y$.

Pb. 7.17 Find the equation of the line that contains the point (3, 2, 1) and is perpendicular to the plane $x + 2y - 2z = 2$. Write the parametric equation for this line.

Pb. 7.18 Find the point of intersection of the plane $2x - 3y + z = 6$ and the line

$$\frac{x-1}{3} = \frac{y+1}{1} = \frac{z-2}{2}$$

Pb. 7.19 Show that the points (1, 5), (3, 11), and (5, 17) are collinear.

Pb. 7.20 Show that the three vectors $\vec{u}$, $\vec{v}$ and $\vec{w}$ are coplanar

$$\vec{u} = (2, 3, 5); \quad \vec{v} = (2, 8, 1); \quad \vec{w} = (8, 22, 12)$$

Pb. 7.21 Find the unit vector normal to the plane determined by the points (0, 0, 1), (0, 1, 0), and (1, 0, 0).

Homework Problems

Pb. 7.22 Determine the tetrahedron with the largest surface area whose vertices P_0, P_1, P_2, and P_3 are on the unit sphere $x^2 + y^2 + z^2 = 1$.

[*Hints*: (1) Designate the point P_0 as north pole and confine P_1 to the zero meridian. With this choice, the coordinates of the vertices are given by

$$P_0 = (\theta_0 = \pi/2, \phi_0 = 0)$$

$$P_1 = (\theta_1, \phi_1 = 0)$$

$$P_2 = (\theta_2, \phi_2)$$

$$P_3 = (\theta_3, \phi_3)$$

(2) From symmetry, the optimal tetrahedron will have a base (P_1, P_2, P_3) that is an equilateral triangle in a plane parallel to the equatorial plane. The latitude of (P_1, P_2, P_3) is θ, while their longitudes are (0, $2\pi/3$, $-2\pi/3$), respectively. (3) The area of the tetrahedron is the sum of the areas of the four triangles (012), (023), (031), (123), where we are indicating each point by its subscript. (4) Express the area as function of θ. Find the value of θ that maximizes this quantity.]

7.6 Tangent, Normal, and Curvature

As you may recall, in Chapter 1, we described curves in 2-D and 3-D by parametric equations. Essentially, we gave each of the coordinates as a function of a parameter. In effect, we generated a vector-valued function because the position of the point describing the curve can be written as

$$\vec{R}(t) = x(t)\hat{e}_1 + y(t)\hat{e}_2 + z(t)\hat{e}_3 \tag{7.53}$$

If the parameter t was chosen to be time, then the tip of the vector $\vec{R}(t)$ would be the position of a point on that curve as a function of time. In mechanics, finding $\vec{R}(t)$ is ultimately the goal of any problem in the dynamics of a point particle.

As pointed out in Chapter 1, the dynamics of a particle can be better visualized if in addition to plotting its trajectory, we also plot its velocity vector. We introduce in this section, the tangent vector and the normal vector to the curve and the curvature of the curve.

The velocity vector field associated with the above position vector is defined through

$$\frac{d\vec{R}(t)}{dt} = \frac{dx(t)}{dt}\hat{e}_1 + \frac{dy(t)}{dt}\hat{e}_2 + \frac{dz(t)}{dt}\hat{e}_3 \tag{7.54}$$

and the unit vector tangent to the curve is given by

$$\hat{T}(t) = \frac{\dfrac{d\vec{R}(t)}{dt}}{\left\|\dfrac{d\vec{R}(t)}{dt}\right\|} \tag{7.55}$$

This is, of course, the unit vector that is always in the direction of the velocity of the particle.

LEMMA If a vector-valued function $\vec{V}(t)$ has a constant value, then its derivative $\dfrac{d\vec{V}(t)}{dt}$ is orthogonal to it.

PROOF The proof of this lemma is straightforward. If the length of the vector is constant, then its dot product with itself is a constant; that is, $\vec{V}(t) \cdot \vec{V}(t) = C$.

Differentiating both sides of this equation gives $\frac{d\vec{V}(t)}{dt} \cdot \vec{V}(t) = 0$ and the orthogonality between the two vectors is thus established.

The tangential unit vector $\hat{T}(t)$ is, by definition, constructed to have unit length. We construct the norm to the curve by taking the unit vector in the direction of the time derivative of the tangential vector; that is,

$$\vec{N}(t) = \frac{\frac{d\hat{T}(t)}{dt}}{\left\| \frac{d\hat{T}(t)}{dt} \right\|} \tag{7.56}$$

The curvature of the curve is

$$\kappa = \frac{\left\| \frac{d\hat{T}(t)}{dt} \right\|}{\left\| \frac{d\vec{R}(t)}{dt} \right\|} \tag{7.57}$$

Example 7.9

Find the tangent, normal, and curvature of the trajectory of a particle moving in uniform circular motion of radius a and with angular frequency ω.

Solution: The parametric equation of motion is

$$\vec{R}(t) = a\cos(\omega t)\hat{e}_1 + a\sin(\omega t)\hat{e}_2 \tag{7.58}$$

The velocity vector is

$$\frac{d\vec{R}(t)}{dt} = -a\omega\sin(\omega t)\hat{e}_1 + a\omega\cos(\omega t)\hat{e}_2 \tag{7.59}$$

and its magnitude is $a\omega$.

The tangent vector is therefore

$$\hat{T}(t) = -\sin(\omega t)\hat{e}_1 + \cos(\omega t)\hat{e}_2 \tag{7.60}$$

The normal vector is

$$\hat{N}(t) = -\cos(\omega t)\hat{e}_1 - \sin(\omega t)\hat{e}_2 \tag{7.61}$$

The radius of curvature is

$$\kappa(t) = \frac{\left\| \dfrac{d\hat{T}(t)}{dt} \right\|}{\left\| \dfrac{d\vec{R}(t)}{dt} \right\|} = \frac{\left\| -\omega \cos(\omega t)\hat{e}_1 - \omega \sin(\omega t)\hat{e}_2 \right\|}{\left\| -a\omega \sin(\omega t)\hat{e}_1 + a\omega \cos(\omega t)\hat{e}_2 \right\|} = \frac{1}{a} = \text{constant} \quad (7.62)$$

□

Homework Problems

Pb. 7.23 Show that in 2-D the radius of curvature can be written as

$$\kappa = \frac{|x'y'' - y'x''|}{((x')^2 + (y')^2)^{3/2}}$$

where the prime refers to the first derivative with respect to time, and the double prime refers to the second derivative with respect to time.

Pb. 7.24 Using the parametric equations for an ellipse given in Example 1.22, find the curvature of the ellipse as function of t.

a. At what points is the curvature a minimum, and at what points is it a maximum?
b. What does the velocity do at the points of minimum and maximum curvature?
c. On what dates of the year does the planet Earth pass through these points on its trajectory around the sun?

7.7 Velocity and Acceleration Vectors in Polar Coordinates*

In the previous section, we found the expressions for the unit tangent and normal vectors and curvature of a curve. We wrote the results for the different vector-valued functions in Cartesian coordinates.

In many dynamical problems of general interest, especially those concerned with the motion of bodies attracted by point-like sources, it is more convenient to analyze the problem's dynamics in polar coordinates. In this section, we will derive the expressions for the velocity and acceleration in polar coordinates and derive some of the most fundamental results of planetary motion physics.

We will limit our discussion here to 2-D. A point is specified by the variables (r, θ).

* The asterisk indicates more advanced material that may be skipped in a first reading.

The basis vectors in polar coordinates are given respectively by

$$\begin{aligned} \hat{e}_r &= \cos(\theta)\,\hat{e}_1 + \sin(\theta)\,\hat{e}_2 \\ \hat{e}_\theta &= -\sin(\theta)\,\hat{e}_1 + \cos(\theta)\,\hat{e}_2 \end{aligned} \tag{7.63}$$

The above unit vectors are referred to respectively as the radial and tangential unit vectors. Geometrically, $\hat{e}_r$ points along the position vector $\vec{r}$, while $\hat{e}_\theta$ is perpendicular to it and is pointing in the anticlockwise direction.

Since both $\hat{e}_r$ and $\hat{e}_\theta$ are dependent on the polar angle, in contrast to the Cartesian unit vectors which are the same everywhere, the time derivatives of these vectors are nonzero, specifically

$$\frac{d\hat{e}_r}{dt} = \frac{d\theta}{dt}(-\sin(\theta)\,\hat{e}_1 + \cos(\theta)\,\hat{e}_2) = \frac{d\theta}{dt}\hat{e}_\theta \tag{7.64}$$

$$\frac{d\hat{e}_\theta}{dt} = \frac{d\theta}{dt}(-\cos(\theta)\,\hat{e}_1 - \sin(\theta)\,\hat{e}_2) = -\frac{d\theta}{dt}\hat{e}_r \tag{7.65}$$

Consequently, the velocity and the acceleration vectors are given in polar coordinates by

$$\vec{v} = \frac{d\vec{r}}{dt} = \frac{dr}{dt}\hat{e}_r + r\frac{d\theta}{dt}\hat{e}_\theta \tag{7.66}$$

$$\vec{a} = \frac{d\vec{v}}{dt} = \left(\frac{d^2r}{dt^2} - r\left(\frac{d\theta}{dt}\right)^2\right)\hat{e}_r + \left(r\frac{d^2\theta}{dt^2} + 2\frac{dr}{dt}\frac{d\theta}{dt}\right)\hat{e}_\theta \tag{7.67}$$

Example 7.10

Using Newton's second law of motion, show that if the force acting on a particle is radial, then the angular momentum of the particle is conserved.

Solution:

- Newton's second law of classical motion states that the force on a point particle is equal to its mass multiplied by its acceleration:

$$\vec{F} = m\vec{a} \tag{7.68}$$

- The angular momentum of a particle is equal to the cross product of its position vector and its momentum ($=m\vec{v}$). In polar coordinates, the angular momentum for a particle moving in the x–y plane is given by

$$\vec{L} = m\vec{r} \times \vec{v} = mr^2 \frac{d\theta}{dt} \hat{e}_3 = ml\hat{e}_3 \tag{7.69}$$

- A radial forces means that

$$\vec{F} = f(r)\, \hat{e}_r \tag{7.70}$$

Combining Newton's second law, with the expression for the acceleration in polar coordinates, and noting that the force is radial, we deduce that

$$r\frac{d^2\theta}{dt^2} + 2\frac{dr}{dt}\frac{d\theta}{dt} = 0 \tag{7.71}$$

Multiplying this equation by r, we obtain

$$r\left(r\frac{d^2\theta}{dt^2} + 2\frac{dr}{dt}\frac{d\theta}{dt}\right) = \frac{dl}{dt} = 0 \tag{7.72}$$

i.e., the angular momentum is conserved. □

EXAMPLE 7.11

Using Newton's law of gravitation, derive Kepler's laws for planetary motion.

Solution:
Some background first:

- The gravitational force exerted on a planet by the sun is given by

$$\vec{F} = -\frac{GMm}{r^2}\hat{e}_r \tag{7.73}$$

where M is the mass of the sun, m the mass of the planet, r the distance between the sun and the planet, and G the gravitational constant. The values of these quantities are

$$M = 1.9891 \times 10^{30}\,\text{kg}$$

$$m_{\text{earth}} = 5.9742 \times 10^{24}\,\text{kg}$$

$$G = 6.673 \times 10^{-11}\,\text{m}^3\,\text{kg}^{-1}\,\text{s}^{-2}$$

- Kepler's laws of planetary motion, formulated in the early seventeenth century (1609–1618) by Johannes Kepler based on observational astronomy state the following:
  - (i) The planetary orbits are ellipses with the sun located at one of the foci.
  - (ii) Each planet sweeps out an equal area in equal time interval.
  - (iii) The square of the orbital period of a planet is proportional to the cube of the major (long) radius of its elliptic trajectory.

To prove Kepler's laws, let us start by using the results of Example 9.10 to establish the second of these laws:

(i) The infinitesimal area swept by the planet in its trajectory around the sun is given by the area of the triangle, with base equal to the arc swept by the planet in time δt, and height equal to r, the distance of the planet to the sun:

$$\delta A = \frac{1}{2} r^2 \delta\theta \tag{7.74}$$

$$\Rightarrow \quad \frac{dA}{dt} = \frac{1}{2} r^2 \frac{d\theta}{dt} = \frac{l}{2} \tag{7.75}$$

As shown in the previous example, this quantity is proportional to the angular momentum and is conserved for particles moving in the fields of radial forces.

(ii) Equating the radial component of the acceleration with the gravitational force per unit mass, and using the conservation of angular momentum to replace the time derivative of the angle by l/r^2 gives us an ODE for r:

$$\frac{d^2r}{dt^2} - \frac{l^2}{r^3} = -\frac{GM}{r^2} \tag{7.76}$$

Introducing the new variable $s = 1/r$, we can write for the time derivaive of r:

$$\frac{dr}{dt} = -\frac{1}{s^2}\frac{ds}{dt} \tag{7.77}$$

To obtain the particle's trajectory, i.e., a functional relation of r and θ, we use the chain rule for derivatives and the conservation of angular momentum, to obtain

$$\frac{dr}{dt} = -r^2 \frac{ds}{d\theta}\frac{d\theta}{dt} = -l\frac{ds}{d\theta} \tag{7.78}$$

Similarly, we can differentiate this quantity again and obtain

$$\frac{d^2r}{dt^2} = -s^2 l^2 \frac{d^2s}{d\theta^2} \tag{7.79}$$

The radial time-dependent equation of motion can now be written as an ODE relating s and θ, solving it gives the planet trajectory. This ODE is

$$\frac{d^2s}{d\theta^2} + s = \frac{GM}{l^2} \tag{7.80}$$

The general solution of this equation is the sum of the homogeneous solution and the particular solution and is given by

$$s = \frac{GM}{l^2}[1 - \varepsilon\cos(\theta)] \tag{7.81}$$

where ε is a constant of integration. This gives for r, the expression

$$r = \frac{r_0}{1 - \varepsilon\cos(\theta)} \qquad \text{where } r_0 = \frac{l^2}{GM} \tag{7.82}$$

Referring back to Pb. 1.16, we know that for $0 < \varepsilon < 1$ this is the equation of an ellipse, which in Cartesian coordinates can be written as

$$\frac{(x - x_0)^2}{a^2} + \frac{y^2}{b^2} = 1 \tag{7.83}$$

where

$$x_0 = \frac{\varepsilon r_0}{1 - \varepsilon^2} \tag{7.84}$$

$$a = \frac{r_0}{1 - \varepsilon^2} \tag{7.85}$$

$$b = \frac{r_0}{(1 - \varepsilon^2)^{1/2}} \tag{7.86}$$

[For the planet Earth, the length a of the semi-major axis (the long radius) of its trajectory around the sun is 149.6×10^9 m, and the eccentricity of this trajectory is 0.0167]. Finally, we need to derive Kepler's third law:

(iii) The period is the time for the planet to sweep a polar angle of 2π, or equivalently the area of the ellipse ($A = \pi ab$). Because of Kepler's second law, the ratio of the ellipse area to the period is equal to dA/dt. This gives

$$T = \frac{2\pi ab}{l} \tag{7.87}$$

If we now substitute the above value for b, the short radius of the ellipse, in the expression of T, we obtain

$$T^2 = \frac{4\pi^2 a^2 b^2}{l^2} = \left(\frac{4\pi^2}{GM}\right) a^3 \tag{7.88}$$

This is Kepler's third law. [Substituting the numerical values for the different quantities, we find that the period of rotation of the planet Earth around the sun is 3.155815×10^7 s.] Numerically, the trajectory can be obtained by solving Newton's equations of motion for the coupled system of Cartesian coordinates describing the motion in a central force.

Recalling that the radial unit vector can be written as

$$\vec{r} = \frac{x}{(x^2 + y^2)^{1/2}}\hat{e}_x + \frac{y}{(x^2 + y^2)^{1/2}}\hat{e}_y \tag{7.89}$$

Newton's equations of motion then reduce to

$$\ddot{x} = -\frac{GMx}{(x^2 + y^2)^{3/2}} \tag{7.90}$$

$$\ddot{y} = -\frac{GMy}{(x^2 + y^2)^{3/2}} \tag{7.91}$$

Introducing the 4-D vector $z(t)$, with components $x(t)$, $y(t)$, $\dot{x}(t)$, and $\dot{y}(t)$ will satisfy the ODE:

$$\dot{z}(t) = \begin{bmatrix} x(t) \\ y(t) \\ -\dfrac{\alpha x(t)}{r^3(t)} \\ -\dfrac{\alpha y(t)}{r^3(t)} \end{bmatrix} \quad \text{where } r(t) = \sqrt{x^2(t) + y^2(t)} \text{ and } \alpha = GM \tag{7.92}$$

The *function m-file* representing this system of equations is, in normalized units:

```
function zdot=Kepler(t,z)
r=sqrt(z(1)^2+z(2)^2);
zdot=[z(3);z(4);-z(1)/r^3;-z(2)/r^3];
```

Using the MATLAB solver **ode23**, we can solve this system of equations. Entering the appropriate values for the initial conditions, we can obtain the trajectory of the planet by executing:

```
tspan=[0 2*pi];
z0=?;      %insert the initial conditions for position and
           %velocity
tol=0.0001;
[t,z]=ode23(@Kepler,tspan,z0,tol);
plot(z(:,1),z(:,2),'k')
axis equal
```
□

Homework Problems

Pb. 7.25 Find analytically the central force acting on a particle that would make its trajectory a spiral of the form

$$r = r_0\theta^2$$

7.8 Line Integral

As you may have already learned in your elementary physics course: if a force $\vec{F}$ is applied to a particle that moves by an infinitesimal distance $\Delta\vec{l}$, then the infinitesimal work done by the force on the particle is the scalar product of the force and the displacement; that is

$$\Delta W = \vec{F} \cdot \Delta\vec{l} \tag{7.93}$$

Now, to calculate the work done when the particle moves along a curve C, located in a plane, we need to define the concept of a line integral.

Suppose that the curve is described parametrically [i.e., $x(t)$ and $y(t)$ are given]. Furthermore, suppose that the vector field representing the force is given by

$$\vec{F} = P(x, y)\hat{e}_x + Q(x, y)\hat{e}_y \tag{7.94}$$

The displacement element is given by

$$\Delta l = \Delta x \hat{e}_x + \Delta y \hat{e}_y \tag{7.95}$$

The infinitesimal element of work, which is the dot product of the above two quantities, can then be written as

$$\Delta W = P\Delta x + Q\Delta y \tag{7.96}$$

This expression can be simplified if the curve is written in parametric form. Assuming the parameter is t, then ΔW can be written as a function of the single parameter t:

$$\Delta W = P(t)\frac{dx}{dt}\Delta t + Q(t)\frac{dy}{dt}\Delta t = \left(P(t)\frac{dx}{dt} + Q(t)\frac{dy}{dt} \right)\Delta t \tag{7.97}$$

and the total work can be written as an integral over the single variable t:

$$W = \int_{t_0}^{t_1} \left(P(t)\frac{dx}{dt} + Q(t)\frac{dy}{dt} \right) dt \tag{7.98}$$

Homework Problems

Pb. 7.26 How much work is done in moving the particle from the point (0, 0) to the point (3, 9) in the presence of the force $\vec{F}$ along the following two different paths?

a. The parabola $y = x^2$.
b. The line $y = 3x$.

The force is given by

$$\vec{F} = xy\hat{e}_x + (x^2 + y^2)\hat{e}_y$$

Pb. 7.27 Let $\vec{F} = y\hat{e}_x + x\hat{e}_y$. Calculate the work moving from (0, 0) to (1, 1) along each of the following curves:

a. The straight line $y = x$.
b. The parabola $y = x^2$.
c. The curve C described by the parametric equations:

$$x(t) = t^{3/2} \quad \text{and} \quad y(t) = t^5$$

(cont'd.)

Homework Problems *(cont'd.)*

A vector field such as the present one, whose line integral is independent of the path chosen between fixed initial and final points, is said to be conservative. In your vector calculus course, you will establish the necessary and sufficient conditions for a vector field to be conservative. The importance of conservative fields lies in the ability of their derivation from a scalar potential. More about this topic will be discussed in electromagnetic courses.

7.9 Infinite-Dimensional Vector Spaces*

This section introduces some preliminary ideas on infinite-dimensional vector spaces. We assume that the components of this vector space are complex numbers rather than real numbers, as we have restricted ourselves thus far. Using these ideas, we discuss, in a very preliminary fashion, Fourier series and expansion of functions in Legendre polynomials.

We use the Dirac notation to stress the commonalties that unite the finite- and infinite-dimensional vector spaces. We, at this level, sacrifice the mathematical rigor for simplicity, and even commit a few sins in our treatment of limits. A more formal and rigorous treatment of this subject can be found in many books on functional analysis, to which we refer the interested reader for further details.

A Hilbert space is much the same type of mathematical object as the vector spaces that you have been introduced to in the preceding sections of this chapter. Its elements are functions, instead of n-dimensional vectors. It is infinite-dimensional because the function has a value, say a component, at each point in space, and space is continuous with an infinite number of points.

The Hilbert space has the following properties:

1. The space is linear under the two conditions that:
   a. If a is a constant and $|\varphi\rangle$ is any element in the space, then $a|\psi\rangle$ is also an element of the space; and
   b. If a and b are constants, and $|\varphi\rangle$ and $|\psi\rangle$ are elements belonging to the space, then $a|\varphi\rangle + b|\psi\rangle$ is also an element of the space.
2. There is an inner (dot) product for any two elements in the space. The definition adopted here for this inner product for functions defined in the interval $t_{\min} \leq t \leq t_{\max}$ is:

$$\langle\psi|\varphi\rangle = \int_{t_{\min}}^{t_{\max}} \bar{\psi}(t)\varphi(t)\,dt \tag{7.99}$$

*The asterisk indicates more advanced material that may be skipped in a first reading.

3. Any element of the space has a norm ("length") that is positive and related to the inner product as follows:

$$\|\varphi\|^2 = \langle \varphi \mid \varphi \rangle = \int_{t_{\min}}^{t_{\max}} \bar{\varphi}(t)\varphi(t)\, dt \tag{7.100}$$

Note that the requirement for the positivity of a norm is that which necessitated the complex conjugation in the definition of the bra-vector.

4. The Hilbert space is complete; or loosely speaking, the Hilbert space contains all its limit points. This condition is too technical and will not be further discussed here.

In this Hilbert space, we define similar concepts to those in finite-dimensional vector spaces:

- *Orthogonality*. Two vectors are orthogonal if

$$\langle \psi \mid \varphi \rangle = \int_{t_{\min}}^{t_{\max}} \bar{\psi}(t)\varphi(t)\, dt = 0 \tag{7.101}$$

- *Basis vectors*. Any function in Hilbert space can be expanded in a linear combination of the basis vectors $\{u_n\}$, such that

$$|\varphi\rangle = \sum_n c_n |u_n\rangle \tag{7.102}$$

and such that the elements of the basis vectors obey the orthonormality relations

$$\langle u_m \mid u_n \rangle = \delta_{m,n} \tag{7.103}$$

- *Decomposition rule*. To find the c_n's, we follow the same procedure adopted for finite-dimensional vector spaces; that is, take the inner product of the expansion in Eq. (7.102) with the bra $\langle u_m|$ .We obtain, using the orthonormality relations [Eq. (7.103)], the following:

$$\langle u_m \mid \varphi \rangle = \sum_n c_n \langle u_m \mid u_n \rangle = \sum_n c_n \delta_{m,n} = c_m \tag{7.104}$$

Said differently, c_m is the projection of the ket $|\varphi\rangle$ onto the bra $\langle u_m|$.

- *The norm as a function of the components*. The norm of a vector can be expressed as a function of its components. Using Eq. (7.102) and Eq. (7.103), we obtain

$$\|\varphi\|^2 = \langle \varphi \mid \varphi \rangle = \sum_n \sum_m \bar{c}_n c_m \langle u_n \mid u_m \rangle = \sum_n \sum_m \bar{c}_n c_m \delta_{n,m} = \sum_n |c_n|^2 \tag{7.105}$$

Said differently, the norm square of a vector is equal to the sum of the magnitude square of the components.

Application 1: The Fourier Series

The theory of Fourier series, as covered in your calculus course, states that a function that is periodic, with period equal to 1, in some normalized units can be expanded as a linear combination of the sequence $\{\exp(j2\pi nt)\}$, where n is an integer that goes from minus infinity to plus infinity. The purpose here is to recast the familiar Fourier series results within the language and notations of the above formalism.

Basis:

$$|u_n\rangle = \exp(j2\pi nt) \quad \text{and} \quad \langle u_n| = \exp(-j2\pi nt) \tag{7.106}$$

Orthonormality of the basis vectors:

$$\langle u_m | u_n\rangle = \int_{-1/2}^{1/2} \exp(-j2\pi mt)\exp(j2\pi nt)\,dt = \begin{cases} 1 & \text{if } m = n \\ 0 & \text{if } m \neq n \end{cases} \tag{7.107}$$

Decomposition rule:

$$|\varphi\rangle = \sum_{n=-\infty}^{\infty} c_n |u_n\rangle = \sum_{n=-\infty}^{\infty} c_n \exp(j2\pi nt) \tag{7.108}$$

where

$$c_n = \langle u_n | \varphi\rangle = \int_{-1/2}^{1/2} \exp(-j2\pi nt)\,\varphi(t)\,dt \tag{7.109}$$

Parseval's identity:

$$\|\varphi\|^2 = \langle \varphi | \varphi \rangle = \int_{-1/2}^{1/2} \overline{\varphi}(t)\varphi(t)dt = \int_{-1/2}^{1/2} |\varphi(t)|^2\,dt = \sum_{n=-\infty}^{\infty} |c_n|^2 \tag{7.110}$$

Example 7.12

Derive the analytic expression for the potential difference across the capacitor in the *RLC* circuit of Figure 4.5 if the temporal profile of the source potential is a periodic function, of period 1, in some normalized units.

Solution:

1. Because the potential is periodic with period 1, it can be expanded using Eq. (7.108) in a Fourier series with basis functions $\{e^{j2\pi nt}\}$:

$$V_s(t) = \text{Re}\left\{\sum_n \tilde{V}_s^n e^{j2\pi nt}\right\} \tag{7.111}$$

where $\tilde{V}_s^n$ is the phasor associated with the frequency mode $(2\pi n)$. (Note that n in the expressions for the phasors is a superscript and not a power.)

2. We find $\tilde{V}_c^n$ the capacitor response phasor associated with the $\tilde{V}_s^n$ excitation. This can be found by noting that the voltage across the capacitor is equal to the capacitor impedance multiplied by the current phasor, giving

$$\tilde{V}_c^n = Z_c^n \tilde{I}^n = \frac{Z_c^n \tilde{V}_s^n}{Z_c^n + Z_R^n + Z_L^n} \tag{7.112}$$

where from the results of Section 6.8, particularly Eq. (6.96) through Eq. (6.98), we have

$$Z_c^n = \frac{1}{j2\pi nC} \tag{7.113}$$

$$Z_L^n = j2\pi nL \tag{7.114}$$

$$Z_R^n = R \tag{7.115}$$

3. Finally, we use the linearity of the ODE system and write the solution as the linear superposition of the solutions corresponding to the response to each of the basis functions; that is,

$$V_c(t) = \text{Re}\left\{\sum_n \frac{Z_c^n \tilde{V}_s^n}{Z_c^n + Z_R^n + Z_L^n} e^{j2\pi nt}\right\} \tag{7.116}$$

leading to the expression

$$V_c(t) = \text{Re}\left\{\sum_n \frac{\tilde{V}_s^n}{1 - (2\pi n)^2 LC + j(2\pi n)RC} e^{j2\pi nt}\right\} \tag{7.117}$$

□

Homework Problem

Pb. 7.28 Consider the *RLC* circuit. Assuming the same notation as in Section 6.5.4, but now assume that the source potential is given by

$$V_s = V_0 \cos^6(\omega t)$$

a. Find analytically the potential difference across the capacitance. (*Hint*: Write the power of the trigonometric function as function of the different multiples of the angle.)

b. Find numerically the steady-state solution to this problem using the techniques of Chapter 4, and assume for some normalized units the following values for the parameters:

$$LC = 1, \quad RC = 1, \quad \omega = 2\pi$$

c. Compare your numerical results with the analytical results.

Application 2: Expansion in Legendre Polynomials

We propose to show that the Legendre polynomials are an orthonormal basis for all functions of compact support over the interval $-1 \le x \le 1$. Thus far, we have encountered the Legendre polynomials twice before. They were defined through their recursion relations in Pb. 2.25, and in Section 4.7.1 through their defining ODE. In this application, we define the Legendre polynomials through their generating function; show how their definitions through their recursion relation, or through their ODE, can be deduced from their definition through their generating function; and show that they constitute an orthonormal basis for functions defined on the interval $-1 \le x \le 1$.

1. The generating function for the Legendre polynomials is given by the simple form

$$G(x,t) = \frac{1}{\sqrt{1-2xt+t^2}} = \sum_{l=0}^{\infty} P_l(x)\, t^l \tag{7.118}$$

2. The lowest orders of $P_l(x)$ can be obtained from the small t-expansion of $G(x, t)$; therefore, expanding Eq. (7.118) to first order in t gives

$$1 + xt + O(t^2) = P_0(x) + tP_1(x) + O(t^2) \tag{7.119}$$

from which, we can deduce that

$$P_0(x) = 1 \tag{7.120}$$

$$P_1(x) = x \tag{7.121}$$

3. By inspection, it is straightforward to verify by substitution that the generating function satisfies the equation

$$(1-2xt+t^2)\frac{\partial G}{\partial t}+(t-x)G=0 \tag{7.122}$$

Because power series can be differentiated term by term, Eq. (7.122) gives

$$(1-2xt+t^2)\sum_{l=0}^{\infty} lP_l(x)t^{l-1}+(t-x)\sum_{l=0}^{\infty} P_l(x)t^l=0 \tag{7.123}$$

Since this equation should hold true for all values of t, this means that all coefficients of any power of t should be zero; therefore;

$$(l+1)P_l(x)-2lxP_l(x)+(l-1)P_{l-1}(x)+(P_{l-1}(x)-xP_l(x)=0 \tag{7.124}$$

or collecting terms, this can be written as

$$(l+1)P_l(x)-(2l+1)xP_l(x)+lP_{l-1}(x)=0 \tag{7.125}$$

This is the recursion relation of Pb. 2.25.

4. By substitution in the explicit expression of the generating function, we can also verify that

$$(1-2xt+t^2)\frac{\partial G}{\partial x}-tG=0 \tag{7.126}$$

which leads to

$$(1-2xt+t^2)\sum_{l=0}^{\infty}\frac{dP_l(x)}{dx}-\sum_{l=0}^{\infty}P_l(x)t^{l+1}=0 \tag{7.127}$$

Again, looking at the coefficients of the same power of t permits us to obtain another recursion relation

$$\frac{dP_{l+1}(x)}{dx}-2x\frac{dP_l(x)}{dx}+\frac{dP_{l-1}(x)}{dx}-P_l(x)=0 \tag{7.128}$$

Differentiating Eq. (7.125), we first eliminate $\frac{dP_{l-1}(x)}{dx}$ and then $\frac{dP_l(x)}{dx}$ from the resulting equation, and use Eq. (7.128) to obtain two new recursion relations:

$$\frac{dP_{l+1}(x)}{dx}-x\frac{dP_l(x)}{dx}=(l+1)P_l(x) \tag{7.129}$$

and

$$x\frac{dP_l(x)}{dx} - \frac{dP_{l-1}(x)}{dx} = lP_l(x) \tag{7.130}$$

Adding Eq. (7.129) and Eq. (7.130), we obtain the more symmetric formula:

$$\frac{dP_{l+1}(x)}{dx} - \frac{dP_{l-1}(x)}{dx} = (2l+1)P_l(x) \tag{7.131}$$

Replacing l by $l-1$ in Eq. (7.129) and eliminating $P'_{l-1}(x)$ from Eq. (7.130), we find that

$$(1-x^2)\frac{dP_l(x)}{dx} = lP_{l-1}(x) - lxP_l(x) \tag{7.132}$$

Differentiating Eq. (7.132) and using Eq. (7.130), we obtain

$$\frac{d}{dx}\left[(1-x^2)\frac{dP_l(x)}{dx}\right] + l(l+1)P_l(x) = 0 \tag{7.133a}$$

which can be written in the equivalent form

$$(1-x^2)\frac{d^2P_l(x)}{dx^2} - 2x\frac{dP_l(x)}{dx} + l(l+1)P_l(x) = 0 \tag{7.133b}$$

which is the ODE for the Legendre polynomial, as previously pointed out in Section 4.7.1.

5. Next, we want to show that if $l \neq m$, we have the orthogonality between any two elements (with different indices) of the basis; that is

$$\int_{-1}^{1} P_l(x)P_m(x)dx = 0 \tag{7.134}$$

To show this relation, we multiply Eq. (7.133) on the LHS by $P_m(x)$ and integrate to obtain

$$\int_{-1}^{1} P_m(x)\left\{\frac{d}{dx}\left[(1-x^2)\frac{dP_l(x)}{dx}\right] + l(l+1)P_l(x)\right\}dx = 0 \tag{7.135}$$

Integrating the first term by parts, we obtain

$$\int_{-1}^{1}\left\{(x^2-1)\frac{dP_m(x)}{dx}\frac{dP_l(x)}{dx} + l(l+1)P_m(x)P_l(x)\right\}dx = 0 \tag{7.136}$$

Similarly, we can write the ODE for $P_m(x)$, and multiply on the LHS by $P_l(x)$; this results in the equation

$$\int_{-1}^{1}\left\{(x^2-1)\frac{dP_l(x)}{dx}\frac{dP_m(x)}{dx}+m(m+1)P_l(x)P_m(x)\right\}dx=0 \tag{7.137}$$

Now, subtracting Eq. (7.137) from Eq. (7.136), we obtain

$$[m(m+1)-l(l+1)]\int_{-1}^{1}P_l(x)P_m(x)\,dx=0 \tag{7.138}$$

But because $l \neq m$, this can only be satisfied if the integral is zero, which is the result that we are after.

6. Finally, we compute the normalization of the basis functions; that is, compute:

$$\int_{-1}^{1}P_l(x)P_l(x)\,dx=N_l^2 \tag{7.139}$$

From Eq. (7.125), we can write

$$P_l(x)-(2l-1)xP_{l-1}(x)+(l-1)P_{l-2}(x)=0 \tag{7.140}$$

If we multiply this equation by $(2l + 1)P_l(x)$ and subtract from it Eq. (7.125), which we multiplied by $(2l + 1)P_{l-1}(x)$, we obtain

$$\begin{aligned}&l(2l+1)P_l^2(x)+(2l-1)(l-1)P_{l-1}(x)P_{l-2}(x)\\&\quad-(l+1)(2l-1)P_{l-1}(x)P_{l+1}(x)-l(2l-1)P_{l-1}^2(x)=0\end{aligned} \tag{7.141}$$

Now integrate over the interval $[-1, 1]$ and using Eq. (7.138), we obtain, for $l = 2, 3, \ldots$:

$$\int_{-1}^{1}P_l^2(x)\,dx=\frac{(2l-1)}{(2l+1)}\int_{-1}^{1}P_{l-1}^2(x)\,dx \tag{7.142}$$

Repeated applications of this formula and the use of Eq. (7.121) yields

$$\int_{-1}^{1}P_l^2(x)dx=\frac{3}{(2l+1)}\int_{-1}^{1}P_\ell^2(x)dx=\frac{2}{(2l+1)} \tag{7.143}$$

Direct calculations show that this is also valid for $l = 0$ and $l = 1$. Therefore, the orthonormal basis functions are given by

$$|u_l\rangle=\sqrt{l+\frac{1}{2}}P_l(x) \tag{7.144}$$

The general theorem that summarizes the decomposition of a function into the Legendre polynomials basis states

THEOREM If the real function $f(x)$ defined over the interval $[-1, 1]$ is piecewise smooth and if the integral $\int_{-1}^{1} f^2(x)\,dx < \infty$ then the series

$$f(x) = \sum_{l=0}^{\infty} c_l P_l(x) \tag{7.145}$$

where

$$c_l = \left(l + \frac{1}{2}\right) \int_{-1}^{1} f(x) P_l(x)\,dx \tag{7.146}$$

converges to $f(x)$ at every continuity point of the function.

The proof of this theorem is not given here.

EXAMPLE 7.13

Find the decomposition into Legendre polynomials of the following function:

$$f(x) = \begin{cases} 0 & \text{for } -1 \leq x \leq a \\ 1 & \text{for } a < x \leq 1 \end{cases} \tag{7.147}$$

Solution: The conditions for the above theorem are satisfied, and

$$c_l = \left(l + \frac{1}{2}\right) \int_{a}^{1} P_l(x)\,dx \tag{7.148}$$

From Eq. (7.131), and noting that $P_l(1) = 1$, we find that

$$c_0 = \frac{1}{2}(1 - a) \tag{7.149}$$

and

$$c_l = -\frac{1}{2}[P_{l+1}(a) - P_{l-1}(a)] \tag{7.150}$$

We show in Figure 7.4 the sum of the truncated decomposition for Example 7.10 for different values of l_{max}. □

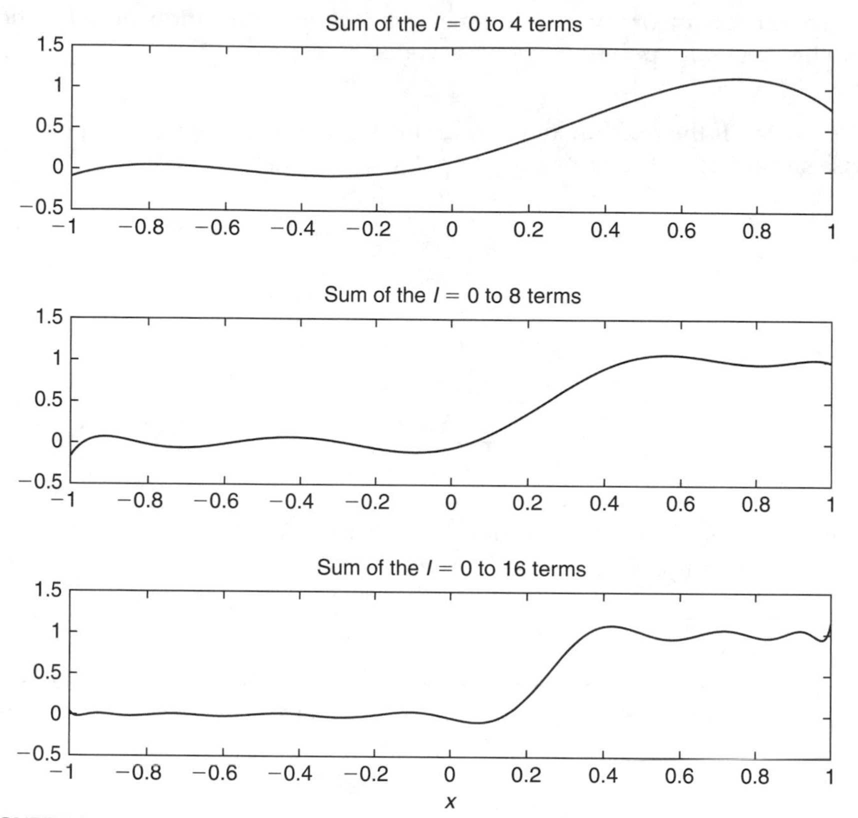

FIGURE 7.4
The plot of the truncated Legendre polynomials expansion of the discontinuous function given by Eq. (7.112), for $a = 0.25$. Top panel: $l_{max} = 4$. Middle panel: $l_{max} = 8$. Bottom panel: $l_{max} = 16$.

7.10 MATLAB Commands Review

`'` Transposition (i.e., for vectors with real components, this changes a row into a column).

`norm` Computes the Euclidean length of a vector.

`cross` Calculates the cross product of two 3-D vectors.

`dot` Calculates the dot product of two 3-D vectors.

8

Matrices

Matrices play the central role in computer-aided engineering design. The emphasis in this chapter is on the methods for constructing matrices and in illustrating their applications in diverse fields. Special matrices and their main properties are also summarized.

8.1 Setting Up Matrices

DEFINITION A matrix is a collection of numbers arranged in a 2-D array structure. Each element of the matrix, call it $M_{i,j}$, occupies the ith row and jth column.

$$\mathbf{M} = \begin{bmatrix} M_{11} & M_{12} & M_{13} & \cdots & M_{1n} \\ M_{21} & M_{22} & M_{23} & \cdots & M_{2n} \\ \vdots & \vdots & \vdots & \ddots & \vdots \\ M_{m1} & M_{m2} & M_{m3} & \cdots & M_{mn} \end{bmatrix} \tag{8.1}$$

We say that $\mathbf{M}$ is an $(m \otimes n)$ matrix, which means that it has m rows and n columns. If $m = n$, we call the matrix square. If $m = 1$, the matrix is a row vector; and if $n = 1$, the matrix is a column vector.

8.1.1 Creating a Matrix by Keying in the Elements

In this method, the different elements of the matrix are keyed in; entering

```
M=[1 3 5 7 11; 13 17 19 23 29; 31 37 41 47 53]
```

returns

```
M =

     1   3   5   7  11
    13  17  19  23  29
    31  37  41  47  53
```

To find the size of the matrix (i.e., the number of rows and columns), enter

```
size(M)
```

returns

```
ans =

      3 5
```

To view a particular element, for example, the (2, 4) element, enter

```
M(2,4)
```

returns

```
ans =

      23
```

To view a particular row such as the 3rd row, enter

```
M(3,:)
```

returns

```
ans =

      31 37 41 47 53
```

To view a particular column such as the 4th column, enter

```
M(:,4)
```

returns

```
ans =

          7
         23
         47
```

If we want to construct a submatrix of the original matrix, for example, one that includes the block from the 2nd to 3rd row (included) and from the 2nd to the 4th column (included), entering

```
M(2:3,2:4)
```

returns

```
ans =

     17  19  23
     37  41  47
```

8.1.2 Retrieving Special Matrices from the MATLAB Library

MATLAB has some commonly used specialized matrices in its library that can be called when needed. For example,

- The matrix of size ($m \otimes n$) with all elements being 0 is **M=zeros(m,n)**

For example, entering

```
M=zeros(3,4)
```

returns

```
 M =

     0  0  0  0
     0  0  0  0
     0  0  0  0
```

- The matrix of size ($m \otimes n$) with all elements equal to 1 is **N=ones(m,n)**

For example, entering

```
N=ones(4,3)
```

returns

```
 N =

     1  1  1
     1  1  1
     1  1  1
     1  1  1
```

- The matrix of size ($n \otimes n$) with only the diagonal elements equal to 1, otherwise 0, is **P=eye(n,n)**

For example, entering

```
P=eye(4,4)
```

returns

```
P =

    1 0 0 0
    0 1 0 0
    0 0 1 0
    0 0 0 1
```

- The matrix of size ($n \otimes n$) with elements randomly chosen from the interval [0, 1], entering

```
Q=rand(4,4)
```

returns, in one particular run

```
Q =

    0.9708  0.4983  0.9601  0.2679
    0.9901  0.2140  0.7266  0.4399
    0.7889  0.6435  0.4120  0.9334
    0.4387  0.3200  0.7446  0.6833
```

- We can select to extract the upper triangular part of the **Q** matrix, but assign to all the lower triangle elements the value 0, entering

```
upQ=triu(Q)
```

returns

```
upQ =

    0.9708  0.4983  0.9601  0.2679
    0       0.2140  0.7266  0.4399
    0       0       0.4120  0.9334
    0       0       0       0.6833
```

or extract the lower triangular part of the **Q** matrix, but assign to all the upper triangle elements the value 0, entering

```
loQ=tril(Q)
```

returns

```
loQ =

      0.9708  0        0        0
      0.9901  0.2140   0        0
      0.7889  0.6435   0.4120   0
      0.4387  0.3200   0.7446   0.6833
```

- The single quotation mark (') after the name of a matrix changes the matrix rows into becoming its columns, and vice versa, if the elements are all real. If the matrix has complex numbers as elements, it also takes their complex conjugate in addition to the transposition.
- Other specialized matrices, including the whole family of sparse matrices, are also included in the MATLAB library. You can find more information about them in the **help** documentation.

8.1.3 Functional Construction of Matrices

The third method for generating matrices is to give, if it exists, an algorithm that generates each element of the matrix. For example, suppose we want to generate the Hilbert matrix of size $(n \otimes n)$, where $n = 4$ and the functional form of the elements are $M_{mn} = 1/(m + n)$. The routine for generating this matrix will be as follows:

```
  M=zeros(4,4);
    for m=1:4
      for n=1:4
        M(m,n)=1/(m+n);
      end
    end
M
```

- We can also create new matrices by appending known matrices. For example,

Let the matrices **A** and **B** be given by

```
A=[1 2 3 4];
B=[5 6 7 8];
```

We want to expand matrix **A** by matrix **B** along the horizontal (this is allowed only if both matrices have the same number of rows). Enter

```
C=[A B]
```

returns

```
C =
      1   2   3   4   5   6   7   8
```

Or, we may want to expand **A** by stacking it on top of **B** (this is allowed only if both matrices have the same number of columns). Enter

```
D=[A;B]
```

produces

```
D =
      1   2   3   4
      5   6   7   8
```

We illustrate the appending operations for larger matrices: define **E** as the (2 ⊗ 3) matrix with 1 for all its elements, and we desire to append it horizontally to **D**. This is allowed because both have the same number of rows (= 2). Enter

```
E=ones(2,3)
```

returns

```
E =
      1   1   1
      1   1   1
```

Enter

```
F =[D E]
```

returns

```
F =
      1   2   3   4   1   1   1
      5   6   7   8   1   1   1
```

Or, we may want to stack two matrices in a vertical configuration. This requires that the two matrices have the same number of columns. Entering

```
G=ones(2,4)
```

returns

```
G =
      1   1   1   1
      1   1   1   1
```

Entering

```
H=[D;G]
```

returns

```
H =
      1   2   3   4
      5   6   7   8
      1   1   1   1
      1   1   1   1
```

The command sum applied to a matrix gives a row in which the m element is the sum of all the elements of the mth column in the original matrix. For example, entering

```
sum(H)
```

returns

```
ans =
      8   10   12   14
```

Finally, we point out that you may also create and save *function M-files* for certain matrices, using a combination of one or more of the above techniques. For example, for the circulant matrices defined as follows,

$$C = \begin{bmatrix} a & b & \cdots & n \\ n & a & \cdots & m \\ \vdots & & \ddots & \vdots \\ b & c & \cdots & a \end{bmatrix}$$

we can save the following *function M-file*:

```
function C=circulantj(V)
n=length(V);
C=zeros(n,n);
C(1,:)=V;
     for k=2:n
            C(k,:)=[C(k-1,n) C(k-1,1:n-1)];
     end
```

Now, entering in the command window

```
C=circulantj([1 3 5 7 9 11])
```

returns the matrix

```
C =
      1     3     5     7     9    11
     11     1     3     5     7     9
      9    11     1     3     5     7
      7     9    11     1     3     5
      5     7     9    11     1     3
      3     5     7     9    11     1
```

8.2 Adding Matrices

Adding two matrices is only possible if they have equal numbers of rows and equal numbers of columns; or in other words, they both have the same size.

The addition operation is the obvious one. That is, the (m, n) element of $(\mathbf{A} + \mathbf{B})$ is the sum of the (m, n) elements of $\mathbf{A}$ and $\mathbf{B}$ respectively:

$$(\mathbf{A} + \mathbf{B})_{mn} = A_{mn} + B_{mn} \tag{8.2}$$

Entering

```
A=[1 2 3 4];
B=[5 6 7 8];
A+B
```

returns

```
ans =
      6   8   10   12
```

If we had subtraction of two matrices, it would be the same syntax as above but using the minus sign between the matrices.

8.3 Multiplying a Matrix by a Scalar

If we multiply a matrix by a number, each element of the matrix is multiplied by that number.

Entering

```
3*A
```

returns

```
ans =
     3   6   9   12
```

Entering

```
3*(A+B)
```

returns

```
ans =
     18   24   30   36
```

8.4 Multiplying Matrices

Two matrices $\mathbf{A}(m \otimes n)$ and $\mathbf{B}(r \otimes s)$ can be multiplied only if $n = r$. The size of the product matrix is $(m \otimes s)$. An element of the product matrix is obtained from those of the constitutent matrices through the following rule:

$$(\mathbf{AB})_{kl} = \sum_h A_{kh} B_{hl} \tag{8.3}$$

This result can be also interpreted by observing that the (k, l) element of the product is the dot product of the k row of **A** and the l column of **B**.

In MATLAB, we denote the product of the matrices **A** and **B** by **A*B**.

Example 8.1

Write the different routines for performing the matrix multiplication from the different definitions of the matrix product.

Solution: Edit and execute the following *script M-file*:

```
D=[1  2  3;  4  5  6];
E=[3  6  9  12;  4  8  12  16;  5  10  15  20];

F=D*E

F1=zeros(2,4);
for i=1:2
```

```
    for j=1:4
      for k=1:3
        F1(i,j)=F1(i,j)+D(i,k)*E(k,j);
      end
    end
end
F1

F2=zeros(2,4);
  for i=1:2
    for j=1:4
      F2(i,j)=D(i,:)*E(:,j);
    end
  end
F2
```

The result **F** is the one obtained using the MATLAB built-in matrix multiplication; the result **F1** is that obtained from Eq. (8.3) and **F2** is the answer obtained by performing, for each element of the matrix product, the dot product of the appropriate row from the first matrix with the appropriate column from the second matrix. Of course, all three results should give the same answer, which they do. □

8.5 Inverse of a Matrix

In this section, we assume that we are dealing with square matrices ($n \otimes n$) because these are the only class of matrices for which we can define an inverse.

DEFINITION A matrix $\mathbf{M}^{-1}$ is called the inverse of matrix **M** if the following conditions are satisfied:

$$\mathbf{M}\mathbf{M}^{-1} = \mathbf{M}^{-1}\mathbf{M} = \mathbf{I} \tag{8.4}$$

(The identity matrix is the ($n \otimes n$) matrix with 1's on the diagonal and 0 everywhere else; the matrix `eye(n,n)` in MATLAB.)

EXISTENCE The existence of an inverse of a matrix hinges on the condition that the determinant of this matrix is nonzero [`det(M)` in MATLAB].

DEFINITION The determinant of a square matrix **M**, of size ($N \otimes N$), is a number equal to

$$\det(\mathbf{M}) = \sum_{P} \varepsilon_{abc\ldots n}\, M_{1a}M_{2b}M_{3c}\ldots M_{Nn} \tag{8.5}$$

where $\varepsilon_{abc...n}$ is the antisymmetric symbol over N objects and its value is given by

$$\varepsilon_{abc\cdots n} = \begin{cases} 1 & \text{if } (a,\ b,\ c, \ldots n) \text{ is an even permutation of } (1, 2, 3, \ldots, N) \\ -1 & \text{if } (a,\ b,\ c, \ldots n) \text{ is an odd permutation of } (1, 2, 3, \ldots, N) \\ 0 & \text{otherwise} \end{cases}$$

1. An even (odd) permutation consists of an even (odd) number of transpositions. (A transposition is a permutation where only two elements exchange positions.)
2. The total number of permutations of N objects is $N!$, and the determinant has this number of terms in the sum.
3. Each term in the sum consists of the product of one element from each row and each column. Products containing more than a single term from a given column or row do not contribute to the determinant because the corresponding antisymmetric symbol is zero.

Example 8.2

Using the definition of a determinant, as given in Eq. (8.5), find the expression for the determinant of a $(2 \otimes 2)$ and a $(3 \otimes 3)$ matrix.

Solution:

a. If $n = 2$, there are only two possibilities for permuting these two numbers, giving the following sequences: (1, 2) and (2, 1). In the first permutation, no transposition was necessary; that is, the multiplying factor in Eq. (8.5) is 1. In the second term, one transposition is needed; that is, the multiplying factor in Eq. (8.5) is -1, giving for the determinant the value:

$$\Delta = M_{11}M_{22} - M_{12}M_{21} \tag{8.6}$$

b. If $n = 3$, there are only six permutations for the sequence (1, 2, 3), namely, (1, 2, 3), (2, 3, 1), and (3, 1, 2), each of which is an even permutation and (3, 2, 1), (2, 1, 3), and (1, 3, 2), which are odd permutations, thereby giving for the determinant the value:

$$\Delta = M_{11}M_{22}M_{33} + M_{12}M_{23}M_{31} + M_{13}M_{21}M_{32} \\ -(M_{13}M_{22}M_{31} + M_{12}M_{21}M_{33} + M_{11}M_{23}M_{32}) \tag{8.7}$$

□

MATLAB Representation

Compute the determinant and the inverse of the matrices **M** and **N**. Entering

```
M=[1 3 5; 7 11 13; 17 19 23];
det_M=det(M)
inv_M=inv(M)
```

returns

```
det_M=
     -84
inv_M=
     -0.0714   -0.3095   -0.1905
     -0.7143   -0.7381   -0.2619
     -0.6429   -0.3810   -0.1190
```

Compare this result with that obtained in the example in Appendix B.
On the other hand, entering

```
N=[2 4 6; 3 5 7; 5 9 13];
det_N=det(N)
inv_N=inv(N)
```

returns

```
det_N =
     0
inv_N
     Warning: Matrix is close to singular or badly
     scaled.
```

Homework Problems

Use the theorems summarized in Appendix B, to solve the following problems:

Pb. 8.1 As defined earlier, a square matrix in which all elements above (below) the diagonal are zeros is called a lower (upper) triangular matrix. Show that the determinant of a triangular $n \otimes n$ matrix is

$$\det(\mathbf{T}) = T_{11}T_{22}T_{33} \ldots T_{nn}$$

Pb. 8.2 If $\mathbf{M}$ is an $n \otimes n$ matrix and k is a constant, show that

$$\det(k\mathbf{M}) = k^n \det(\mathbf{M})$$

Pb. 8.3 Prove that if the inverse of the matrix $\mathbf{M}$ exists, then

$$\det(\mathbf{M}^{-1}) = \frac{1}{\det(\mathbf{M})}$$

8.6 Solving a System of Linear Equations

Let us assume that we have a system of n linear equations in n unknowns that we want to solve

$$\begin{aligned}
M_{11}x_1 + M_{12}x_2 + M_{13}x_3 + \cdots + M_{1n}x_n &= b_1 \\
M_{21}x_1 + M_{22}x_2 + M_{23}x_3 + \cdots + M_{2n}x_n &= b_2 \\
&\vdots \\
M_{n1}x_1 + M_{n2}x_2 + M_{n3}x_3 + \cdots + M_{nn}x_n &= b_n
\end{aligned} \tag{8.8}$$

The above equations can be readily written in matrix notation

$$\begin{bmatrix} M_{11} & M_{12} & M_{13} & \cdots & M_{1n} \\ M_{21} & M_{22} & M_{23} & \cdots & M_{2n} \\ \vdots & \vdots & \vdots & \ddots & \vdots \\ \vdots & \vdots & \vdots & \cdots & \vdots \\ M_{n1} & M_{n2} & M_{n3} & \cdots & M_{nn} \end{bmatrix} \begin{bmatrix} x_1 \\ x_2 \\ \vdots \\ \vdots \\ x_n \end{bmatrix} = \begin{bmatrix} b_1 \\ b_2 \\ \vdots \\ \vdots \\ b_n \end{bmatrix} \tag{8.9}$$

or

$$\mathbf{MX} = \mathbf{B} \tag{8.10}$$

where the column of b's and x's are denoted by **B** and **X**. Multiplying, on the left, both sides of this matrix equation by $\mathbf{M}^{-1}$, we find that

$$\mathbf{X} = \mathbf{M}^{-1}\mathbf{B} \tag{8.11}$$

As pointed out previously, remember that the condition for the existence of solutions is a nonzero value for the determinant of **M**.

Example 8.3

Use MATLAB to solve the system of equations given by

$$\begin{aligned}
x_1 + 3x_2 + 5x_3 &= 22 \\
7x_1 + 11x_2 - 13x_3 &= -10 \\
17x_1 + 19x_2 - 23x_3 &= -14
\end{aligned}$$

Solution: Edit and execute the following *script M-file*:

```
M=[1 3 5; 7 11 -13; 17 19 -23];
B=[22;-10;-14];
det_M=det(M)
inv_M=inv(M)
X=inv(M)*B
```

Verify that the vector **X** could also have been obtained using the left slash notation `X=M\B`. □

NOTE In this and the immediately preceding sections, we said very little about the algorithm used for computing essentially the inverse of a matrix. This is a subject that will be amply covered in your linear algebra courses. What the interested reader needs to know at this stage is that the Gaussian elimination technique (and its different refinements) is essentially the numerical method of choice for the built-in algorithms of numerical softwares. The following two examples are essential building blocks in such constructions.

EXAMPLE 8.4

Without using the MATLAB inverse command, solve the system of equations:

$$\mathbf{LX} = \mathbf{B} \tag{8.12}$$

where **L** is a lower triangular matrix.

Solution: In matrix form, the system of equations to be solved is

$$\begin{bmatrix} L_{11} & 0 & 0 & \cdots & 0 \\ L_{21} & L_{22} & 0 & \cdots & 0 \\ \vdots & \vdots & \vdots & \ddots & \vdots \\ \vdots & \vdots & \vdots & \cdots & \vdots \\ L_{n1} & L_{n2} & L_{n3} & \cdots & L_{nn} \end{bmatrix} \begin{bmatrix} x_1 \\ x_2 \\ \vdots \\ \vdots \\ x_n \end{bmatrix} = \begin{bmatrix} b_1 \\ b_2 \\ \vdots \\ \vdots \\ b_n \end{bmatrix} \tag{8.13}$$

The solution of this system can be directly obtained if we proceed iteratively. That is, we find in the following order: $x_1, x_2, \ldots, x_n$, obtaining

$$\begin{aligned} x_1 &= \frac{b_1}{L_{11}} \\ x_2 &= \frac{(b_2 - L_{21}x_1)}{L_{22}} \\ &\vdots \\ x_k &= \frac{\left(b_k - \sum_{j=1}^{k-1} L_{kj}x_j\right)}{L_{kk}} \end{aligned} \tag{8.14}$$

The above solution can be implemented by executing the following *script M-file*:

```
L=[?];                    % enter the L matrix
B=[?];                    % enter the B column
```

```
n=length(B);
X=zeros(n,1);
X(1)=B(1)/L(1,1);
   for k=2:n
     X(k)=(B(k)-L(k,1:k-1)*X(1:k-1))/L(k,k);
   end
X □
```

Example 8.5

Solve the system of equations, **UX** = **B**, where **U** is an upper triangular matrix.

Solution: The matrix form of the problem becomes

$$\begin{bmatrix} U_{11} & U_{12} & U_{13} & \cdots & U_{1n} \\ 0 & U_{22} & U_{23} & \cdots & U_{2n} \\ \vdots & \vdots & \ddots & \vdots & \vdots \\ 0 & 0 & \cdots & U_{n-1\,n-1} & U_{n-1\,n} \\ 0 & 0 & 0 & \cdots & U_{nn} \end{bmatrix} \begin{bmatrix} x_1 \\ x_2 \\ \vdots \\ x_{n-1} \\ x_n \end{bmatrix} = \begin{bmatrix} b_1 \\ b_2 \\ \vdots \\ b_{n-1} \\ b_n \end{bmatrix} \tag{8.15}$$

In this case, the solution of this system can also be directly obtained if we proceed iteratively, but this time in the backward order $x_n, x_{n-1}, \ldots, x_1$, obtaining

$$\begin{aligned} x_n &= \frac{b_n}{U_{nn}} \\ x_{n-1} &= \frac{(b_{n-1} - U_{n-1\,n}\; x_n)}{U_{n-1\,n-1}} \\ &\vdots \\ x_k &= \frac{\left(b_k - \sum_{j=k+1}^{n} U_{kj} x_j\right)}{U_{kk}} \end{aligned} \tag{8.16}$$

The corresponding *script M-file* is

```
U=[?];                  % enter the U matrix
B=[?];                  % enter the B column
n=length(B);
X=zeros(n,1);
X(n)=B(n)/U(n,n);
         for k=n-1:-1:1
           X(k)=(B(k)-U(k,k+1:n)*X(k+1:n))/U(k,k);
         end
X □
```

8.7 Application of Matrix Methods

This section provides seven representative applications that illustrate the immense power that matrix formulation and tools can provide to diverse problems of common interest in electrical engineering.

8.7.1 dc Circuit Analysis

EXAMPLE 8.6

Find the voltages and currents for the circuit given in Figure 8.1.

Solution: Using Kirchoff's current and voltage laws and Ohm's law, we can write the following equations for the voltages and currents in the circuit, assuming that $R_L = 2\Omega$:

$$\begin{aligned}
&V_1 = 5 \\
&V_1 - V_2 = 50I_1 \\
&V_2 - V_3 = 100I_2 \\
&V_2 = 300I_3 \\
&V_3 = 2I_2 \\
&I_1 = I_2 + I_3
\end{aligned}$$

NOTE These equations can be greatly simplified if we use the method of elimination of variables. This is essentially the method of nodes analysis covered in circuit theory courses. At this time, our purpose is to show a direct numerical method for obtaining the solutions.

If we form column vector **VI**, the top three components referring to the voltages V_1, V_2, and V_3, and the bottom three components referring to the currents

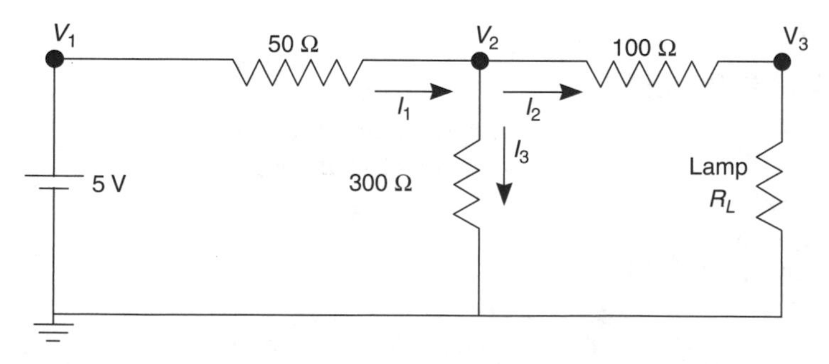

FIGURE 8.1
Circuit of Example 8.6.

I_1, I_2, and I_3, then the following *script M-file* provides the solution to the above circuit:

```
M=[1 0 0 0 0 0;1 -1 0 -50 0 0;0 1 -1 0 -100 0;...
   0 1 0 0 0 -300;0 0 1 0 -2 0;0 0 0 1 -1 -1];
Vs=[5;0;0;0;0;0];
VI=M\Vs
```
□

IN-CLASS EXERCISE

Pb. 8.4 Use the same technique as shown in Example 8.6 to solve for the potentials and currents in the circuit given in Figure 8.2.

8.7.2 dc Circuit Design

In design problems, we are usually faced with the reverse problem of the direct analysis problem, such as the one solved in Section 8.7.1.

EXAMPLE 8.7

Find the value of the lamp resistor in Figure 8.1, so that the current flowing through it is given, *a priori*.

Solution: We approach this problem by first defining a function file for the relevant current. In this case, it is

```
function ilamp=circuit872(RL)
M=[1 0 0 0 0 0;1 -1 0 -50 0 0;0 1 -1 0 -100 0;...
   0 1 0 0 0 -300;0 0 1 0 -RL 0;0 0 0 1 -1 -1];
```

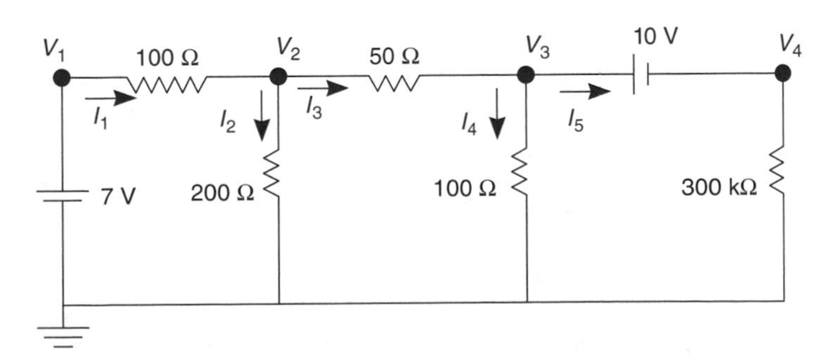

FIGURE 8.2
Circuit of Pb. 8.4.

```
Vs=[5;0;0;0;0;0];
VI=M\Vs;
ilamp=VI(5);
```

Second, from the command window, we proceed by calling this function and plotting the current in the lamp as a function of the resistance. Third, using the **ginput** command we graphically read for the value of R_L, which gives the desired current value. □

IN-CLASS EXERCISE

Pb. 8.5 For the circuit of Figure 8.1, find R_L that gives a 22 mA current in the lamp. (*Hint*: Plot the current as function of the load resistor.)

8.7.3 ac Circuit Analysis

Conceptually, there is no difference between performing an ac steady-state analysis of a circuit with purely resistive elements, as was done in Section 8.7.1, and performing the analysis for a circuit that includes capacitors and inductors, if we adopt the tool of impedance introduced in Section 6.8, and we write the circuit equations instead with phasors. The only modification from an all-resistors circuit is that matrices now have complex numbers as elements, and the impedances have frequency dependence. For convenience, we illustrate again the relationships of the voltage–current phasors across resistors, inductors, and capacitors:

$$\tilde{V}_R = \tilde{I}R \tag{8.17}$$

$$\tilde{V}_L = \tilde{I}(j\omega L) \tag{8.18}$$

$$\tilde{V}_C = \frac{\tilde{I}}{(j\omega C)} \tag{8.19}$$

and restate Kirchoff's laws again.

- Kirchoff's voltage law: The sum of all voltage drops around a closed loop is balanced by the sum of all voltage sources around the same loop.

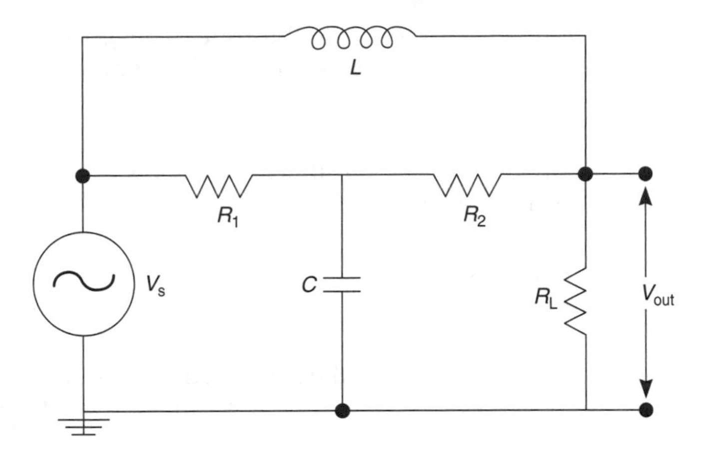

FIGURE 8.3
Bridged-T filter. Circuit of Pb. 8.6.

- Kirchoff's current law: The algebraic sum of all currents entering (exiting) a circuit node must be zero.

IN-CLASS EXERCISE

Pb. 8.6 In a bridged-T filter, the voltage $V_s(t)$ is the input voltage, and the output voltage is that across the load resistor R_L. The circuit is given in Figure 8.3.
Assuming that $R_1 = R_2 = 3\,\Omega$, $R_L = 2\,\Omega$, $C = 0.25\,\text{F}$, and $\text{L} = 1\,\text{H}$

a. Write the equations for the phasors of the voltages and currents.
b. Form the matrix representation for the equations found in part (a).
c. Plot the magnitude and phase of $\dfrac{\tilde{V}_{\text{out}}}{\tilde{V}_S}$ as a function of the frequency.

8.7.4 Accuracy of a Truncated Taylor Series

In this subsection and Section 8.7.5, we illustrate the use of matrices as a convenient constructional tool to state and manipulate problems with two indices. In this application, we desire to verify the accuracy of the truncated Taylor series $S = \sum_{n=0}^{N} \dfrac{x^n}{n!}$ as an approximation to the function $y = \exp(x)$, over the interval $0 \le x < 1$.

Because this application's purpose is to illustrate a constructional scheme, we write the code lines as we are proceeding with the different computational steps:

1. We start by dividing the (0, 1) interval into equally spaced segments. This array is given by

```
x=[0:0.01:1];
M=length(x);
```

2. Assume that we are truncating the series at the value $N = 10$

```
N=10;
```

3. Construct the matrix **W** having the following form:

$$\mathbf{W} = \begin{bmatrix} 1 & x_1 & \frac{x_1^2}{2!} & \frac{x_1^3}{3!} & \cdots & \frac{x_1^N}{N!} \\ 1 & x_2 & \frac{x_2^2}{2!} & \frac{x_2^3}{3!} & \cdots & \frac{x_2^N}{N!} \\ 1 & x_3 & \frac{x_3^2}{2!} & \frac{x_3^3}{3!} & \cdots & \frac{x_3^N}{N!} \\ \vdots & & \vdots & & \cdots & \\ \vdots & & \vdots & & \ddots & \vdots \\ 1 & x_M & \frac{x_M^2}{2!} & \frac{x_M^3}{3!} & \cdots & \frac{x_M^N}{N!} \end{bmatrix} \tag{8.20}$$

Specify the size of **W**, and then give the induction rule to go from one column to the next

$$W(i, j) = x(i) * \frac{W(i, j-1)}{j-1} \tag{8.21}$$

This is implemented in the code as follows:

```
W=ones(M,N);
for i=1:M
  for j=2:N
    W(i,j)=x(i)*W(i,j-1)/(j-1);
  end
end
```

4. The value of the truncated series at a specific point is the sum of the row elements corresponding to its index; however, since MATLAB command **sum** acting on a matrix adds the column elements, we take the sum of the adjoint (the matrix obtained, for real elements, by changing the rows to columns and vice versa) of **W** to obtain our result. Consequently, add to the code

```
serexp=sum(W');
```

5. Finally, compare the values of the truncated series with that of the exponential function

```
y=exp(x);
plot(x,serexp,x,y,'--')
```

In examining the plot resulting from executing the above instructions, we observe that the truncated series give a very good approximation to the exponential over the whole interval.

If you would also like to check the error of the approximation as a function of x, enter

```
dy=abs(y-serexp);
semilogy(x,dy)
```

Examining the output graph, you will find, as expected, that the error increases with an increase in the value of x. However, the approximation of the exponential by the partial sum of the first 10 elements of the truncated Taylor series is accurate over the whole domain considered, to an accuracy of better than one part per million.

Question: Could you have estimated the maximum error in the above computed value of **dy** by evaluating the first neglected term in the Taylor's series at $x = 1$?

IN-CLASS EXERCISE

Pb. 8.7 Verify the accuracy of truncating at the fifth element the following Taylor series, in a domain that you need to specify, so the error is everywhere less than one part in 10,000

a. $\ln(1+x) = \sum_{n=1}^{\infty} (-1)^{n+1} \frac{x^n}{n}$

b. $\sin(x) = \sum_{n=0}^{\infty} (-1)^n \frac{x^{2n+1}}{(2n+1)!}$

c. $\cos(x) = \sum_{n=0}^{\infty} (-1)^n \frac{x^{2n}}{(2n)!}$

8.7.5 Reconstructing a Function from Its Fourier Components

From the results of Section 7.9, where we discussed the Fourier series, it is easy to show that any even periodic function with period 2π can be written in the form of a cosine series, and that an odd periodic function can be written in the form of a sine series of the fundamental frequency and its higher harmonics.

Knowing the coefficients of its Fourier series, we would like to plot the function over a period. The purpose of the following example is twofold:

1. On the mechanistic side, to illustrate again the setting up of a two indices problem in a matrix form.
2. On the mathematical contents side, examining the effects of truncating a Fourier series on the resulting curve.

EXAMPLE 8.8

Plot $y(x) = \sum_{k=1}^{M} C_k \cos(kx)$, if $C_k = \frac{(-1)^k}{k^2+1}$. Choose successively for **M** the values 5, 20, and 40.

Solution: Edit and execute the following *script M-file*:

```
M=? ;                           %Enter a value for M
p=500;
k=1:M;
n=0:p;
x=(2*pi/p)*n;
a=cos((2*pi/p)*n'*k);
c=((-1).^k)./(k.^2+1);
y=a*c';
plot(x,y)
axis([0 2*pi -1 1.2]) □
```

Plot the shape of the resulting curve for different values of **M** and compare the results.

Alternate to the above solution, we could, of course, have created the following *function M-file*:

```
function y=Example_88(x,M)
k=1:M;
y=sum((((-1).^k)./(k.^2+1)).*cos(k.*x));
```

and called from the command window this function for the purpose of plotting it for different values of **M** as follows:

```
fplot(@(x)Example_88(x,?),[0 2*pi])       %input M
```

In-Class Exercises

Pb. 8.8 For different values of the cutoff, plot the resulting curves for the functions given by the following Fourier series:

$$y_1(x) = \frac{8}{\pi^2}\sum_{k=1}^{\infty}\left(\frac{1}{(2k-1)^2}\right)\cos((2k-1)x)$$

$$y_2(x) = \frac{4}{\pi}\sum_{k=1}^{\infty}\left(\frac{(-1)^{k-1}}{(2k-1)}\right)\cos((2k-1)x)$$

$$y_3(x) = \frac{2}{\pi}\sum_{k=1}^{\infty}\frac{1}{(2k-1)}\sin((2k-1)x)$$

Pb. 8.9 The purpose of this problem is to explore the Gibbs phenomenon. This phenomenon occurs as a result of truncating the Fourier series of a discontinuous function. Examine, for example, this phenomenon in detail for the function $y_3(x)$ given in Pb. 8.8.

The function under consideration is given analytically by

$$y_3(x) = \begin{cases} 0.5 & \text{for} \quad 0 < x < \pi \\ -0.5 & \text{for} \quad \pi < x < 2\pi \end{cases}$$

a. Verify graphically that the truncated Fourier series in the neighborhood of $x = 0$ oscillates. In particular, show that for large values of the summation upper bound, it overshoots the value of 0.5, and the limiting value of this first maximum is 0.58949, while the limiting value of the first local minimum is 0.45142.

b. Find analytically the locations of the extrema of the partial sum of the first M terms of the series.

$$\textit{Hint: } y_3^M(x) = \begin{cases} = \dfrac{2}{\pi}\displaystyle\sum_{k=1}^{M}\frac{1}{(2k-1)}\sin((2k-1)x) = \frac{2}{\pi}\sum_{k=1}^{M}\int_0^x \cos((2k-1)t)dt \\ = \dfrac{2}{\pi}\displaystyle\int_0^x\sum_{k=1}^{M}\cos((2k-1)t)dt = \frac{1}{\pi}\int_0^x \frac{\sin(2Mt)}{\sin(t)}dt \end{cases}$$

c. Evaluate numerically the values of $y_3^M(x)$ for large M at the positions of the first two extrema derived in (b), and compare these results with what you found in (a).

It is worth noting at this point that an important goal of filter theory is to find methods to smooth these kinds of oscillations, near discontinuities.

8.7.6 Interpolating the Coefficients of an $(n - 1)$-Degree Polynomial from n Points

The problem at hand can be posed as follows:

Given the coordinates of n points: $(x_1, y_1), (x_2, y_2), \ldots, (x_n, y_n)$, we want to find the polynomial of degree $(n - 1)$, denoted by $p_{n-1}(x)$, whose curve passes through these points.

Let us assume that the polynomial has the following form:

$$p_{n-1}(x) = a_0 + a_1x + a_2x^2 + \cdots + a_{n-1}x^{n-1} \tag{8.22}$$

From a knowledge of the column vectors **X** and **Y**, we can formulate this problem in the standard linear system form. In particular, in matrix form, we can write

$$\mathbf{V} * \mathbf{A} = \begin{bmatrix} 1 & x_1 & x_1^2 & \cdots & x_1^{n-1} \\ 1 & x_2 & x_2^2 & \cdots & x_2^{n-1} \\ \vdots & \vdots & \vdots & & \vdots \\ \vdots & \vdots & \vdots & & \vdots \\ 1 & x_n & x_n^2 & \cdots & x_n^{n-1} \end{bmatrix} \begin{bmatrix} a_0 \\ a_1 \\ a_2 \\ \vdots \\ a_{n-1} \end{bmatrix} = \begin{bmatrix} y_1 \\ y_2 \\ \vdots \\ \vdots \\ y_n \end{bmatrix} = \mathbf{Y} \tag{8.23}$$

Knowing the matrix **V** and the column **Y**, it is then a trivial matter to deduce the column **A**:

$$\mathbf{A} = \mathbf{V}^{-1} * \mathbf{Y} \tag{8.24}$$

What remains to be done is to generate in an efficient manner the matrix **V** using the column vector **X** as input. We note the following recursion relation for the elements of **V**:

$$V(k, j) = x(k) * V(k, j - 1) \tag{8.25}$$

Furthermore, the first column of **V** has all its elements equal to 1.

The following routine computes **A**:

```
X=[x1;x2;x3;.......;xn];
Y=[y1;y2;y3;.......;yn];
n=length(X);
V=ones(n,n);
  for j=2:n
    V(:,j)=X.*V(:,j-1);
  end
A=V\Y
```

In-Class Exercise

Pb. 8.10 Find the polynomials that are uniquely defined through

a. The points (1, 5), (2, 11), and (3, 19).
b. The points (1, 8), (2, 39), (3, 130), (4, 341), and (5, 756).

8.7.7 Least-Squares Fit of Data

In Section 8.7.6, we found the polynomial of degree $(n - 1)$ that was uniquely determined by the coordinates of n points on its curve. However, when data fitting is the tool used by experimentalists to verify a theoretical prediction, many more points than the minimum are measured in order to minimize the effects of random errors generated in the acquisition of the data. But this overdetermination in the system parameters faces us with the dilemma of what confidence level one gives to the accuracy of specific data points, and which data points to accept or reject. *A priori*, one takes all data points, and resorts to a determination of the vector **A** whose corresponding polynomial comes closest to all the experimental points. Closeness is defined through the Euclidean distance between the experimental points and the predicted curve. This method for minimizing the sum of the square of the Euclidean distance between the optimal curve and the experimental points is referred to as the least-squares fit of the data (refer back to Example 5.8).

To have a geometrical understanding of what we are attempting to do, consider the conceptually analogous problem in 3-D of having to find the plane with the least total square distance from five given data points. So what do we do? Using the projection procedure derived in Chapter 7, we deduce each point's distance from the plane; then we go ahead and adjust the parameters of the plane equation to obtain the smallest total square distance between the points and the plane. In linear algebra courses, using generalized optimization techniques, you will be shown that the best fit to **A** (i.e., the one called least-square fit) is given (using the notation of the previous subsection) by

$$\mathbf{A}_N = (\mathbf{V}^T\mathbf{V})^{-1}\,\mathbf{V}^T\mathbf{Y} \tag{8.26}$$

A MATLAB routine to fit a number of (n) points to a polynomial of order $(m - 1)$ now reads:

```
X=[x1;x2;x3;.......;xn];
Y=[y1;y2;y3;.......;yn];
n=length(X);
m=?            %(m-1) is the degree of the polynomial
V=ones(n,m);
  for j=2:m
    V(:,j)=X.*V(:,j-1);
  end
AN=inv(V'*V)*(V'*Y)
```

MATLAB also has a built-in command to achieve the least-squares fit of data. Look up the **polyfit** function in your help documentation, and learn its use and point out what difference exists between its notation and that of the above routine.

IN-CLASS EXERCISE

Pb. 8.11 Find the second-degree polynomials that best fit the data points: (1, 8.1), (2, 24.8), (3, 52.5), (4, 88.5), (5, 135.8), and (6, 193.4).

8.7.8 Numerical Solution of Fredholm Equations

In Section 4.8, we introduced integral equations and showed that an ODE can be written as an integral equation. Here, we illustrate the method for solving this class of equations.

We consider the Fredholm integral equation given by

$$y(x) = 1 + \int_0^1 (1 - 4x\xi)\, y(\xi)\, d\xi \tag{8.27}$$

The key to solving this problem is to rewrite the integral in a summation form. This reduces to

$$y(i) = 1 + d\xi \sum_{j=1}^{2N+1} (1 - 4x(i)\,\xi(j))\, w(j)\, y(j) \tag{8.28}$$

where $d\xi$ is the width of the infinitesimal subinterval and the w's are the particular integration algorithm weight functions.

In the program below, we use the weight functions of Simpson's rule in evaluating the integral and discretize the integral equation to a matrix form.

```
xin=0;
xfin=1;
N=500;
x=linspace(xin,xfin,2*N+1);
xi=linspace(xin,xfin,2*N+1);
dxi=(xfin-xin)/(2*N+1);
K=zeros(2*N+1,2*N+1);
y=zeros(2*N+1,1);
F=ones(2*N+1,1);
M=eye(2*N+1,2*N+1);
      for i=1:2*N+1
            for j=1:2*N+1
```

```
                    if j==1 | j==2*N+1
                          K(i,j)=(1-4*x(i)*xi(j))*1;
                    elseif rem(j,2)==0
                          K(i,j)=(1-4*x(i)*xi(j))*4;
                    else
                          K(i,j)=(1-4*x(i)*xi(j))*2;
                    end
              end
        end
    y=(M-(dxi/3)*K)\F;
    plot(x,y,'k')
```

Homework Problem

Pb. 8.12 Solve the above integral equation analytically. (*Hint*: Assume that the solution is a polynomial and solve for its coefficients.)

a. Compare the analytical solution with the numerical solution. In particular, find the numerical errors at $x = 0$ and $x = 1$.
b. Plot the value of the error in the numerical solution at $x = 0$ as a function of the number of subintervals.

8.8 Eigenvalues and Eigenvectors*

DEFINITION If **M** is a square $n \otimes n$ matrix, then a vector $|v\rangle$ is called an eigenvector and λ, a scalar, is called an eigenvalue, if they satisfy the relation

$$\mathbf{M}|v\rangle = \lambda|v\rangle \tag{8.29}$$

that is, the vector $\mathbf{M}|v\rangle$ is a scalar multiplied by the vector $|v\rangle$.

8.8.1 Finding the Eigenvalues of a Matrix

To find the eigenvalues, note that the above definition of eigenvectors and eigenvalues can be rewritten in the following form:

$$(\mathbf{M} - \lambda\mathbf{I})|v\rangle = 0 \tag{8.30}$$

where **I** is the identity $n \otimes n$ matrix. The above set of homogeneous equations admits a solution only if the determinant of the matrix multiplying the

*The asterisk indicates more advanced material that may be skipped in a first reading.

vector $|v\rangle$ is zero. Therefore, the eigenvalues are the roots of the polynomial $p(\lambda)$, defined as follows:

$$p(\lambda) = \det(\mathbf{M} - \lambda\mathbf{I}) \tag{8.31}$$

This equation is called the characteristic equation of the matrix **M**. It is of degree n in λ. (This last assertion can be proven by noting that the contribution to the determinant of $(\mathbf{M} - \lambda\mathbf{I})$, coming from the product of the diagonal elements of this matrix, contributes a factor of λ^n to the expression of the determinant.)

Example 8.9

Find the eigenvalues and the eigenvectors of the matrix **M**, defined as follows:

$$\mathbf{M} = \begin{pmatrix} 2 & 4 \\ 1/2 & 3 \end{pmatrix} \tag{8.32}$$

Solution: The characteristic polynomial for this matrix is given by

$$p(\lambda) = (2 - \lambda)(3 - \lambda) - (4)(1/2) = \lambda^2 - 5\lambda + 4 \tag{8.33}$$

The roots of this polynomial (i.e., the eigenvalues of the matrix) are, respectively,

$$\lambda_1 = 1 \quad \text{and} \quad \lambda_2 = 4 \tag{8.34}$$

To find the eigenvectors corresponding to the above eigenvalues, which we shall denote respectively by $|v_1\rangle$ and $|v_2\rangle$, we must satisfy the following two equations separately:

$$\begin{pmatrix} 2 & 4 \\ 1/2 & 3 \end{pmatrix}\begin{pmatrix} a \\ b \end{pmatrix} = 1\begin{pmatrix} a \\ b \end{pmatrix} \tag{8.35}$$

and

$$\begin{pmatrix} 2 & 4 \\ 1/2 & 3 \end{pmatrix}\begin{pmatrix} c \\ d \end{pmatrix} = 4\begin{pmatrix} c \\ d \end{pmatrix} \tag{8.36}$$

From the first set of equations, we deduce that $b = -a/4$; and from the second set of equations that $d = c/2$, thus giving for the eigenvectors $|v_1\rangle$ and $|v_2\rangle$, the following expressions:

$$|v_1\rangle = a\begin{pmatrix} -1 \\ 1/4 \end{pmatrix} \tag{8.37}$$

$$|v_2\rangle = c\begin{pmatrix} -1 \\ -1/2 \end{pmatrix} \tag{8.38}$$

It is common to give the eigenvectors in the normalized form, that is, fix a and c to make $\langle v_1|v_1\rangle = \langle v_2|v_2\rangle = 1$, thus giving for $|v_1\rangle$ and $|v_2\rangle$, the normalized values:

$$|v_1\rangle = \sqrt{\frac{16}{17}}\begin{pmatrix} -1 \\ 1/4 \end{pmatrix} = \begin{pmatrix} -0.9701 \\ 0.2425 \end{pmatrix} \tag{8.39}$$

$$|v_2\rangle = \sqrt{\frac{4}{5}}\begin{pmatrix} -1 \\ -1/2 \end{pmatrix} = \begin{pmatrix} -0.8944 \\ -0.4472 \end{pmatrix} \tag{8.40}$$

□

Example 8.10

Find the eigenvalues of a $(N \otimes N)$ matrix having all its elements equal.

Solution: The matrix is given by

$$M = \begin{pmatrix} a & a & \cdots & a & a \\ a & a & & & \vdots \\ \vdots & & \ddots & & \vdots \\ \vdots & & & \ddots & \vdots \\ a & a & \cdots & \cdots & a \end{pmatrix} \tag{8.41}$$

and the eigenvalues are obtained by solving the polynomial equation:

$$\det(\mathbf{M} - \lambda\mathbf{I}) = \det\begin{pmatrix} a-\lambda & a & \cdots & a & a \\ a & a-\lambda & & & \vdots \\ \vdots & & \ddots & & \vdots \\ \vdots & & & \ddots & \vdots \\ a & a & \cdots & \cdots & a-\lambda \end{pmatrix} = 0 \tag{8.42}$$

Using Theorem 8 of the determinants summarized in Appendix B, we do not change the value of the determinant if we subtract the first column from all other columns, we obtain

$$\begin{vmatrix} a-\lambda & \lambda & \cdots & \lambda & \lambda \\ a & -\lambda & & & 0 \\ \vdots & & \ddots & & \vdots \\ \vdots & & & -\lambda & 0 \\ a & 0 & \cdots & 0 & -\lambda \end{vmatrix} = 0 \tag{8.43}$$

Next, we add rows 2 through N to the first row. All elements above the diagonal now vanish, specifically

$$\begin{vmatrix} Na-\lambda & 0 & \cdots & 0 & 0 \\ a & -\lambda & & & 0 \\ \vdots & & \ddots & & \vdots \\ \vdots & & & -\lambda & 0 \\ a & 0 & \cdots & 0 & -\lambda \end{vmatrix} = 0 \tag{8.44}$$

From the result of Pb. 8.1 on the determinant of triangular matrices, we know that the above determinant reduces to

$$(Na - \lambda)\lambda^{n-1} = 0 \tag{8.45}$$

i.e., we have a single-fold root $\lambda = Na$, and $(N - 1)$-fold roots $\lambda = 0$. □

8.8.2 Finding the Eigenvalues and Eigenvectors Using MATLAB

Given a matrix **M**, the MATLAB command to find the eigenvectors and eigenvalues is given by `[V,D]=eig(M)`; the columns of `V` are the eigenvectors and `D` is a diagonal matrix whose elements are the eigenvalues. Entering the matrix `M` and the eigensystem commands gives

```
V =
       -0.9701 -0.8944
       -0.2425 -0.4472
D =
       1 0
       0 4
```

Finding the matrices **V** and **D** is referred to as diagonalizing the matrix **M**. It should be noted that this is not always possible. For example, the matrix is not diagonalizable when one or more of the roots of the characteristic polynomial is zero. In courses of linear algebra, you will study the necessary and sufficient conditions for **M** to be diagonalizable.

IN-CLASS EXERCISES

Pb. 8.13 Show that if $\mathbf{M}|v\rangle = \lambda|v\rangle$, then $\mathbf{M}^n|v\rangle = \lambda^n|v\rangle$, that is, the eigenvalues of $\mathbf{M}^n$ are λ^n; however, the eigenvectors $|v\rangle$'s remain the same as those of **M**.
Verify this theorem using the choice in Example 8.9 for the matrix **M**.

Pb. 8.14 Find the eigenvalues of the upper triangular matrix

$$\mathbf{T} = \begin{pmatrix} 1/4 & 0 & 0 \\ -1 & 1/2 & 0 \\ 2 & -3 & 1 \end{pmatrix}$$

Generalize your result to prove analytically that the eigenvalues of any triangular matrix are its diagonal elements. (*Hint:* Use the previously derived result in Pb. 8.1 for the expression of the determinant of a triangular matrix.)

Pb. 8.15 A general theorem, which will be proven to you in linear algebra courses, states that if a matrix is diagonalizable, then, using the above notation

$$\mathbf{V}\mathbf{D}\mathbf{V}^{-1} = \mathbf{M}$$

Verify this theorem for the matrix **M** of Example 8.9.

a. Using this theorem, show that

$$\det(\mathbf{M}) = \det(\mathbf{D}) = \prod_i^n \lambda_i$$

b. Also show that

$$\mathbf{V}\mathbf{D}^n\mathbf{V}^{-1} = \mathbf{M}^n$$

c. Apply this theorem to compute the matrix $\mathbf{M}^5$, for the matrix **M** of Example 8.9.

Pb. 8.16 Find the nonzero eigenvalues of the $2 \otimes 2$ matrix **A** that satisfies the equation

$$\mathbf{A} = \mathbf{A}^3$$

Homework Problems

The function of a matrix can formally be defined through a Taylor series expansion. For example, the exponential of a matrix **M** can be defined through

$$\exp(\mathbf{M}) = \sum_{n=0}^{\infty} \frac{\mathbf{M}^n}{n!}$$

Pb. 8.17 Use the results from Pb. 8.15 to show that

$$\exp(\mathbf{M}) = \mathbf{V}\exp(\mathbf{D})\mathbf{V}^{-1}$$

where, for any diagonal matrix

$$\exp\begin{pmatrix} \lambda_1 & 0 & \cdots & \cdots & 0 \\ 0 & \lambda_2 & & & \vdots \\ \vdots & & \ddots & & \vdots \\ \vdots & & & \lambda_{n-1} & 0 \\ 0 & 0 & \cdots & 0 & \lambda_n \end{pmatrix} = \begin{pmatrix} \exp(\lambda_1) & 0 & \cdots & \cdots & 0 \\ 0 & \exp(\lambda_2) & & & \vdots \\ \vdots & & \ddots & & \vdots \\ \vdots & & & \exp(\lambda_{n-1}) & 0 \\ 0 & 0 & \cdots & 0 & \exp(\lambda_n) \end{pmatrix}$$

Pb. 8.18 Using the results from Pb. 8.17, we deduce a direct technique for solving the initial value problem for any system of coupled linear ODEs with constant coefficients.

Find and plot the solutions in the interval $0 \le t \le 1$ for the following set of ODEs:

$$\frac{dx_1}{dt} = x_1 + 2x_2$$

$$\frac{dx_2}{dt} = 2x_1 - 2x_2$$

with the initial conditions: $x_1(0) = 1$ and $x_2(0) = 3$. (*Hint*: The solution of $\frac{dX}{dt} = \mathbf{AX}$ is $\mathbf{X}(t) = \exp(\mathbf{A}t)\mathbf{X}(0)$, where **X** is a time-dependent vector and **A** a time-independent matrix.)

Pb. 8.19 MATLAB has a shortcut for computing the exponential of a matrix. While the command `exp(M)` takes the exponential of each element of the matrix, the command `expm(M)` computes the matrix exponential. Verify your results for Pb. 8.18 using this built-in function.

8.9 The Cayley–Hamilton and Other Analytical Techniques*

In Section 8.8, we presented the general techniques for computing the eigenvalues and eigenvectors of square matrices, and showed their power in solving systems of coupled linear differential equations. In this section, we add to our analytical tools arsenal some techniques that are particularly powerful when elegant solutions are desired in low-dimensional problems. We start with the Cayley–Hamilton theorem.

8.9.1 Cayley–Hamilton Theorem

The matrix **M** satisfies its own characteristic equation.

PROOF As per Eq. (8.31), the characteristic equation for a matrix is given by

$$p(\lambda) = \det(\mathbf{M} - \lambda\mathbf{I}) = 0 \tag{8.46}$$

Let us now form the polynomial of the matrix **M** having the same coefficients as that of the characteristic equation, $p(\mathbf{M})$. Using the result from Pb. 8.15, and assuming that the matix is diagonalizable, we can write for this polynomial,

$$p(\mathbf{M}) = \mathbf{V}p(\mathbf{D})\mathbf{V}^{-1} \tag{8.47}$$

where

$$p(\mathbf{D}) = \begin{pmatrix} p(\lambda_1) & 0 & \cdots & \cdots & 0 \\ 0 & p(\lambda_2) & & & 0 \\ \vdots & & \ddots & & \\ \vdots & & & p(\lambda_{n-1}) & 0 \\ 0 & 0 & \cdots & 0 & p(\lambda_n) \end{pmatrix} \tag{8.48}$$

However, we know that $\lambda_1, \lambda_2, \ldots, \lambda_{n-1}, \lambda_n$ are all roots of the characteristic equation. Therefore,

$$p(\lambda_1) = p(\lambda_2) = \cdots = p(\lambda_{n-1}) = p(\lambda_n) = 0 \tag{8.49}$$

thus giving

$$p(\mathbf{D}) = 0 \tag{8.50}$$

$$\Rightarrow p(\mathbf{M}) = 0 \tag{8.51}$$

---

* The asterisk indicates more advanced material that may be skipped in a first reading.

Example 8.11

Using the Cayley–Hamilton theorem, find the inverse of the matrix **M** given in Example 8.9.

Solution: The characteristic equation for this matrix is given by

$$p(\mathbf{M}) = \mathbf{M}^2 - 5\mathbf{M} + 4\mathbf{I} = 0 \tag{8.52}$$

Now multiply this equation by $\mathbf{M}^{-1}$ to obtain

$$\mathbf{M} - 5\mathbf{I} + 4\mathbf{M}^{-1} = 0 \tag{8.53}$$

and

$$\Rightarrow \mathbf{M}^{-1} = 0.25(5\mathbf{I} - \mathbf{M}) = \begin{pmatrix} \frac{3}{4} & -1 \\ -\frac{1}{8} & \frac{1}{2} \end{pmatrix} \tag{8.54}$$

□

Example 8.12

Reduce the following fourth-order polynomial in **M**, where **M** is given in Example 8.9, to a first-order polynomial in **M**:

$$P(\mathbf{M}) = \mathbf{M}^4 + \mathbf{M}^3 + \mathbf{M}^2 + \mathbf{M} + \mathbf{I} \tag{8.55}$$

Solution: From the results of Example 8.10, we have

$$\begin{gathered} \mathbf{M}^2 = 5\mathbf{M} - 4\mathbf{I} \\ \mathbf{M}^3 = 5\mathbf{M}^2 - 4\mathbf{M} = 5(5\mathbf{M} - 4\mathbf{I}) - 4\mathbf{M} = 21\mathbf{M} - 20\mathbf{I} \\ \mathbf{M}^4 = 21\mathbf{M}^2 - 20\mathbf{M} = 21(5\mathbf{M} - 4\mathbf{I}) - 20\mathbf{M} = 85\mathbf{M} - 84\mathbf{I} \\ \Rightarrow P(\mathbf{M}) = 112\mathbf{M} - 107\mathbf{I} \end{gathered} \tag{8.56}$$

Verify the answer numerically using MATLAB. □

8.9.2 Solution of Equations of the Form $\frac{dX}{dt} = AX$

We sketched a technique in Pb. 8.17 that uses the eigenvectors matrix and solves this equation. In Example 8.13, we solve the same problem using the Cayley–Hamilton technique.

Example 8.13

Using the Cayley–Hamilton technique, solve the system of equations

$$\frac{dx_1}{dt} = x_1 + 2x_2 \tag{8.57}$$

$$\frac{dx_2}{dt} = 2x_1 - 2x_2 \tag{8.58}$$

with the initial conditions: $x_1(0) = 1$ and $x_2(0) = 3$

Solution: The matrix **A** for this system is given by

$$\mathbf{A} = \begin{pmatrix} 1 & 2 \\ 2 & -2 \end{pmatrix} \tag{8.59}$$

and the solution of this system is given by

$$\mathbf{X}(t) = e^{\mathbf{A}t}\mathbf{X}(0) \tag{8.60}$$

Given that **A** is a 2 ⊗ 2 matrix, we know from the Cayley–Hamilton result that the exponential function of **A** can be written as a first-order polynomial in **A**; thus

$$P(\mathbf{A}) = e^{\mathbf{A}t} = a\mathbf{I} + b\mathbf{A} \tag{8.61}$$

To determine a and b, we note that the polynomial equation holds as well for the eigenvalues of **A**, which are equal to -3 and 2; therefore

$$e^{-3t} = a - 3b \tag{8.62}$$

$$e^{2t} = a + 2b \tag{8.63}$$

giving

$$a = \frac{2}{5}e^{-3t} + \frac{3}{5}e^{2t} \tag{8.64}$$

$$b = \frac{1}{5}e^{2t} - \frac{1}{5}e^{-3t} \tag{8.65}$$

and

$$e^{\mathbf{A}t} = \begin{pmatrix} \frac{1}{5}e^{-3t} + \frac{4}{5}e^{2t} & \frac{2}{5}e^{2t} - \frac{2}{5}e^{-3t} \\ \frac{2}{5}e^{2t} - \frac{2}{5}e^{-3t} & \frac{4}{5}e^{-3t} + \frac{1}{5}e^{2t} \end{pmatrix} \tag{8.66}$$

Therefore, the solution of the system of equations is

$$X(t) = \begin{pmatrix} 2e^{2t} - e^{-3t} \\ e^{2t} + 2e^{-3t} \end{pmatrix} \tag{8.67}$$

□

8.9.3 Solution of Equations of the Form $\frac{dX}{dt} = AX + B(t)$

Multiplying this equation on the LHS, by $e^{-\mathbf{A}t}$, we obtain

$$e^{-\mathbf{A}t}\frac{d\mathbf{X}}{dt} = e^{-\mathbf{A}t}\mathbf{AX} + e^{-\mathbf{A}t}\mathbf{B}(t) \tag{8.68}$$

Rearranging terms, we write this equation as

$$e^{-\mathbf{A}t}\frac{d\mathbf{X}}{dt} - e^{-\mathbf{A}t}\mathbf{AX} = e^{-\mathbf{A}t}\mathbf{B}(t) \tag{8.69}$$

We note that the LHS of this equation is the derivative of $e^{-\mathbf{A}t}\mathbf{X}$. Therefore, we can now write Eq. (8.69) as

$$\frac{d}{dt}[e^{-\mathbf{A}t}\mathbf{X}(t)] = e^{-\mathbf{A}t}\mathbf{B}(t) \tag{8.70}$$

This can be directly integrated to give

$$[e^{-\mathbf{A}t}\mathbf{X}(t)]\big|_0^t = \int_0^t e^{-\mathbf{A}\tau}\mathbf{B}(\tau)\,d\tau \tag{8.71}$$

or, written differently as

$$e^{-\mathbf{A}t}\,\mathbf{X}(t) - \mathbf{X}(0) = \int_0^t e^{-\mathbf{A}\tau}\mathbf{B}(\tau)\,d\tau \tag{8.72a}$$

which leads to the standard form of the solution

$$\mathbf{X}(t) = e^{\mathbf{A}t}\,\mathbf{X}(0) + \int_0^t e^{\mathbf{A}(t-\tau)}\mathbf{B}(\tau)\,d\tau \tag{8.72b}$$

We illustrate the use of this solution in finding the classical motion of an electron in the presence of both an electric field and a magnetic flux density.

Example 8.14

Find the motion of an electron in the presence of a constant electric field and a constant magnetic flux density that are parallel.

Solution: Let the electric field and the magnetic flux density be given by

$$\vec{E} = E_0 \hat{e}_3 \tag{8.73}$$

$$\vec{B} = B_0 \hat{e}_3 \tag{8.74}$$

Newton's equation of motion in the presence of both an electric field and a magnetic flux density is written as

$$m \frac{d\vec{v}}{dt} = q(\vec{E} + \vec{v} \times \vec{B}) \tag{8.75}$$

where $\vec{v}$ is the velocity of the electron, and m and q are its mass and charge, respectively. Writing this equation in component form, it reduces to the following matrix equation:

$$\frac{d}{dt}\begin{pmatrix} v_1 \\ v_2 \\ v_3 \end{pmatrix} = \alpha \begin{pmatrix} 0 & 1 & 0 \\ -1 & 0 & 0 \\ 0 & 0 & 0 \end{pmatrix} \begin{pmatrix} v_1 \\ v_2 \\ v_3 \end{pmatrix} + \beta \begin{pmatrix} 0 \\ 0 \\ 1 \end{pmatrix} \tag{8.76}$$

where $\alpha = \dfrac{qB_0}{m}$ and $\beta = \dfrac{qE_0}{m}$.

This equation can be put in the above standard form for an inhomogeneous first-order equation if we make the following identifications:

$$\mathbf{A} = \alpha \begin{pmatrix} 0 & 1 & 0 \\ -1 & 0 & 0 \\ 0 & 0 & 0 \end{pmatrix} \quad \text{and} \quad \mathbf{B} = \beta \begin{pmatrix} 0 \\ 0 \\ 1 \end{pmatrix} \tag{8.77}$$

First, we note that the matrix $\mathbf{A}$ is block-diagonalizable; that is, all off-diagonal elements with 3 as either the row or column index are zero, and therefore we can separately do the exponentiation of the third component giving $e^0 = 1$; the exponentiation of the top block can be performed along the same steps, using the Cayley–Hamilton techniques from Example 8.12 , giving finally

$$e^{\mathbf{A}t} = \begin{pmatrix} \cos(\alpha t) & \sin(\alpha t) & 0 \\ -\sin(\alpha t) & \cos(\alpha t) & 0 \\ 0 & 0 & 1 \end{pmatrix} \tag{8.78}$$

Therefore, we can write the solutions for the electron's velocity components as follows:

$$\begin{pmatrix} v_1(t) \\ v_2(t) \\ v_3(t) \end{pmatrix} = \begin{pmatrix} \cos(\alpha t) & \sin(\alpha t) & 0 \\ -\sin(\alpha t) & \cos(\alpha t) & 0 \\ 0 & 0 & 1 \end{pmatrix} \begin{pmatrix} v_1(0) \\ v_2(0) \\ v_3(0) \end{pmatrix} + \beta \begin{pmatrix} 0 \\ 0 \\ t \end{pmatrix} \tag{8.79}$$

or equivalently

$$\begin{aligned} v_1(t) &= v_1(0)\cos(\alpha t) + v_2(0)\sin(\alpha t) \\ v_2(t) &= -v_1(0)\sin(\alpha t) + v_2(0)\cos(\alpha t) \\ v_3(t) &= v_3(0) + \beta t \end{aligned} \tag{8.80}$$

□

In-Class Exercises

Pb. 8.20 Plot the 3-D curve, with time as parameter, for the tip of the velocity vector of an electron with an initial velocity $v = v_0\hat{e}_1$, where $v_0 = 10^5$ m/s, entering a region of space where a constant electric field and a constant magnetic flux density are present and are described by $\vec{E} = E_0\hat{e}_3$, where $E_0 = -10^4$ V/m, and $\vec{B} = B_0\hat{e}_3$, where $B_0 = 10^{-2}$ Wb/m^2. The mass of the electron is $m_e = 9.1094 \times 10^{-31}$ kg, and the magnitude of the electron charge is $e = 1.6022 \times 10^{-19}$ C.

Pb. 8.21 Integrate the expression of the velocity vector in Pb. 8.20 to find the parametric equations of the electron position vector for the preceding problem configuration, and plot its 3-D curve. Let the origin of the axis be fixed to where the electron enters the region of the electric and magnetic fields.

Pb. 8.22 Find the parametric equations for the electron velocity if the electric field and the magnetic flux density are still parallel, the magnetic flux density is still constant, but the electric field is now described by $\vec{E} = E_0\cos(\omega t)\hat{e}_3$.

EXAMPLE 8.15

Find the motion of an electron in the presence of a constant electric field and a constant magnetic flux density perpendicular to it.

Solution: Let the electric field and the magnetic flux density be given by

$$\vec{E} = E_0\hat{e}_3 \tag{8.81}$$

$$\vec{B} = B_0\hat{e}_1 \tag{8.82}$$

The matrix **A** is given in this instance by

$$\mathbf{A} = \alpha\begin{pmatrix} 0 & 0 & 0 \\ 0 & 0 & 1 \\ 0 & -1 & 0 \end{pmatrix} \tag{8.83}$$

while the vector **B** is still given by

$$\mathbf{B} = \beta\begin{pmatrix} 0 \\ 0 \\ 1 \end{pmatrix} \tag{8.84}$$

The matrix $e^{\mathbf{A}t}$ is now given by

$$e^{\mathbf{A}t} = \begin{pmatrix} 1 & 0 & 0 \\ 0 & \cos(\alpha t) & \sin(\alpha t) \\ 0 & -\sin(\alpha t) & \cos(\alpha t) \end{pmatrix} \tag{8.85}$$

and the solution for the velocity vector for this configuration is given, using Eq. (8.72), by

$$\begin{aligned} \begin{pmatrix} v_1(t) \\ v_2(t) \\ v_3(t) \end{pmatrix} &= \begin{pmatrix} 1 & 0 & 0 \\ 0 & \cos(\alpha t) & \sin(\alpha t) \\ 0 & -\sin(\alpha t) & \cos(\alpha t) \end{pmatrix}\begin{pmatrix} v_1(0) \\ v_2(0) \\ v_3(0) \end{pmatrix} \\ &+ \int_0^t \begin{pmatrix} 1 & 0 & 0 \\ 0 & \cos[\alpha(t-\tau)] & \sin[\alpha(t-\tau)] \\ 0 & -\sin[\alpha(t-\tau)] & \cos[\alpha(t-\tau)] \end{pmatrix}\begin{pmatrix} 0 \\ 0 \\ \beta \end{pmatrix} d\tau \end{aligned} \tag{8.86}$$

leading to the following parametric representation for the velocity vector:

$$\begin{aligned} v_1(t) &= v_1(0) \\ v_2(t) &= v_2(0)\cos(\alpha t) + v_3(0)\sin(\alpha t) + \frac{\beta}{\alpha}[1-\cos(\alpha t)] \\ v_3(t) &= -v_2(0)\sin(\alpha t) + v_3(0)\cos(\alpha t) + \frac{\beta}{\alpha}\sin(\alpha t) \end{aligned} \tag{8.87}$$

□

Homework Problems

Pb. 8.23 Plot the 3-D curve, with time as parameter, for the tip of the velocity vector of an electron with an initial velocity $\vec{v}(0) = \frac{v_0}{\sqrt{3}}(\hat{e}_1 + \hat{e}_2 + \hat{e}_3)$, where $v_0 = 10^5$ m/s, entering a region of space where the electric field and the magnetic flux density are constant and described by $\vec{E} = E_0\hat{e}_3$, where $E_0 = -10^4$ V/m; and $\vec{B} = B_0\hat{e}_1$, where $B_0 = 10^{-2}$ Wb/m^2.

Pb. 8.24 Find the parametric equations for the position vector for Pb. 8.23, assuming that the origin of the axis is where the electron enters the region of the force fields. Plot the 3-D curve that describes the position of the electron.

8.9.4 Pauli Spinors

We have shown thus far in this section the power of the Cayley–Hamilton theorem in helping us avoid the explicit computation of the eigenvectors while still analytically solving a number of problems of linear algebra where the dimension of the matrices was essentially $2 \otimes 2$, or in some special cases $3 \otimes 3$. In this subsection, we discuss another analytical technique for matrix manipulation, one that is based on a generalized underlying abstract algebraic structure: the Pauli spin matrices. This is the prototype and precursor to more advanced computational techniques from a field of mathematics called group theory. The Pauli matrices are $2 \otimes 2$ matrices given by

$$\boldsymbol{\sigma}_1 = \begin{pmatrix} 0 & 1 \\ 1 & 0 \end{pmatrix} \tag{8.88a}$$

$$\boldsymbol{\sigma}_2 = j\begin{pmatrix} 0 & -1 \\ 1 & 0 \end{pmatrix} \tag{8.88b}$$

$$\boldsymbol{\sigma}_3 = \begin{pmatrix} 1 & 0 \\ 0 & -1 \end{pmatrix} \tag{8.88c}$$

These matrices have the following properties, which can be easily verified by inspection

Property 1:
$$\boldsymbol{\sigma}_1^2 = \boldsymbol{\sigma}_2^2 = \boldsymbol{\sigma}_3^2 = \mathbf{I} \tag{8.89}$$

where $\mathbf{I}$ is the $2 \otimes 2$ identity matrix.

Property 2:
$$\boldsymbol{\sigma}_1\boldsymbol{\sigma}_2 + \boldsymbol{\sigma}_2\boldsymbol{\sigma}_1 = \boldsymbol{\sigma}_1\boldsymbol{\sigma}_3 + \boldsymbol{\sigma}_3\boldsymbol{\sigma}_1 = \boldsymbol{\sigma}_2\boldsymbol{\sigma}_3 + \boldsymbol{\sigma}_3\boldsymbol{\sigma}_2 = 0 \tag{8.90}$$

Property 3:
$$\boldsymbol{\sigma}_1\boldsymbol{\sigma}_2 = j\boldsymbol{\sigma}_3; \quad \boldsymbol{\sigma}_2\boldsymbol{\sigma}_3 = j\boldsymbol{\sigma}_1; \quad \boldsymbol{\sigma}_3\boldsymbol{\sigma}_1 = j\boldsymbol{\sigma}_2 \tag{8.91}$$

If we define the quantity $\vec{\sigma} \cdot \vec{v}$ to mean

$$\vec{\boldsymbol{\sigma}} \cdot \vec{v} = \boldsymbol{\sigma}_1 v_1 + \boldsymbol{\sigma}_2 v_2 + \boldsymbol{\sigma}_3 v_3 \tag{8.92}$$

that is, $\vec{v} = (v_1, v_2, v_3)$, where the parameters v_1, v_2, v_3 are represented as the components of a vector, the following theorem is valid.

THEOREM

$$(\vec{\boldsymbol{\sigma}} \cdot \vec{v})(\vec{\boldsymbol{\sigma}} \cdot \vec{w}) = (\vec{v} \cdot \vec{w})\mathbf{I} + j\vec{\boldsymbol{\sigma}} \cdot (\vec{v} \times \vec{w}) \tag{8.93}$$

where the vectors' dot and cross products have the standard definition.

PROOF The LHS of this equation can be expanded as follows:

$$\begin{aligned}(\vec{\boldsymbol{\sigma}} \cdot \vec{v})(\vec{\boldsymbol{\sigma}} \cdot \vec{w}) &= (\boldsymbol{\sigma}_1 v_1 + \boldsymbol{\sigma}_2 v_2 + \boldsymbol{\sigma}_3 v_3)(\boldsymbol{\sigma}_1 w_1 + \boldsymbol{\sigma}_2 w_2 + \boldsymbol{\sigma}_3 w_3) \\ &= (\boldsymbol{\sigma}_1^2 v_1 w_1 + \boldsymbol{\sigma}_2^2 v_2 w_2 + \boldsymbol{\sigma}_3^2 v_3 w_3) + (\boldsymbol{\sigma}_1\boldsymbol{\sigma}_2 v_1 w_2 + \boldsymbol{\sigma}_2\boldsymbol{\sigma}_1 v_2 w_1) \\ &\quad + (\boldsymbol{\sigma}_1\boldsymbol{\sigma}_3 v_1 w_3 + \boldsymbol{\sigma}_3\boldsymbol{\sigma}_1 v_3 w_1) + (\boldsymbol{\sigma}_2\boldsymbol{\sigma}_3 v_2 w_3 + \boldsymbol{\sigma}_3\boldsymbol{\sigma}_2 v_3 w_2)\end{aligned} \tag{8.94}$$

Using Property 1 of the Pauli's matrices, the first parenthesis on the RHS of Eq. (8.94) can be written as

$$(\boldsymbol{\sigma}_1^2 v_1 w_1 + \boldsymbol{\sigma}_2^2 v_2 w_2 + \boldsymbol{\sigma}_3^2 v_3 w_3) = (v_1 w_1 + v_2 w_2 + v_3 w_3)\mathbf{I} = (\vec{v} \cdot \vec{w})\mathbf{I} \tag{8.95}$$

Using Properties 2 and 3 of the Pauli's matrices, the second, third, and fourth parentheses on the RHS of Eq. (8.94) can respectively be written as

$$(\boldsymbol{\sigma}_1\boldsymbol{\sigma}_2 v_1 w_2 + \boldsymbol{\sigma}_2\boldsymbol{\sigma}_1 v_2 w_1) = j\boldsymbol{\sigma}_3(v_1 w_2 - v_2 w_1) \tag{8.96}$$

$$(\boldsymbol{\sigma}_1\boldsymbol{\sigma}_3 v_1 w_3 + \boldsymbol{\sigma}_3\boldsymbol{\sigma}_1 v_3 w_1) = j\boldsymbol{\sigma}_2(-v_1 w_3 + v_3 w_1) \tag{8.97}$$

$$(\boldsymbol{\sigma}_2\boldsymbol{\sigma}_3 v_2 w_3 + \boldsymbol{\sigma}_3\boldsymbol{\sigma}_2 v_3 w_2) = j\boldsymbol{\sigma}_1(v_2 w_3 - v_3 w_2) \tag{8.98}$$

Recalling that the cross product of two vectors $(\vec{v} \times \vec{w})$ can be written from Eq. (7.49) in components form as

$$(\vec{v} \times \vec{w}) = (v_2 w_3 - v_3 w_2, -v_1 w_3 + v_3 w_1, v_1 w_2 - v_2 w_1) \tag{7.49}$$

the second, third, and fourth parentheses on the RHS of Eq. (8.94) can be combined to give $j\vec{\boldsymbol{\sigma}} \cdot (\vec{v} \times \vec{w})$, thus completing the proof of the theorem.

COROLLARY If $\hat{e}$ is a unit vector, then

$$(\vec{\boldsymbol{\sigma}} \cdot \hat{e})^2 = \mathbf{I} \tag{8.99}$$

PROOF Using Eq. (8.93), we have

$$(\vec{\boldsymbol{\sigma}} \cdot \hat{e})^2 = (\hat{e} \cdot \hat{e})\mathbf{I} + j\vec{\boldsymbol{\sigma}} \cdot (\hat{e} \times \hat{e}) = \mathbf{I}$$

where, in the last step, we used the fact that the norm of a unit vector is one and that the cross product of any vector with itself is zero.

A direct result of this corollary is that

$$(\vec{\boldsymbol{\sigma}} \cdot \hat{e})^{2m} = \mathbf{I} \tag{8.100}$$

and

$$(\vec{\boldsymbol{\sigma}} \cdot \hat{e})^{2m+1} = (\vec{\boldsymbol{\sigma}} \cdot \hat{e}) \tag{8.101}$$

From the above results, we are led to the theorem:

THEOREM

$$\exp(j\vec{\boldsymbol{\sigma}} \cdot \hat{e}\phi) = \cos(\phi) + j\vec{\boldsymbol{\sigma}} \cdot \hat{e}\sin(\phi) \tag{8.102}$$

PROOF If we Taylor expand the exponential function, we obtain

$$\exp(j\vec{\boldsymbol{\sigma}} \cdot \hat{e}\phi) = \sum_m \frac{[j\phi(\vec{\boldsymbol{\sigma}} \cdot \hat{e})]^m}{m!} \tag{8.103}$$

Now separating the even- and odd-power terms, using the just-derived result for the odd and even powers of $(\vec{\sigma} \cdot \hat{e})$, and Taylor expansions of the cosine and sine functions, we obtain the desired result.

EXAMPLE 8.16

Find the time development of the spin state of an electron in a constant magnetic flux density.

Solution: [Readers not interested in the physical background of this problem can immediately jump to the paragraph following Eq. (8.106).]

Physical Background: In addition to the spatio-temporal dynamics, the electron and all other elementary particles of nature also have internal degrees of freedom; which means that even if the particle has no translational motion, its state may still evolve in time. The spin of a particle is such an internal degree of freedom. The electron spin internal degree of freedom requires for its representation a 2-D vector, that is, two fundamental states are possible. As it may be familiar to you from your elementary chemistry courses, the up and down states of the electron are required to satisfactorily describe the number of electrons in the different orbitals of the atoms. For the up state, the eigenvalue of the spin matrix is positive, while for the down state, the eigenvalue is negative ($h/2$ and $h/2$, respectively where $h = 1.0546 \times 10^{-34}\,\mathrm{J\,s} = h/(2\pi)$, and h is Planck's constant).

Due to spin, the quantum mechanical dynamics of an electron in a magnetic flux density does not only include quantum mechanically the time development equivalent to the classical motion that we described in Example 8.13 and Example 8.14; it also includes precession of the spin around the external magnetic flux density, similar to that experienced by a small magnet dipole in the presence of a magnetic flux density.

The magnetic dipole moment due to the spin internal degree of freedom of an electron is proportional to the Pauli's spin matrix; specifically

$$\vec{\mu} = -\mu_B \vec{\sigma} \tag{8.104}$$

where $\mu_B = 0.927 \times 10^{-23}\,\mathrm{J/Tesla}$.

In the same notation, the electron spin angular momentum is given by

$$\vec{\mathbf{S}} = \frac{\hbar}{2}\vec{\sigma} \tag{8.105}$$

The electron magnetic dipole, due to spin, interaction with the magnetic flux density is described by the potential

$$\mathbf{V} = \mu_B \vec{\sigma} \cdot \vec{B} \tag{8.106}$$

and the dynamics of the electron spin state in the magnetic flux density is described by Schrodinger's equation

$$j\hbar \frac{d}{dt} | \psi\rangle = \mu_B \vec{\boldsymbol{\sigma}} \cdot \vec{B} | \psi\rangle \tag{8.107}$$

where, as previously mentioned, the Dirac ket-vector is two-dimensional.

Mathematical Problem: To put the problem in purely mathematical form, we are asked to find the time development of the 2-D vector $|\Psi\rangle$ if this vector obeys the system of equations:

$$\frac{d}{dt}\begin{pmatrix} a(t) \\ b(t) \end{pmatrix} = -j\frac{\Omega}{2}(\vec{\boldsymbol{\sigma}} \cdot \hat{e})\begin{pmatrix} a(t) \\ b(t) \end{pmatrix} \tag{8.108}$$

where $\Omega/2 = \mu_B B_0/\hbar$, and is called the Larmor frequency, and the magnetic flux density is given by $\vec{B} = B_0\hat{e}$. The solution of Eq. (8.108) can be immediately written because the magnetic flux density is constant. The solution at an arbitrary time is related to the state at the origin of time through

$$\begin{pmatrix} a(t) \\ b(t) \end{pmatrix} = \exp\left[-j\frac{\Omega}{2}(\vec{\boldsymbol{\sigma}} \cdot \hat{e})t\right]\begin{pmatrix} a(0) \\ b(0) \end{pmatrix} \tag{8.109}$$

which from Eq. (8.102) can be simplified to read

$$\begin{pmatrix} a(t) \\ b(t) \end{pmatrix} = \left[\cos\left(\frac{\Omega}{2}t\right)\mathbf{I} - j(\vec{\boldsymbol{\sigma}} \cdot \hat{e})\sin\left(\frac{\Omega}{2}t\right)\right]\begin{pmatrix} a(0) \\ b(0) \end{pmatrix} \tag{8.110}$$

If we choose the magnetic flux density to point in the z-direction, then the solution takes the very simple form

$$\begin{pmatrix} a(t) \\ b(t) \end{pmatrix} = \begin{pmatrix} e^{-j\frac{\Omega}{2}t}a(0) \\ e^{j\frac{\Omega}{2}t}b(0) \end{pmatrix} \tag{8.111}$$

Physically, the above result can be interpreted as the precession of the electron around the direction of the magnetic flux density. To understand this statement, let us find the eigenvectors of the $\boldsymbol{\sigma}_x$ and $\boldsymbol{\sigma}_y$ matrices. These are given by

$$\boldsymbol{\alpha}_x = \frac{1}{\sqrt{2}}\begin{pmatrix}1\\1\end{pmatrix} \text{ and } \boldsymbol{\beta}_x = \frac{1}{\sqrt{2}}\begin{pmatrix}1\\-1\end{pmatrix} \tag{8.112a}$$

$$\boldsymbol{\alpha}_y = \frac{1}{\sqrt{2}}\begin{pmatrix}1\\j\end{pmatrix} \text{ and } \boldsymbol{\beta}_y = \frac{1}{\sqrt{2}}\begin{pmatrix}1\\-j\end{pmatrix} \tag{8.112b}$$

The eigenvalues of $\boldsymbol{\sigma}_x$ and $\boldsymbol{\sigma}_y$ corresponding to the eigenvectors $\boldsymbol{\alpha}$ are equal to 1, while those corresponding to the eigenvectors $\boldsymbol{\beta}$ are equal to -1.

Now, assume that the electron was initially in the state $\boldsymbol{\alpha}_x$

$$\begin{pmatrix}a(0)\\b(0)\end{pmatrix} = \frac{1}{\sqrt{2}}\begin{pmatrix}1\\1\end{pmatrix} = \boldsymbol{\alpha}_x \tag{8.113}$$

By substitution in Eq. (8.111), we can compute the electron spin state at different times. Thus, for the time indicated, the electron spin state is given by the second column in the list below:

$$t = \frac{\pi}{2\Omega} \Rightarrow |\psi\rangle = e^{-j\pi/4}\boldsymbol{\alpha}_y \tag{8.114}$$

$$t = \frac{\pi}{\Omega} \Rightarrow |\psi\rangle = e^{-j\pi/2}\boldsymbol{\beta}_x \tag{8.115}$$

$$t = \frac{3\pi}{2\Omega} \Rightarrow |\psi\rangle = e^{-j3\pi/4}\boldsymbol{\beta}_y \tag{8.116}$$

$$t = \frac{2\pi}{\Omega} \Rightarrow |\psi\rangle = e^{-j\pi}\boldsymbol{\alpha}_x \tag{8.117}$$

In examining the above results, we note that, up to an overall phase, the electron spin state returns to its original state following a cycle. During this cycle, the electron "pointed" successively in the positive x-axis, the positive y-axis, the negative x-axis, and the negative y-axis before returning again to the positive x-axis, thus mimicking the hand of a clock moving in the counterclockwise direction. It is this "motion" that is referred to as the electron spin precession around the direction of the magnetic flux density. □

In-Class Exercises

Pb. 8.25 Find the Larmor frequency for an electron in a magnetic flux density of 100 G (10^{-2} T).

Pb. 8.26 Similar to the electron, the proton and the neutron also have spin as one of their internal degrees of freedom, and similarly attached to this spin, both the proton and the neutron each have a magnetic moment. The magnetic moment attached to the proton and neutron have, respectively, the values $\mu_n = -1.91\,\mu_N$ and $\mu_p = 2.79\,\mu_N$, where μ_N is called the nuclear magneton and is equal to $\mu_N = 0.505 \times 10^{-26}$ J/Tesla.

Find the precession frequency of the proton spin if the proton is in the presence of a magnetic flux density of strength 1 T.

Homework Problem

Pb. 8.27 Magnetic resonance imaging (MRI) is one of the most accurate techniques in biomedical imaging. Its principle of operation is as follows. A strong dc magnetic flux density aligns in one of two possible orientations of the spins of the protons of the hydrogen nuclei in the water of the tissues (we say that it polarizes them). The other molecules in the system have zero magnetic moments and are therefore not affected. In thermal equilibrium and at room temperature, there are slightly more protons aligned parallel to the magnetic flux density because this is the lowest energy level in this case. A weaker rotating ac transverse flux density attempts to flip these aligned spins. The energy of the transverse field absorbed by the biological system, which is proportional to the number of spin flips, is the quantity measured in an MRI scan. It is a function of the density of the polarized particles present in that specific region of the image, and of the frequency of the ac transverse flux density.

In this problem, we want to find the frequency of the transverse field that will induce the maximum number of spin flips.

The ODE describing the spin system dynamics in this case is given by

$$\frac{d}{dt}|\psi\rangle = j[\Omega_\perp \cos(\omega t)\boldsymbol{\sigma}_1 - \Omega_\perp \sin(\omega t)\boldsymbol{\sigma}_2 + \Omega\boldsymbol{\sigma}_3]|\psi\rangle$$

where $\Omega = \mu_p B_0/\hbar$, $\Omega_\perp = \mu_p B_\perp/\hbar$, and μ_p is given in Pb. 8.26. The magnetic flux density is given by

$$\vec{B} = B_\perp \cos(\omega t)\hat{e}_1 - B_\perp \sin(\omega t)\hat{e}_2 + B_0\hat{e}_3$$

(cont'd.)

Homework Problem *(cont'd.)*

Assume for simplicity the initial state $|\psi(t=0)\rangle\rangle = \begin{pmatrix} 1 \\ 0 \end{pmatrix}$, and denote the state of the system at time t by $|\psi(t)\rangle\rangle = \begin{pmatrix} a(t) \\ b(t) \end{pmatrix}$:

a. Find numerically at which frequency ω the magnitude of $b(t)$ is maximum.

b. Once you have determined the optimal ω, go back and examine what strategy you should adopt in the choice of $\Omega_\perp$ to ensure maximum resolution.

c. Verify your numerical answers with the analytical solution of this problem, which is given by

$$|b(t)|^2 = \frac{\Omega_\perp^2}{\tilde{\omega}^2} \sin^2(\tilde{\omega} t)$$

where $\tilde{\omega}^2 = (\Omega - \omega/2)^2 + \Omega_\perp^2$.

8.10 Special Classes of Matrices*

8.10.1 Hermitian Matrices

Hermitian matrices of finite or infinite dimensions (operators) play a key role in quantum mechanics, the primary tool for understanding and solving physical problems at the atomic and subatomic scales. In this section, we define these matrices and find key properties of their eigenvalues and eigenvectors.

DEFINITION The Hermitian adjoint of a matrix **M**, denoted by $\mathbf{M}^\dagger$, is equal to the complex conjugate of its transpose:

$$\mathbf{M}^\dagger = \overline{\mathbf{M}}^T \tag{8.118}$$

For example, in complex vector spaces, the bra-vector will be the Hermitian adjoint of the corresponding ket-vector:

$$\langle v| = (|v\rangle)^\dagger \tag{8.119}$$

*The asterisk indicates more advanced material that may be skipped in a first reading.

LEMMA

$$(\mathbf{AB})^{\dagger} = \mathbf{B}^{\dagger}\mathbf{A}^{\dagger} \tag{8.120}$$

PROOF From the definition of matrix multiplication and Hermitian adjoint, we have

$$\begin{aligned}
[(\mathbf{AB})^{\dagger}]_{ij} &= (\bar{\mathbf{A}}\,\bar{\mathbf{B}})_{ji} \\
&= \sum_k \bar{\mathbf{A}}_{jk}\bar{\mathbf{B}}_{ki} = \sum_k (\mathbf{A}^{\dagger})_{kj}(\mathbf{B}^{\dagger})_{ik} \\
&= \sum_k (\mathbf{B}^{\dagger})_{ik}(\mathbf{A}^{\dagger})_{kj} = (\mathbf{B}^{\dagger}\mathbf{A}^{\dagger})_{ij}
\end{aligned} \tag{8.121}$$

DEFINITION A matrix is Hermitian if it is equal to its Hermitian adjoint, that is,

$$\mathbf{H}^{\dagger} = \mathbf{H} \tag{8.122}$$

THEOREM 1 The eigenvalues of a Hermitian matrix are real.

PROOF Let λ_m be an eigenvalue of $\mathbf{H}$ and let $|v_m\rangle$ be the corresponding eigenvector; then

$$\mathbf{H}|v_m\rangle = \lambda_m |v_m\rangle \tag{8.123}$$

Taking the Hermitian adjoints of both sides, using the above lemma, and remembering that $\mathbf{H}$ is Hermitian, we successively obtain

$$(\mathbf{H}|v_m\rangle)^{\dagger} = \langle v_m|\mathbf{H}^{\dagger} = \langle v_m|\mathbf{H} = \langle v_m|\bar{\lambda}_m \tag{8.124}$$

Now multiply (in an inner-product sense) Eq. (8.123) on the LHS with the bra $\langle v_m|$ and Eq. (8.124) on the RHS by the ket-vector $|v_m\rangle$, we obtain

$$\langle v_m|\mathbf{H}|v_m\rangle = \lambda_m\langle v_m|v_m\rangle = \bar{\lambda}_m\langle v_m|v_m\rangle \Rightarrow \lambda_m = \bar{\lambda}_m \tag{8.125}$$

THEOREM 2 The eigenvectors of a Hermitian matrix corresponding to different eigenvalues are orthogonal; that is, given that

$$\mathbf{H}|v_m\rangle = \lambda_m |v_m\rangle \tag{8.126}$$

$$\mathbf{H}|v_n\rangle = \lambda_n |v_n\rangle \tag{8.127}$$

and

$$\lambda_m \neq \lambda_n \tag{8.128}$$

then

$$\langle v_n | v_m \rangle = \langle v_m | v_n \rangle = 0 \tag{8.129}$$

PROOF Because the eigenvalues are real, we can write

$$\langle v_n | \mathbf{H} = \langle v_n | \lambda_n \tag{8.130}$$

Dot this quantity on the right by the ket $|v_m\rangle$ to obtain

$$\langle v_n | \mathbf{H} | v_m \rangle = \langle v_n | \lambda_n | v_m \rangle = \lambda_n \langle v_n | v_m \rangle \tag{8.131}$$

On the other hand, if we dotted Eq. (8.126) on the left with the bra-vector , $\langle v_n|$, we obtain

$$\langle v_n | \mathbf{H} | v_m \rangle = \langle v_n | \lambda_m | v_m \rangle = \lambda_m \langle v_n | v_m \rangle \tag{8.132}$$

Now compare Eq. (8.131) and Eq. (8.132). They are equal, or that

$$\lambda_m \langle v_n | v_m \rangle = \lambda_n \langle v_n | v_m \rangle \tag{8.133}$$

However, because $\lambda_m \neq \lambda_n$, this equality can only be satisfied if $\langle v_n | v_m \rangle = 0$, which is the desired result.

IN-CLASS EXERCISES

Pb. 8.28 Show that any Hermitian $2 \otimes 2$ matrix has a unique decomposition into the Pauli spin matrices and the identity matrix.

Pb. 8.29 Find the multiplication rule for two $2 \otimes 2$ Hermitian matrices that have been decomposed into the Pauli spin matrices and the identity matrix; that is

if $$\mathbf{M} = a_0\mathbf{I} + a_1\boldsymbol{\sigma}_1 + a_2\boldsymbol{\sigma}_2 + a_3\boldsymbol{\sigma}_3$$

and $$\mathbf{N} = b_0\mathbf{I} + b_1\boldsymbol{\sigma}_1 + b_2\boldsymbol{\sigma}_2 + b_3\boldsymbol{\sigma}_3$$

Find the p-components in $\mathbf{P} = \mathbf{MN} = p_0\mathbf{I} + p_1\boldsymbol{\sigma}_1 + p_2\boldsymbol{\sigma}_2 + p_3\boldsymbol{\sigma}_3$

Homework Problem

Pb. 8.30 For the most general case of dielectric materials, the electric displacement vector and the electric field are related by

$$D_i = \sum_j \varepsilon_{ij} E_j$$

where the dielectric tensor is symmetric for nonabsorbing material, i.e., $\varepsilon_{ij} = \varepsilon_{ji}$, while this matrix is Hermitian for the most general case.

The physical consequence for the nonvanishing off-diagonal elements of the matrix $\boldsymbol{\varepsilon}$ is that the displacement vector and the electric field are not in general parallel inside such materials. The only exception being when the electric field is parallel to one of the "principal axes" (i.e., the axes defined by the eigenvectors of the matrix $\boldsymbol{\varepsilon}$).

As the matrix $\boldsymbol{\varepsilon}$ is Hermitian, we know from Theorem 2 above that its different eigenvectors are orthogonal to each other.

Consider a material such that

$$\varepsilon_{11} \neq \varepsilon_{22} \neq \varepsilon_{33}$$

$$\varepsilon_{13} = \varepsilon_{23} = 0$$

Determine the directions of the principal axes. (Let the indices (1, 2, 3) refer respectively to the (x, y, z) components.) (*Hint:* Let the normalized eigenvectors be written in the form

$$\begin{pmatrix} \cos(\alpha) \\ \cos(\beta) \\ \cos(\gamma) \end{pmatrix}$$

where α is the angle the unit eigenvector subtends with the x-axis, β the angle it subtends with the y-axis, and γ the angle it subtends with the z-axis.)

Answer: The principal axes are the z-axis and the lines that are in the x–y plane and which make an angle α with the x-axis and such that

$$\tan(2\alpha) = \frac{2\varepsilon_{12}}{\varepsilon_{11} - \varepsilon_{22}}$$

8.10.2 Unitary Matrices

DEFINITION A unitary matrix has the property that its Hermitian adjoint is equal to its inverse

$$\mathbf{U}^{\dagger} = \mathbf{U}^{-1} \tag{8.134}$$

An example of a unitary matrix would be the matrix $e^{j\mathbf{H}t}$, if $\mathbf{H}$ was Hermitian.

THEOREM 1 The eigenvalues of a unitary matrix all have magnitude one.

PROOF The eigenvalues and eigenvectors of the unitary matrix satisfy the usual equations for these quantities; that is

$$\mathbf{U}|v_n\rangle = \lambda_n |v_n\rangle \tag{8.135}$$

Taking the Hermitian conjugate of this equation, we obtain

$$\langle v_n|\mathbf{U}^{\dagger} = \langle v_n|\mathbf{U}^{-1} = \langle v_n|\bar{\lambda}_n \tag{8.136}$$

Multiplying Eq. (8.135) on the LHS by Eq. (8.136), we obtain

$$\langle v_n|\mathbf{U}^{-1}\mathbf{U}|v_n\rangle = \langle v_n|v_n\rangle = |\lambda_n|^2 \langle v_n|v_n\rangle \tag{8.137}$$

from which we deduce the desired result that: $|\lambda_n|^2 = 1$.

A direct corollary of the above theorem is that $|\det(\mathbf{U})| = 1$. This can be proven directly if we remember the result of Pb. 8.15, which states that the determinant of any diagonalizable matrix is the product of its eigenvalues, and the above theorem that proved that each of these eigenvalues has unit magnitude.

THEOREM 2 A transformation represented by a unitary matrix keeps invariant the scalar (dot, or inner) product of two vectors.

PROOF The matrix $\mathbf{U}$ acting on the vectors $|\varphi\rangle$ and $|\psi\rangle$ results in two new vectors, denoted by $|\varphi'\rangle$ and $|\psi'\rangle$ and such that

$$|\varphi'\rangle = \mathbf{U}|\varphi\rangle \tag{8.138}$$

$$|\psi'\rangle = \mathbf{U}|\psi\rangle \tag{8.139}$$

Taking the Hermitian adjoint of Eq. (8.138), we obtain

$$\langle\varphi'| = \langle\varphi|\mathbf{U}^{\dagger} = \langle\varphi|\mathbf{U}^{-1} \tag{8.140}$$

Multiplying Eq. (8.139) on the LHS by Eq. (8.140), we obtain

$$\langle \varphi' | \psi' \rangle = \langle \varphi | \mathbf{U}^{-1}\mathbf{U} | \psi \rangle = \langle \varphi | \psi \rangle \tag{8.141}$$

which is the result that we are after. In particular, note that the norm of the vector under this matrix multiplication remains invariant. We will have the opportunity to study a number of examples of such transformations in Chapter 9.

8.10.3 Unimodular Matrices

DEFINITION A unimodular matrix has the defining property that its determinant is equal to one. In the remainder of this section, we restrict our discussion to 2 ⊗ 2 unimodular matrices, as these form the tools for the matrix formulation of ray optics and Gaussian optics, which are two of the major subfields of photonics engineering.

Example 8.17

Find the eigenvalues and eigenvectors of the 2 ⊗ 2 unimodular matrix.

Solution: Let the matrix **M** be given by the following expression:

$$\mathbf{M} = \begin{pmatrix} a & b \\ c & d \end{pmatrix} \tag{8.142}$$

The unimodularity condition is then written as

$$\det(\mathbf{M}) = ad - bc = 1 \tag{8.143}$$

Using Eq. (8.143), the eigenvalues of this matrix are given by

$$\lambda_{\pm} = \frac{1}{2}\left[(a+d) \pm \sqrt{(a+d)^2 - 4}\right] \tag{8.144}$$

Depending on the value of $(a + d)$, these eigenvalues can be parameterized in a simple expression. We choose, here, the range $-2 \leq (a + d) \leq 2$ for illustrative purposes. Under this constraint, the following parameterization is convenient:

$$\cos(\theta) = \frac{1}{2}(a+d) \tag{8.145}$$

(For the ranges below -2 and above 2, the hyperbolic cosine function will be more appropriate and similar steps to the ones that we will follow can be repeated.)

Having found the eigenvalues, which can now be expressed in the simple form:

$$\lambda_{\pm} = e^{\pm j\theta} \tag{8.146}$$

let us proceed to find the matrix **V**, defined as

$$\mathbf{M} = \mathbf{V}\mathbf{D}\mathbf{V}^{-1} \quad \text{or} \quad \mathbf{M}\mathbf{V} = \mathbf{V}\mathbf{D} \tag{8.147}$$

and where **D** is the diagonal matrix of the eigenvalues. By direct substitution, in the matrix equation defining **V**, Eq. (8.147), the following relations can be directly obtained

$$\frac{V_{11}}{V_{21}} = \frac{\lambda_{+} - d}{c} \tag{8.148}$$

and

$$\frac{V_{12}}{V_{22}} = \frac{\lambda_{-} - d}{c} \tag{8.149}$$

If we choose for convenience $V_{11} = V_{22} = c$ (which is always possible because each eigenvector can have the value of one of its components arbitrary chosen with the other components expressed as functions of it), the matrix **V** can be written as

$$\mathbf{V} = \begin{pmatrix} e^{j\theta} - d & e^{-j\theta} - d \\ c & c \end{pmatrix} \tag{8.150}$$

and the matrix **M** can be then written as

$$\mathbf{M} = \frac{\begin{pmatrix} e^{j\theta} - d & e^{-j\theta} - d \\ c & c \end{pmatrix} \begin{pmatrix} e^{j\theta} & 0 \\ 0 & e^{-j\theta} \end{pmatrix} \begin{pmatrix} c & d - e^{-j\theta} \\ -c & e^{j\theta} - d \end{pmatrix}}{(2j\sin(\theta))\, c} \tag{8.151}$$

□

Homework Problem

Pb. 8.31 Use Cayley–Hamilton theorem to prove Sylvester theorem for the unimodular matrix, which states that

$$\mathbf{M}^n = \begin{pmatrix} a & b \\ c & d \end{pmatrix}^n = \begin{pmatrix} \dfrac{a\sin(n\theta) - \sin((n-1)\theta)}{\sin(\theta)} & \dfrac{b\sin(n\theta)}{\sin(\theta)} \\ \dfrac{c\sin(n\theta)}{\sin(\theta)} & \dfrac{d\sin(n\theta) - \sin((n-1)\theta)}{\sin(\theta)} \end{pmatrix}$$

where θ is defined in Eq. (8.145).

Application: Dynamics of the Trapping of an Optical Ray in an Optical Fiber

Optical fibers, the main waveguides of land-based optical broadband networks are hair-thin glass fibers that transmit light pulses over very long distances with very small losses. Their waveguiding property is due to a quadratic index of refraction radial profile built into the fiber. This profile is implemented in the fiber manufacturing process, through doping the glass with different concentrations of impurities at different radial distances.

The purpose of this application is to explain how waveguiding can be achieved if the index of refraction inside the fiber has the following profile:

$$n = n_0\left(1 - \frac{n_2^2}{2} r^2\right) \tag{8.152}$$

where r is the radial distance from the fiber axis and $n_2^2 r^2$ a number smaller than 0.01 everywhere inside the fiber.

This problem can, of course, be solved by finding the solution of Maxwell equations, or the differential equation of geometrical optics for ray propagation in a nonuniform medium. However, we will not do this in this application. Here, we use only Snell's law of refraction (see Figure 8.4), which states that at the boundary between two transparent materials with two different indices of refraction, light refracts such that the product of the index of refraction of each medium multiplied by the sine of the angle that the ray makes with the normal to the interface in each medium is constant, and Sylvester's theorem derived in Pb. 8.31.

Let us describe a light ray going through the fiber at any point z along its length, by the distance r that the ray is displaced from the fiber axis, and by the small angle α that the ray's direction makes with the fiber axis. Now consider two points on the fiber axis separated by the small distance δz. We want to

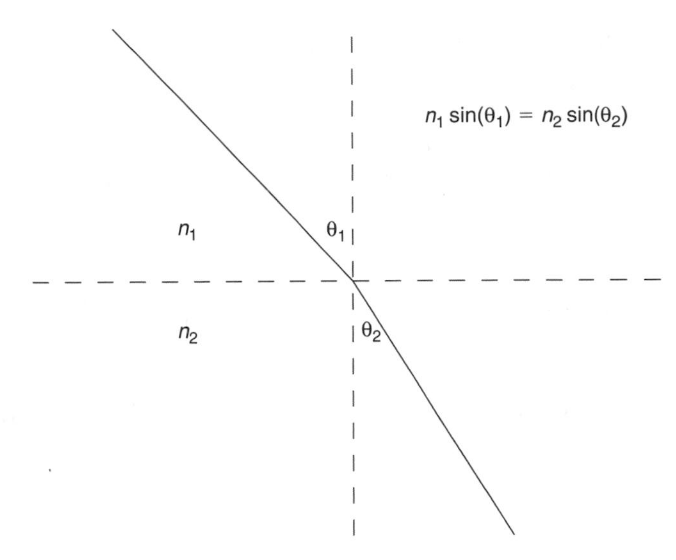

FIGURE 8.4
Parameters of Snell's law of refraction.

find $r(z + \delta z)$ and $\alpha(z + \delta z)$, knowing $r(z)$ and $\alpha(z)$. We are looking for the iteration relation that successive applications will permit us to find the ray displacement r and α slope at any point inside the fiber if we knew their values at the fiber entrance plane.

We solve the problem in two steps. We first assume that there was no bending in the ray, and then find the ray transverse displacement following a small longitudinal displacement. This is straightforward from the definition of the slope of the ray

$$\delta r = \alpha(z)\,\delta z \tag{8.153}$$

Because the angle α is small, we approximated the tangent of the angle by the value of the angle in radians.

Therefore, if we represent the position and slope of the ray as a column matrix, Eq. (8.153) can be represented by the following matrix representation:

$$\begin{pmatrix} r(z+\delta z) \\ \alpha(z+\delta z) \end{pmatrix} = \begin{pmatrix} 1 & \delta z \\ 0 & 1 \end{pmatrix} \begin{pmatrix} r(z) \\ \alpha(z) \end{pmatrix} \tag{8.154}$$

Next, we want to find the bending experienced by the ray in advancing through the distance δz. Because the angles that should be used in Snell's law are the complementary angles to those that the ray forms with the axis of the fiber, and recalling that the glass index of refraction is changing only in the radial direction, we deduce from Snell's law that

$$n(r + \delta r)\cos(\alpha + \delta\alpha) = n(r)\cos(\alpha) \tag{8.155}$$

Now, taking the leading terms of a Taylor expansion of the LHS of this equation leads us to

$$\left[n(r) + \frac{dn(r)}{dr}\delta r\right]\left[1 - \frac{(\alpha + \delta\alpha)^2}{2}\right] \approx n(r)\left(1 - \frac{\alpha^2}{2}\right) \tag{8.156}$$

Further simplification of this equation gives to first order in the variations

$$\delta\alpha \approx \frac{1}{\alpha n(r)}\frac{dn(r)}{dr}\delta r \approx \frac{1}{n_0}(-n_0 n_2^2 r)\,\delta z = -(n_2^2\,\delta z)\,r \tag{8.157}$$

which can be expressed in matrix form as

$$\begin{pmatrix} r(z+\delta z) \\ \alpha(z+\delta z) \end{pmatrix} = \begin{pmatrix} 1 & 0 \\ -n_2^2\,\delta z & 1 \end{pmatrix}\begin{pmatrix} r(z) \\ \alpha(z) \end{pmatrix} \tag{8.158}$$

The total variation in the values of the position and slope of the ray can be obtained by taking the product of the two matrices in Eq. (8.154) and Eq. (8.158), giving

$$\begin{pmatrix} r(z+\delta z) \\ \alpha(z+\delta z) \end{pmatrix} = \begin{pmatrix} 1-(n_2\delta z)^2 & \delta z \\ -n_2^2\delta z & 1 \end{pmatrix}\begin{pmatrix} r(z) \\ \alpha(z) \end{pmatrix} \tag{8.159}$$

Equation (8.159) provides us with the required recursion relation to numerically iterate the progress of the ray inside the fiber. Thus, the ray distance from the fiber axis and the angle that it makes with this axis can be computed at any z in the fiber if we know the values of the ray transverse coordinate and its slope at the entrance plane.

The problem can also be solved analytically if we note that the determinant of this matrix is 1 (the matrix is unimodular). Sylvester's theorem provides the means to obtain the following result:

$$\begin{pmatrix} r(z) \\ \alpha(z) \end{pmatrix} = \begin{pmatrix} \cos(n_2 z) & \dfrac{\sin(n_2 z)}{n_2} \\ -n_2\sin(n_2 z) & \cos(n_2 z) \end{pmatrix}\begin{pmatrix} r(0) \\ \alpha(0) \end{pmatrix} \tag{8.160}$$

Homework Problems

Pb. 8.32 Consider an optical fiber of radius $a = 30\,\mu\text{m}$, $n_0 = 4/3$, and $n_2 = 10^3\,\text{m}^{-1}$. Three rays enter this fiber parallel to the fiber axis at distances of $5\,\mu\text{m}$, $10\,\mu\text{m}$, and $15\,\mu\text{m}$ from the fiber's axis.

a. Write a MATLAB program to follow the progress of the rays through the fiber, properly choosing the δz increment.
b. Trace these rays going through the fiber.

Figure 8.5 shows the answer that you should obtain for a fiber length of 3 cm.

Pb. 8.33 Using Sylvester's theorem, derive Eq. (8.160). (*Hint*: Define the angle θ, such that $\sin\left(\frac{\theta}{2}\right) = \frac{\alpha\,\delta z}{2}$, and recall that while δz goes to zero, its product with the number of iterations is finite and is equal to the distance of propagation inside the fiber.)

Pb. 8.34 Find the maximum angle that an incoming ray can have so that it does not escape from the fiber. (Remember to include the refraction at the entrance of the fiber.)

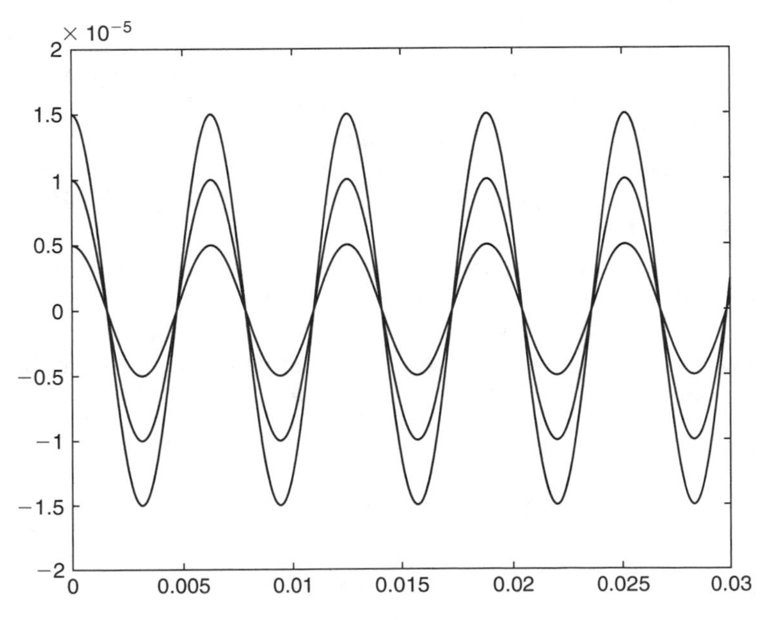

FIGURE 8.5
Traces of rays, originally parallel to the fiber's axis, when propagating inside an optical fiber.

8.11 Transfer Matrices*

In the example of Section 8.10.3 (Dynamics of the trapping of an optical ray in an optical fiber), we actually solved by matrix technique the differential equation of geometrical optics describing the propagation of a ray in a fiber. This technique can be generalized to the solution of many important differential equations of interest to electrical engineers. This technique is often referred to in the literature as the transfer matrix method. In particular, this technique has been very useful when considering the extremely important problems of the propagation of EM fields in media with position-dependent index of refraction. We will illustrate this technique by considering a number of applications for cases with one or more discontinuities.

Maxwell's equation describing the time-independent function representing a cw electric field propagating in a 1-D geometry is given by

$$\frac{d^2E}{dz^2} + \varepsilon\frac{\omega^2}{c^2}E = 0 \tag{8.161}$$

where ω is the angular frequency of the cw field and ε the dielectric constant of the medium (which may be frequency dependent). The dielectric constant and the index of refraction are related through $\varepsilon = n^2$. The complex form of the electrical field is $(E\exp(j\omega t))$.

If the index of refraction is z-independent, the general solution of the above equation is

$$E = A\exp(-jkz) + B\exp(jkz) \tag{8.162}$$

where the wave vector $k = (n\omega/c)$. The first term on the RHS is referred to as the forward wave and the second term as the backward wave. (In this geometry, the incident wave is assumed to be traveling from left to right.)

Our first order of business is to find the relationship between the corresponding (A, B)'s if two materials with unequal indices of refraction with a sharp boundary occupied the domain of space of interest. Let us, for the sake of definiteness, denote the index of refraction of the material on the left by n_1 and that on the right by n_2, and let the interface between the materials be at $z = l$.

For a plane-wave field normally incident on the interface, the continuity equations respectively for the electric and magnetic fields require that both E and dE/dz be continuous at the interface, i.e.,

$$E(z = l^-) = E(z = l^+) \tag{8.163}$$

and

$$\left.\frac{dE}{dz}\right|_{z=l^-} = \left.\frac{dE}{dz}\right|_{z=l^+} \tag{8.164}$$

* The asterisk indicates more advanced material that may be skipped in a first reading.

These boundary conditions lead to the following relations between (A_1, B_1) and (A_2, B_2):

$$A_1 \exp(-jk_1 l) + B_1 \exp(jk_1 l) = A_2 \exp(-jk_2 l) + B_2 \exp(jk_2 l) \tag{8.165a}$$

$$-jk_1 A_1 \exp(-jk_1 l) + jk_1 B_1 \exp(jk_1 l) = -jk_2 A_2 \exp(-jk_2 l) + jk_2 B_2 \exp(jk_2 l) \tag{8.165b}$$

In matrix form, the above two equations can be written as

$$\begin{pmatrix} \exp(-jk_1 l) & \exp(jk_1 l) \\ -jk_1 \exp(-jk_1 l) & jk_1 \exp(jk_1 l) \end{pmatrix} \begin{pmatrix} A_1 \\ B_1 \end{pmatrix} = \begin{pmatrix} \exp(-jk_2 l) & \exp(jk_2 l) \\ -jk_2 \exp(-jk_2 l) & jk_2 \exp(jk_2 l) \end{pmatrix} \begin{pmatrix} A_2 \\ B_2 \end{pmatrix} \tag{8.166}$$

Multiplying both sides by the inverse of the matrix on the left, we get

$$\begin{pmatrix} A_1 \\ B_1 \end{pmatrix} = \frac{1}{2k_1} \begin{pmatrix} (k_1 + k_2)\exp(j(k_1 - k_2)l) & (k_1 - k_2)\exp(j(k_1 + k_2)l) \\ (k_1 - k_2)\exp(-j(k_1 + k_2)l) & (k_1 + k_2)\exp(-j(k_1 - k_2)l) \end{pmatrix} \begin{pmatrix} A_2 \\ B_2 \end{pmatrix} \tag{8.167}$$

If we define the transfer matrix $\mathbf{M}(1 \to 2)$ as

$$\mathbf{M}(1 \to 2) = \frac{1}{2k_1} \begin{pmatrix} (k_1 + k_2)\exp(j(k_1 - k_2)l) & (k_1 - k_2)\exp(j(k_1 + k_2)l) \\ (k_1 - k_2)\exp(-j(k_1 + k_2)l) & (k_1 + k_2)\exp(-j(k_1 - k_2)l) \end{pmatrix} \tag{8.168}$$

using matrix multiplication, we can generalize the matrix relation between coefficients to include as many interfaces as exist in the problem and

$$\mathbf{MT}(1 \to N) = \mathbf{M}(1 \to 2)\mathbf{M}(2 \to 3) \cdots \mathbf{M}(N - 1 \to N) \tag{8.169}$$

where

$$\begin{pmatrix} A_1 \\ B_1 \end{pmatrix} = \mathbf{MT}(1 \to N) \begin{pmatrix} A_N \\ B_N \end{pmatrix} \tag{8.170}$$

and where each of the transfer matrices is evaluated at the coordinate of the corresponding interface.

Example 8.18

Consider the physical configuration where from the left a plane wave propagating in vacuum is normally incident on a semi-infinite glass medium ($n = 1.5$). What is the reflection coefficient and reflectance of this wave due to the presence of the interface?

Solution: Let us choose $z = 0$ at the vacuum–glass interface. On the left is vacuum and on the right is glass. The index of refraction of vacuum is 1.

Let the magnitude of the incoming wave be A, then

$$A_1 = A, \quad B_1 = rA, \quad A_2 = tA, \quad B_2 = 0 \tag{8.171}$$

where the coefficients r and t are respectively called the reflection and transmission coefficients. We assumed that there was no backward wave on the right, i.e., in the semi-infinite glass medium since there was nothing on the right to cause reflection of the wave after it enters the glass medium.

The matrix equation relating the different quantities is then

$$\begin{pmatrix} A \\ rA \end{pmatrix} = \frac{1}{2}\begin{pmatrix} (1+n) & (1-n) \\ (1-n) & (1+n) \end{pmatrix}\begin{pmatrix} tA \\ 0 \end{pmatrix} \tag{8.172}$$

from which we deduce that

$$t = \frac{2}{(1+n)} \quad \text{and} \quad r = \frac{(1-n)}{(1+n)} \tag{8.173}$$

The reflectance is the ratio of the intensities of the reflected wave to the incoming wave and is given by

$$R = |r|^2 = \left(\frac{1-n}{1+n}\right)^2 = 4\% \quad \text{for glass } (n_{\text{glass}} = 1.5) \tag{8.174}$$

□

EXAMPLE 8.19

In a number of applications, one wishes to suppress any reflection from an interface surface for a specific frequency (or color). The technique to achieve this task is the antireflection coating. A thin coat of a transparent material is deposited on the surface of a substrate. Our goal here is to find the index of refraction of the coating material and the thickness of this coat that would make the reflectance from the composite coat–substrate system zero at a given frequency (color).

Solution: We have a situation here with two interfaces, the vacuum–coating interface and the coating–substrate interface. Assume for simplicity that the electromagnetic wave is coming from the left (i.e., from vacuum). Let the vacuum–coat interface be at $z = 0$, then the coat–substrate interface is at d, the thickness of the coat. The two transfer matrices are given by

$$\mathbf{M1} = \frac{1}{2}\begin{pmatrix} (1+n_c) & (1-n_c) \\ (1-n_c) & (1+n_c) \end{pmatrix} \tag{8.175}$$

and

$$\mathbf{M2} = \frac{1}{2n_c}\begin{pmatrix} (n_c + n)\exp(j(n_c - n)\bar{d}) & (n_c - n)\exp(j(n_c + n)\bar{d}) \\ (n_c - n)\exp(-j(n_c + n)\bar{d}) & (n_c + n)\exp(-j(n_c - n)\bar{d}) \end{pmatrix} \quad (8.176)$$

where n_c is the index of refraction of the coating material, n the index of refraction of the substrate, and $\bar{d} = (\omega d/c)$. The total transfer matrix is

$$\mathbf{MT} = \mathbf{M1} * \mathbf{M2} \quad (8.177)$$

The reflection coefficient is obtained in the same manner as in the previous example and is given by

$$r = \frac{(\mathrm{MT})_{21}}{(\mathrm{MT})_{11}} \quad (8.178)$$

Consequently if we find the conditions for which the numerator (i.e., $(\mathrm{MT})_{21}$) is zero, we have found the conditions for which the reflection coefficient is zero. This matrix element is given by

$$(MT)_{21} = \frac{1}{2n_c}\exp(-jn_c\bar{d})[(n_c - n_c n)\cos(n_c\bar{d}) + j(n - n_c^2)\sin(n_c\bar{d})] \quad (8.179)$$

This quantity is identically zero if the square bracket (a complex number) is zero. Equating both the real and imaginary parts of this quantity to zero, the following two conditions are obtained:

$$n_c = \sqrt{n} \quad \text{and} \quad \bar{d} = \frac{(2m+1)\pi}{2n_c} \quad \text{where } m \text{ is an integer} \quad (8.180)$$

□

Homework Problem

Pb. 8.35 The single film coat in the previous example succeeds in reducing reflectance to zero for a specific frequency. In many practical applications, one may desire to produce an antireflecting coating that would reduce reflectance over a wide band of frequencies. This can be usually achieved through multiple layers coating. In this problem, we study a system that succeeds in reducing reflectance over almost all the frequencies of visible light.

The coating consists here of three layers, that we will denote, beginning with the layer closest to the vacuum, by the indices 1, 2, and 3. Calling

(cont'd.)

Homework Problem *(cont'd.)*

the reference wavelength $\lambda_0 = 550\,\text{nm}$, the characteristics of the different layers are as follows:

$$n_1 = 1.38, \quad d_1 = \frac{\lambda_0}{4n_1}, \quad n_2 = 2.2, \quad d_2 = \frac{\lambda_0}{2n_2}, \quad n_3 = 1.7, \quad d_3 = \frac{\lambda_0}{4n_3}$$

where d_1, d_2, and d_3 are respectively the thickness of the different films.

1. Compute the transfer matrices at the four interfaces (assume $n = 1.52$ for the glass substrate here).
2. Plot the reflectance in the range $400\,\text{nm} \le \lambda \le 800\,\text{nm}$ if only one film was present, $n_1 = \sqrt{n}$ and $d_1 = \lambda_0/4n_1$.
3. Plot the reflectance in the same frequency range, in the presence of the three film layers specified above.

8.12 Covariance Matrices*

In Chapter 1, we introduced the probability density function for a single random variable. A random vector is a vector of random variables. For example, the electric conductivity, the specific heat, the tensile strength, the optical reflectivity at a certain wavelength, and the mass density of a certain metallic object found in an excavation, in the absence of other information, can be represented by a random vector of length 5.

Henceforth let us assume a generic random vector including N random variables:

$$\mathbf{X} = \begin{bmatrix} X_1 \\ X_2 \\ \vdots \\ X_N \end{bmatrix} \tag{8.181}$$

The joint density function describing the behavior of such a random vector has the general form

$$f_{\mathbf{X}}(\mathbf{x}) = f_{\mathbf{X}}(x_1, x_2, \ldots, x_N) \tag{8.182}$$

* The asterisk indicates more advanced material that may be skipped in a first reading.

This joint density function satisfies the following conditions

(i) $f_{\mathbf{X}}(x_1, x_2, \ldots, x_N) \geq 0$

(ii) $\int_{-\infty}^{\infty} dx_N \int_{-\infty}^{\infty} dx_{N-1} \cdots \int_{-\infty}^{\infty} dx_1 \, f_{\mathbf{X}}(x_1, x_2, \ldots, x_N) = 1$

(iii) The marginal density function for any of its random variables X_i is obtained by integrating the joint density function over all other variables, i.e.,

$$f_{X_i}(x_i) = \int_{-\infty}^{\infty} dx_N \int_{-\infty}^{\infty} dx_{N-1} \cdots \int_{-\infty}^{\infty} dx_{i+1} \int_{-\infty}^{\infty} dx_{i-1} \cdots \int_{-\infty}^{\infty} dx_1 \, f_{\mathbf{X}}(x_1, x_2, \ldots, x_N) \tag{8.183}$$

In case that the multivariate joint density function is a Gaussian, it takes the form

$$f_{\mathbf{X}}(x_1, x_2, \ldots, x_N) = \frac{1}{(2\pi)^{N/2}(\det(\mathbf{K}))^{1/2}} \exp\left\{-\frac{1}{2}[\mathbf{x}-\mathbf{m}]^T \mathbf{K}^{-1}[\mathbf{x}-\mathbf{m}]\right\} \tag{8.184}$$

where the mean vector $\mathbf{m}$ and the covariance matrix $\mathbf{K}$ are given by

$$m_i = E[X_i] = \int_{-\infty}^{\infty} dx_N \int_{-\infty}^{\infty} dx_{N-1} \cdots \int_{-\infty}^{\infty} dx_1 \, x_i \, f_{\mathbf{X}}(x_1, x_2, \ldots, x_N) \tag{8.185}$$

$$\begin{aligned} K_{ij} &= \mathrm{Cov}(X_i, X_j) = E[(X_i - E[X_i])(X_j - E[X_j])] \\ &= \int_{-\infty}^{\infty} dx_N \int_{-\infty}^{\infty} dx_{N-1} \cdots \int_{-\infty}^{\infty} dx_1 \, (x_i - E[X_i])(x_j - E[X_j]) \, f_{\mathbf{X}}(x_1, x_2, \ldots, x_N) \end{aligned} \tag{8.186}$$

Physically, the covariance between two variables measures the tendency of these variables to vary together. In case that $\mathrm{Cov}(X, Y) = 0$, the variables X and Y are said to be uncorrelated.

The standard deviation of the random variable X_i can be expressed as

$$\sigma_{X_i}^2 = \mathrm{Var}(X_i) = \mathrm{Cov}(X_i, X_i) \tag{8.187}$$

Since the covariance between two random variables is not dimensionless, it is often convenient to use the dimensionless correlation coefficient matrix defined through

$$C_{ij} = C(X_i, X_j) = \frac{\mathrm{Cov}(X_i, X_j)}{\sigma_{X_i}\sigma_{X_j}} \tag{8.188}$$

The magnitude of each of the elements of the matrix $\mathbf{C}$ is smaller or equal to one.

In most cases of practical interest, the mean vector and the covariance matrix are determined from experimental or historical data. In the next two subsections, we seek to construct the algorithm for obtaining these quantities, and find the technique to uncorrelate the components of this random vector.

8.12.1 Parametric Estimation

Let us consider the situation where data for different random variables for each of different times or different samples are given. We desire to find the mean vector and the covariance matrix for the associated random vector.

The algorithm is obtained as follows:

(i) Construct **S** the data matrix, ordered such that its row index corresponds to the sample number, and its column index refers to the random variable index.

(ii) The components of the mean vector are obtained from

$$m_i = \frac{1}{N}\sum_{j=1}^{N} S_{ji} \tag{8.189}$$

where N is the number of samples.

(iii) The deviation from the mean matrix is

$$\mathbf{\Delta} = \mathbf{S} - \mathbf{M} \tag{8.190}$$

where

$$\mathbf{M} = \mathbf{ones}(N, 1) * \mathbf{m} \tag{8.191}$$

(iv) The covariance matrix **K** is then

$$\mathbf{K} = \frac{1}{(N-1)}\mathbf{\Delta}^T\mathbf{\Delta} \tag{8.192}$$

The above normalization is consistent with the definition of the standard deviation given in Chapter 1.

(v) The correlation coefficient matrix elements are given by

$$C_{ij} = \frac{K_{ij}}{\sqrt{K_{ii}\,K_{jj}}} \tag{8.193}$$

EXAMPLE 8.20

The following table gives the percentage net returns from stocks A and B over the last 20 quarters:

Returns	Q1	Q2	Q3	Q4	Q5	Q6	Q7	Q8	Q9	Q10
A	5	4	3	4	5	6	4	5	6	5
B	7	9	8	7	4	−2	8	3	2	4

Returns	Q11	Q12	Q13	Q14	Q15	Q16	Q17	Q18	Q19	Q20
A	7	4	5	6	3	5	4	3	4	3
B	−2	7	4	−3	9	8	6	9	8	9

Compute the average return and standard deviation for each stock, and the covariance matrix, and the correlation coefficient matrix for the system.

Solution: Editing and entering the following program:

```
V1=[5;4;3;4;5;6;4;5;6;5;7;4;5;...
    6;3;5;4;3;4;3];
V2=[7;8;8;7;4;-2;8;3;2;4;-2;7;4;...
    -3;9;8;6;9;8;9];
N=length(V1);
S=[V1 V2];
m=(1/N)*sum(S)
M=ones(N,1)*m;
Dev=S-M;
K=(1/(N-1))*Dev'*Dev;
     for i=1:2
           for j=1:2
                      sigma2(i,j)=sqrt(K(i,i)*K(j,j));
                      C(i,j)=K(i,j)/sigma2(i,j);
           end
     end
sigma=[sqrt(K(1,1))    sqrt(K(2,2))]
K
C
```

returns

```
m =
     4.5500    5.2000
sigma =
     1.1459    3.8607
K =
     1.3132    -3.9053
     -3.9053   14.9053
C =
      1.0000   -0.8827
     -0.8827    1.0000
```

The interpretations of the above results are as follows:

(i) A negative off-diagonal element of the **C** matrix with magnitude close to one indicates that the two stocks have the tendencies to move in opposite directions.
(ii) Although the average (mean) return of stock B is higher than the average (mean) return of stock A, investing in B is far more riskier (i.e., larger standard deviation). □

Homework Problem

Pb. 8.36 Portfolio allocation theory concerns itself with allocating assets among different possible investments. A portfolio analysis consists of computing, using historical data, the anticipated average return on investment and the risk factor associated with different weight factors for the different investments.

a. Assuming the two assets case described below, compute the average return and the standard deviation of the portfolio, for different weights in the allocation. (*Hint*: The sum of the weights is equal to 1.)
Stock A: Average annual return: 10%; $\sigma_A = 15\%$
Stock B: Average annual return: 20%; $\sigma_B = 30\%$
Correlation coefficient $C_{AB} = -0.9$

b. Compute analytically the allocation configuration that is the least risky (i.e., that gives the smallest value for the portfolio standard deviation). Compare your result with that obtained numerically in (a).

Example 8.21

Develop an algorithm that will determine the weight factors, which minimize the risk for a portfolio with n assets.

Solution: The variance of the portfolio's return can be written in the following matrix form:

$$\text{Var}(r_p) = (w_1 \; w_2 \; \cdots \cdots \; w_n) \begin{pmatrix} K_{11} & K_{12} & \cdots & \cdots & K_{1n} \\ K_{21} & K_{22} & \cdots & \cdots & K_{2n} \\ \vdots & & \ddots & & \vdots \\ \vdots & & & \ddots & \vdots \\ K_{n1} & K_{n2} & \cdots & \cdots & K_{nn} \end{pmatrix} \begin{pmatrix} w_1 \\ w_2 \\ \vdots \\ \vdots \\ w_n \end{pmatrix} \tag{8.194}$$

where r_p is the return on the portfolio, the w's are the weights of the different assets and $\mathbf{K}$ is the covariance matrix for the different assets. Since the covariance matrix is symmetric, we can write the above expression in the form

$$\mathrm{Var}(r_p) = \sum_{i=1}^{n} K_{ii} w_i^2 + 2 \sum_{i=1}^{n}{}' \sum_{j=1}^{n} K_{ij} w_i w_j \tag{8.195}$$

where the prime in the summation sign in the second term indicates that we are excluding the $i = j$ term from this summation.

Minimizing the risk factor subject to the condition that

$$\sum_{i=1}^{n} w_i = 1 \tag{8.196}$$

can be solved by the Lagrange multiplier technique described in Section 5.5. This leads to the system of equations:

$$\begin{aligned}
&K_{11}\, w_1 + K_{12}\, w_2 + K_{13}\, w_3 + \cdots + K_{1n}\, w_n = \lambda \\
&K_{21}\, w_1 + K_{22}\, w_2 + K_{23}\, w_3 + \cdots + K_{2n}\, w_n = \lambda \\
&\vdots \\
&K_{n1}\, w_1 + K_{n2}\, w_2 + K_{n3}\, w_3 + \cdots + K_{nn}\, w_n = \lambda \\
&w_1 + w_2 + w_3 + \cdots + w_n = 1
\end{aligned} \tag{8.197}$$

The solution of this system is

$$w_i = \frac{\sum_{j=1}^{n} (\mathbf{K}^{-1})_{ij}\, (\mathbf{ones(n, 1)})_j}{\sum_{i=1}^{n} \sum_{j=1}^{n} (\mathbf{K}^{-1})_{ij}\, (\mathbf{ones(n, 1)})_j} \tag{8.198}$$

where the matrix **ones (n, 1)** is the column vector with n rows and 1 column and all elements are equal to 1.

In MATLAB the commands to obtain the w array, having previously computed the matrix **K**, will simply be

```
non_norm_w=K\ones(n,1);
w=non_norm_w/sum(non_norm_w)
```
□

8.12.2 Karhunen–Loeve Transform

The purpose of the Karhunen–Loeve transform is to form from the random vector components new variables which covariance matrix off-diagonal elements are 0, i.e., the new variables obtained through an orthogonal transformation over the old variables will be uncorrelated.

To find these transforms, recall that the covariance matrix is Hermitian. Therefore, its eigenvalues are real and its eigenvectors are mutually orthogonal and the matrix can be decomposed in the following form:

$$\mathbf{K} = \mathbf{VDV}^{-1} \tag{8.199}$$

where the matrix **V** is formed by the eigenvectors of the matrix **K**, and **D** is the diagonal matrix formed by the eigenvalues of the matrix **K**.

The new variables are obtained by a rotation of the original variables vector. The rotation matrix is the matrix $\mathbf{V}^{-1}$. The covariance matrix of the new variables is diagonal and its elements are the same as those of the diagonal matrix.

Example 8.22

In a series of $N = 20$ experiments, the following data was obtained for the measurement of the variables X_1 and X_2:

#	X_1	X_2
01	0.4295	0.9898
02	2.2615	5.8507
03	1.9964	3.5261
04	4.1779	13.1570
05	0.8081	2.2250
06	2.3042	9.5246
07	−0.1338	0.4045
08	3.5807	7.9035
09	0.1166	0.3666
10	2.2093	4.1771
11	1.2087	1.5689
12	4.5514	11.2349
13	3.5091	9.1240
14	4.1300	11.9860
15	2.2164	2.9652
16	0.4517	1.2261
17	5.3787	13.4331
18	2.8916	8.5232
19	2.4861	4.7120
20	3.7007	6.9165

1. Compute the covariance matrix for the vector **X**.
2. Compute the eigen **V** and **D** matrices of the covariance matrix.
3. Verify that the covariance matrix for the transformed variables **Y** is diagonal and that its elements, which represent the variance of the transformed uncorrelated variables, are equal to those of the elements of the matrix **D**.

Solution: Entering the following program:

```
X1 =[0.4295;2.2615;1.9964;4.1779;0.8081; ...
     2.3042;-0.1338;3.5807;0.1166;2.2093; ...
     1.2087;4.5514;3.5091;4.1300;2.2164; ...
     0.4517;5.3787;2.8916;2.4861;3.7007];
X2 =[0.9898;5.8507;3.5261;13.1570;2.2250; ...
     9.5246;0.4045;7.9035;0.3666;4.1771; ...
     1.5689;11.2349;9.1240;11.9860;2.9652; ...
     1.2261;13.4331;8.5232;4.7120;6.9165];
N=length(X1);
SX=[X1 X2];
mX=(1/N)*sum(SX);
MX=ones(N,1)*mX;
DevX=SX-MX;
KX=(1/(N-1))*DevX'*DevX
[V,D]=eig(KX)
y=inv(V)*SX';
SY=y';
mY=(1/N)*sum(SY);
MY=ones(N,1)*mY;
DevY=SY-MY;
KY=(1/(N-1))*DevY'*DevY
```

returns

```
KX =
      2.5128      6.5058
      6.5058     19.3619
V =
     -0.9464     -0.3229
      0.3229     -0.9464
D =
      0.2932           0
           0     21.5815
KY =
      0.2932     -0.0000
     -0.0000     21.5815
```

□

8.13 MATLAB Commands Review

det Compute the determinant of a matrix.
expm Computes the matrix exponential.
eye Identity matrix.

inv	Find the inverse of a matrix.
ones	Matrix with all elements equal to 1.
polyfit	Fit polynomial to data.
triu	Extract upper triangle of a matrix.
tril	Extract lower triangle of a matrix.
zeros	Matrix with all elements equal to 0.
[V,D]=eig(M)	Finds the eigenvalues and eigenvectors of a matrix.
.'	Applied to a matrix gives its transpose.
'	Applied to a matrix gives its complex conjugate transpose.

9

Transformations

The theory of transformations concerns itself with changes in the coordinates and shapes of objects upon the action of geometrical operations, dynamical boosts, or other operators. In this chapter, we deal only with linear transformations, using examples from both plane geometry and relativistic dynamics (space–time geometry). We also show how transformation techniques play an important role in image processing, and in generating iterative constructs. We formulate both the problems and their solutions in the language of matrices. Matrices are still denoted by boldface type and matrix multiplication by an asterisk.

9.1 Two-Dimensional Geometric Transformations

We first concern ourselves with the operations of inversion about the origin of axes, reflection about the coordinate axes, rotation around the origin, scaling, and translation. But prior to going into the details of these transformations, we need to learn how to draw closed polygonal figures in MATLAB so that we can implement and graph the different cases.

9.1.1 Construction of Polygonal Figures

Consider a polygonal figure whose vertices are located at the points:

$$(x_1, y_1),\ (x_2, y_2),\ \ldots,\ (x_n, y_n)$$

The polygonal figure can then be thought of as line segments (edges) connecting the vertices in a given order, including the edge connecting the last point to the initial point to ensure that we obtain a closed figure. The implementation of the steps leading to the drawing of the figure are as follows:

1. Label all vertex points.
2. Label the path you follow.

3. Construct a $(2 \otimes (n + 1))$ matrix, the **G** matrix, where the elements of the first row consist of the ordered $(n + 1)$-tuplet, $(x_1, x_2, x_3, \ldots, x_n, x_1)$, and those of the second row consist of the corresponding y-coordinates $(n + 1)$-tuplet.
4. Plot the second row of **G** as function of its first row.

EXAMPLE 9.1

Plot the trapezoid whose vertices are located at points (2, 1), (6, 1), (5, 3), and (3, 3).

Solution: Enter and execute the following commands:

```
G=[2 6 5 3 2; 1 1 3 3 1];
plot(G(1,:),G(2,:))
```

To ensure that the exact geometrical shape is properly reproduced, remember to instruct your computer to choose the axes such that you have equal x- and y-ranges and an aspect ratio of 1. If you would like to add any text anywhere in the figure, use the command **gtext**. □

9.1.2 Inversion about the Origin and Reflection about the Coordinate Axes

We concern ourselves here with inversion with respect to the origin and with reflection about the x- or y-axis. Inversion about other points or reflection about other than the coordinate axes can be deduced from a composition of the present transformations and those discussed later.

- The inversion about the origin changes the coordinates as follows:

$$\begin{aligned} x' &= -x \\ y' &= -y \end{aligned} \tag{9.1}$$

In the matrix form, this transformation can be represented by

$$\mathbf{P} = \begin{bmatrix} -1 & 0 \\ 0 & -1 \end{bmatrix} \tag{9.2}$$

- For the reflection about the x-axis, denoted by $\mathbf{P}_x$, and the reflection about the y-axis, denoted by $\mathbf{P}_y$, the transformation matrices are given by

$$\mathbf{P}_x = \begin{bmatrix} 1 & 0 \\ 0 & -1 \end{bmatrix} \tag{9.3}$$

$$\mathbf{P}_y = \begin{bmatrix} -1 & 0 \\ 0 & 1 \end{bmatrix} \tag{9.4}$$

IN-CLASS EXERCISES

Pb. 9.1 Using the trapezoid of Example 9.1, obtain all the transformed **G**'s as a result of the action of each of the three transformations defined in Eq. (9.2) through Eq. (9.4), and plot the transformed figures on the same graph.

Pb. 9.2 In drawing the original trapezoid, we followed the counterclockwise direction in the sequencing of the different vertices. What is the sequencing of the respective points in each of the transformed **G**'s?

Pb. 9.3 Show that the quantity $(x^2 + y^2)$ is invariant under the action of $\mathbf{P}_x$, $\mathbf{P}_y$, or $\mathbf{P}$.

9.1.3 Rotation around the Origin

The new coordinates of a point in the x–y plane rotated by an angle θ around the z-axis can be directly derived through some elementary trigonometry. Here, instead, we derive the new coordinates using results from the complex numbers chapter (Chapter 6). Recall that every point in a 2-D plane represents a complex number, and multiplication by a complex number of modulus 1 and argument θ results in a rotation of angle θ of the original point. Therefore

$$z' = ze^{j\theta}$$

$$x' + jy' = (x + jy)(\cos(\theta) + j\sin(\theta)) \tag{9.5}$$

$$= (x\cos(\theta) - y\sin(\theta)) + j(x\sin(\theta) + y\cos(\theta))$$

Equating separately the real and the imaginary parts, we deduce the action of rotation on the coordinates of a point

$$\begin{aligned} x' &= x\cos(\theta) - y\sin(\theta) \\ y' &= x\sin(\theta) + y\cos(\theta) \end{aligned} \tag{9.6}$$

The above transformation can also be written in matrix form. That is, if the point is represented by a size 2 column vector, then the new vector is related to the old one through the following transformation:

$$\begin{bmatrix} x' \\ y' \end{bmatrix} = \begin{bmatrix} \cos(\theta) & -\sin(\theta) \\ \sin(\theta) & \cos(\theta) \end{bmatrix} \begin{bmatrix} x \\ y \end{bmatrix} = \mathbf{R}(\theta) \begin{bmatrix} x \\ y \end{bmatrix} \tag{9.7}$$

The convention for the sign of the angle is the same as that used in Chapter 6, namely, that it is measured positive when in the counterclockwise direction.

PREPARATORY EXERCISES

Using the above form for the rotation matrix, verify the following properties:

Pb. 9.4 Its determinant is equal to 1.

Pb. 9.5 $\mathbf{R}(-\theta) = [\mathbf{R}(\theta)]^{-1} = [\mathbf{R}(\theta)]^T$

Pb. 9.6 $\mathbf{R}(\theta_1) * \mathbf{R}(\theta_2) = \mathbf{R}(\theta_1 + \theta_2) = \mathbf{R}(\theta_2) * \mathbf{R}(\theta_1)$

Pb. 9.7 $(x')^2 + (y')^2 = x^2 + y^2$

Pb. 9.8 Show that $\mathbf{P} = \mathbf{R}(\theta = \pi)$. Also show that there is no rotation that can reproduce $\mathbf{P_x}$ or $\mathbf{P_y}$.

IN-CLASS EXERCISES

Pb. 9.9 Find the coordinates of the image of the point (x, y) obtained by reflection about the line $y = x$. Test your results using MATLAB.

Pb. 9.10 Find the transformation matrix corresponding to a rotation of $-\pi/3$, followed by an inversion around the origin. Solve the problem in two different ways.

Pb. 9.11 By what angle should you rotate the trapezoid so that point (6, 1) of the trapezoid of Example 9.1 is now on the y-axis?

9.1.4 Scaling

If the x-coordinate of each point in the plane is multiplied by a positive constant s_x, then the effect of this transformation is to expand or compress each plane figure in the x-direction. If $0 < s_x < 1$, the result is a compression, and if $s_x > 1$, the result is an expansion. The same can also be done along the y-axis. This class of transformations is called scaling.

The matrices corresponding to these transformations, in 2-D, are respectively,

$$\mathbf{S_x} = \begin{bmatrix} s_x & 0 \\ 0 & 1 \end{bmatrix} \tag{9.8}$$

$$\mathbf{S_y} = \begin{bmatrix} 1 & 0 \\ 0 & s_y \end{bmatrix} \tag{9.9}$$

IN-CLASS EXERCISES

Pb. 9.12 Find the transformation matrix for simultaneously compressing the x-coordinate by a factor of 2, while expanding the y-coordinate by a factor of 2. Apply this transformation to the trapezoid of Example 9.1 and plot the result.

Pb. 9.13 Find the inverse matrices for $\mathbf{S_x}$ and $\mathbf{S_y}$.

9.1.5 Translation

A translation is defined by a vector $\vec{T} = (t_x, t_y)$, and the transformation of the coordinates is given simply by

$$x' = x + t_x$$
$$y' = y + t_y \tag{9.10}$$

or written in matrix form as

$$\begin{bmatrix} x' \\ y' \end{bmatrix} = \begin{bmatrix} x \\ y \end{bmatrix} + \begin{bmatrix} t_x \\ t_y \end{bmatrix} \tag{9.11}$$

The effect of translation over the matrix **G** is described by the relation

$$\mathbf{G_T} = \mathbf{G} + \mathbf{T} \ * \ \mathbf{ones}(1, n+1) \tag{9.12}$$

where n is the number of points being translated.

IN-CLASS EXERCISE

Pb. 9.14 Translate the trapezoid of Example 9.1 by a vector of length 5 that makes an angle of 30° with the x-axis.

9.2 Homogeneous Coordinates

As we have seen in Section 9.1, inversion about the origin, reflection about the coordinate axes, rotation, and scaling are operations that can be represented by a multiplicative matrix, and therefore the composite operation of acting successively on a figure by one or more of these operations can be described by a product of matrices. The translation operation, on the other hand, is represented by an addition, and thus cannot be incorporated, as yet, into the matrix multiplication scheme; and consequently, the expression for composite operations becomes less tractable. We illustrate this situation with the following example:

EXAMPLE 9.2

Find the new **G** that results from rotating the trapezoid of Example 9.1 by a $\pi/4$ angle around the point Q $(-5, 5)$.

Solution: Because we have thus far defined the rotation matrix only around the origin, our task here is to generalize this result. We solve the problem by reducing it to a combination of elementary operations thus far defined. The strategy for solving the problem goes as follows:

1. Perform a translation to place Q at the origin of a new coordinate system.
2. Perform a $\pi/4$ rotation around the new origin using the above form for rotation.
3. Translate back the origin to its initial location.

Written in matrix form, the above operations can be written sequentially as follows:

1.
$$\mathbf{G}_1 = \mathbf{G} + \mathbf{T} \ * \ \mathbf{ones}(1, \ n+1) \tag{9.13}$$

where
$$\mathbf{T} = \begin{bmatrix} 5 \\ -5 \end{bmatrix} \tag{9.14}$$

and $n = 4$.

2.
$$\mathbf{G}_2 = \mathbf{R}(\pi / 4) * \mathbf{G}_1 \tag{9.15}$$

3.
$$\mathbf{G}_3 = \mathbf{G}_2 - \mathbf{T} \ * \ \mathbf{ones}(1, \ n+1) \tag{9.16}$$

and the final result can be written as

$$\mathbf{G}_3 = \mathbf{R}(\pi / 4) * \mathbf{G} + [(\mathbf{R}(\pi / 4) - 1) * \mathbf{T}] * \mathbf{ones}(1, n+1) \tag{9.17}$$

We can implement the above sequence of transformations through the following *script M-file*:

```
plot(-5,5,'*')
hold on
G=[2 6 5 3 2; 1 1 3 3 1];
plot(G(1,:),G(2,:),'b')
T=[5;-5];
G1=G+T*ones(1,5);
plot(G1(1,:),G1(2,:), 'r')
```

```
R=[cos(pi/4) -sin(pi/4);sin(pi/4) cos(pi/4)];
G2=R*G1;
plot(G2(1,:),G2(2,:),'g')
G3=G2-T*ones(1,5);
plot(G3(1,:),G3(2,:),'k')
axis([-12 12 -12 12])
axis square
hold off □
```

Although the above formulation of the problem is absolutely correct, the number of terms in the final expression for the image can wind up, in more involved problems, being large and cumbersome because of the existence of sums and products in the intermediate steps. Thus, the question arises: can we incorporate all the transformations discussed thus far into only multiplicative matrices?

The answer comes from an old trick that mapmakers have used successfully; namely, the technique of homogeneous coordinates. In this technique, as applied to the present case, we append to any column vector the row with value 1, that is, the point (x_m, y_m) is now represented by the column vector:

$$\begin{bmatrix} x_m \\ y_m \\ 1 \end{bmatrix} \tag{9.18}$$

Similarly in the definition of **G**, we should append to the old definition, a row with all elements being 1.

In this coordinate representation, the different transformations thus far discussed are now multiplicative and take the following forms:

$$\mathbf{P} = \begin{bmatrix} -1 & 0 & 0 \\ 0 & -1 & 0 \\ 0 & 0 & 1 \end{bmatrix} \tag{9.19}$$

$$\mathbf{P_x} = \begin{bmatrix} 1 & 0 & 0 \\ 0 & -1 & 0 \\ 0 & 0 & 1 \end{bmatrix} \tag{9.20}$$

$$\mathbf{P_y} = \begin{bmatrix} -1 & 0 & 0 \\ 0 & 1 & 0 \\ 0 & 0 & 1 \end{bmatrix} \tag{9.21}$$

$$\mathbf{S} = \begin{bmatrix} s_x & 0 & 0 \\ 0 & s_y & 0 \\ 0 & 0 & 1 \end{bmatrix} \tag{9.22}$$

$$\mathbf{R}(\theta) = \begin{bmatrix} \cos(\theta) & -\sin(\theta) & 0 \\ \sin(\theta) & \cos(\theta) & 0 \\ 0 & 0 & 1 \end{bmatrix} \tag{9.23}$$

$$\mathbf{T} = \begin{bmatrix} 1 & 0 & t_x \\ 0 & 1 & t_y \\ 0 & 0 & 1 \end{bmatrix} \tag{9.24}$$

The composite matrix of any two transformations can now be written as the product of the matrices representing the constituent transformations. Of course, this economizes on the writing of expressions and makes the calculations less prone to trivial errors originating in the expansion of products of sums.

Example 9.3

Repeat Example 9.2, but now use the homogeneous coordinates.

Solution: The following *script M-file* implements the required task:

```
plot(-5,5,'*')
hold on
G=[2 6 5 3 2; 1 1 3 3 1;1 1 1 1 1];
plot(G(1,:),G(2,:),'b')
T=[1 0 5;0 1 -5;0 0 1];
G1=T*G;
plot(G1(1,:),G1(2,:), 'r')
R=[cos(pi/4) -sin(pi/4) 0;sin(pi/4) cos(pi/4) 0;...
  0 0 1];
G2=R*G1;
plot(G2(1,:),G2(2,:),'g')
G3=inv(T)*G2;
plot(G3(1,:),G3(2,:),'k')
axis([-12 12 -12 12])
axis square
hold off
```
□

9.3 Manipulation of 2-D Images

Currently, more and more images are being stored or transmitted in digital form. What does this mean?

To simplify the discussion, consider a black and white image and assume that it has a square boundary. The digital image is constructed by the optics of the detecting system (i.e., the camera) to form on a plane containing a 2-D array of detectors, instead of the traditional photographic film. Each of these detectors, called a pixel (picture element), measures the intensity of light falling on it. The image is then represented by a matrix having the same size as the detectors' 2-D array structure, and such that the value of each of the matrix elements is proportional to the intensity of the light falling on the associated detector element. Of course, the resolution of the picture increases as the number of arrays increases.

9.3.1 Geometrical Manipulation of Images

Having the image represented by a matrix, it is now possible to perform all kinds of manipulations on it in MATLAB. For example, we could flip it in the left/right direction (**`fliplr`**), or in the up/down direction (**`flipud`**), or rotate it by 90° (**`rot90`**), or for that matter transform it by any matrix transformation. In the remainder of this section, we explore some of the techniques commonly employed in the handling and manipulation of digital images.

Let us explore and observe the structure of a matrix subjected to the above elementary transformations. For this purpose, execute and observe the outputs from each of the following commands:

```
M=(1/25)*[1 2 3 4 5;6 7 8 9 10;11 12 13 14 15;...
  16 17 18 19 20;21 22 23 24 25]
lrM=fliplr(M)
udM=flipud(M)
Mr90=rot90(M)
```

A careful examination of the resulting matrix elements will indicate the general features of each of these transformations. You can also see in a visually more suggestive form how each of the transformations changed the image of the original matrix, if we render the image of **M** and its transform in false colors, that is, we assign a color to each number.

To perform this task, choose the **`colormap(hot)`** command to obtain the images. In this mapping, the program assigns a color to each pixel, varying from black-red-yellow-white, depending on the magnitude of the intensity at the corresponding detector.

Enter, in the following sequence, each of the following commands and at each step note the color distributions of the image:

```
colormap(hot)
imagesc(M, [0 1])
imagesc(lrM, [0 1])
imagesc(udM, [0 1])
imagesc(Mr90, [0 1])
```

The command **imagesc** produces an intensity image of a data matrix that spans a given range of values.

9.3.2 Digital Image Processing

A typical problem in digital image processing involves the analysis of the raw data of an image that was subject, during acquisition, to a blur due to the movement of the camera or to other sources of noise. An example of this situation occurs in the analysis of aerial images; the images are blurred due, *inter alia*, to the motion of the plane while the camera shutter is open. The question is, can we do anything to obtain a crisper image from the raw data if we know the speed and altitude of the plane when it took the photograph?

The answer is affirmative. We consider for our example the photograph of a rectangular board. Construct this image by entering

```
N=64;
A=zeros(N,N);
A(15:35,15:45)=1;
colormap(gray);
imagesc(A, [0 1])
```

where (N, N) is the size of the image (here, $N = 64$).

Now assume that the camera that took the image had moved while the shutter was open by a distance that would correspond in the image plane to L pixels. What will the image look like now? (See Figure 9.1.)

The blurring operation was modeled here by the matrix **B**. The blurred image is simulated through the matrix product:

$$\mathbf{A1} = \mathbf{A} * \mathbf{B} \tag{9.25}$$

where **B**, the blurring matrix, is given by the following Toeplitz matrix:

```
L=9;
B=toeplitz([ones(L,1);zeros(N-L,1)],...
  [1;zeros(N-1,1)])/L;
```

Here, the blur length was $L = 9$, and the blurred image **A1** was obtained by executing the following commands:

```
A1=A*B;
imagesc(A1, [0 1])
```

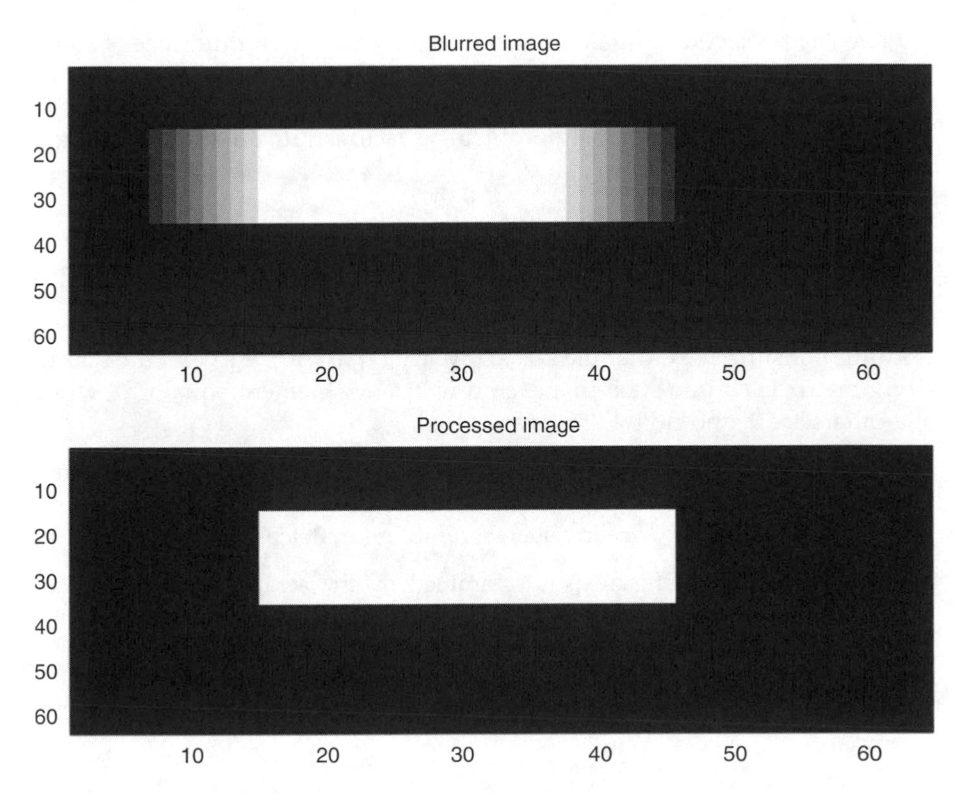

FIGURE 9.1
The raw and processed images of a rectangular board photographed from a moving plane. Top panel: raw (blurred) image. Bottom panel: processed image.

To bring back the unblurred picture, simply multiply the matrix **A1** on the right by **inv(B)** and obtain the original image.

In practice, one is given the blurred image and asked to reconstruct it while correcting for the blur. What should one do?

1. Compute the blur length from the plane speed and height.
2. Construct the Toeplitz matrix, and take its inverse.
3. Apply the inverse of the Toeplitz matrix to the blurred image matrix, obtaining the processed image.

9.3.3 Encrypting an Image

If for any reason, two individuals desire to exchange an image but want to keep its contents only to themselves, they may agree beforehand on a scrambling matrix that the first individual applies to scramble the sent image, while the second individual applies the inverse of the scramble matrix to unscramble the received image.

Given that an average quality image currently has a minimum size of about (1000×1000) pixels, reconstructing the scrambling matrix, if chosen cleverly, would be inaccessible except to the most powerful and specialized computers.

The purpose of the following problems is to illustrate an efficient method for building a scrambling matrix.

In-Class Exercises

Assume, for simplicity, that the 2-D array size is (10×10), and that the scrambling matrix is chosen such that each row has one element equal to 1, while the others are 0, and no two rows are equal.

Pb. 9.15 For the (10×10) matrix dimension, how many possible scrambling matrices **S**, constructed as per the above prescription, are there? If the matrix size is (1000×1000), how many such scrambling matrices will there be?

Pb. 9.16 An original figure was scrambled by the scrambling matrix **S** to obtain the image shown in Figure 9.2. The matrix **S** is (10×10) and has all its elements equal to zero, except $S(1, 6) = S(2, 3) = S(3, 2) = S(4, 1) = S(5, 9) = S(6, 4) = S(7, 10) = S(8, 7) = S(9, 8) = S(10, 5) = 1$. Find the original image.

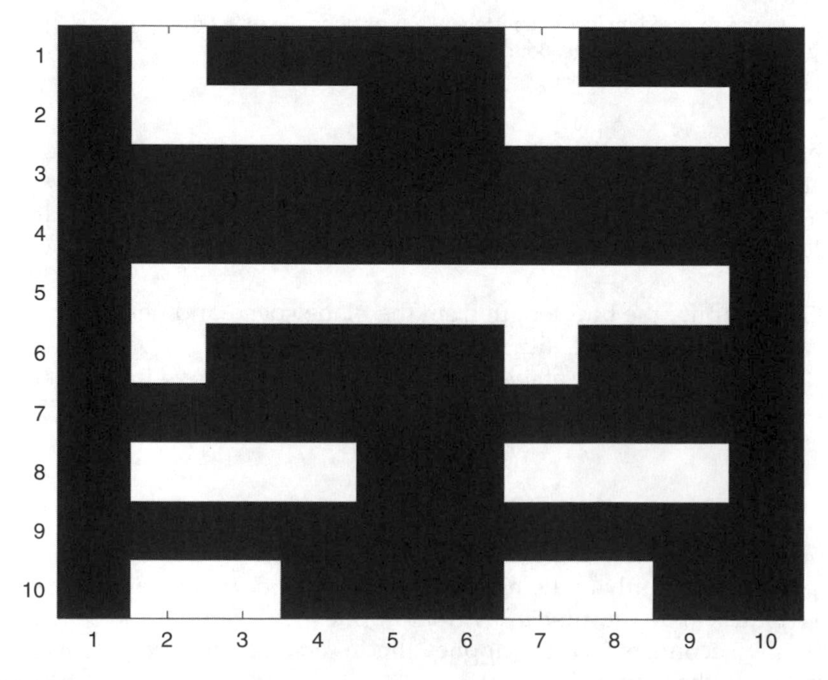

FIGURE 9.2
Scrambled image of Pb. 9.16.

9.4 Lorentz Transformation*

9.4.1 Space–Time Coordinates

Einstein's theory of special relativity studies the relationship of the dynamics of a system, if described in two coordinate systems moving with constant speed from each other. The theory of special relativity does not assume, as classical mechanics does, that there exists an absolute time common to all coordinate systems. It associates with each coordinate system a four-dimensional space (three space coordinates and one time coordinate). The theory of special relativity associates a space–time transformation to go between two coordinate systems moving uniformly with respect to each other. Each real point event (e.g., the arrival of a light flash on a screen) will be measured in both systems. If we distinguish by primes the data of the second observer from those of the first, then the first observer will ascribe to the event the coordinates (x, y, z, t), while the second observer will ascribe to it the coordinates (x', y', z', t'); that is, there is no absolute time. The Lorentz transformation gives the rules for going from one coordinate system to the other.

Assuming that the velocity v between the two systems has the same direction as the positive x-axis and where the x-axis direction continuously coincides with that of the x'-axis; and furthermore, that the origin of the spatial coordinates of one system at time $t = 0$ coincides with the origin of the other system at time $t' = 0$, Einstein, on the basis of two postulates, derived the following transformation relating the coordinates of the two systems:

$$x' = \frac{x - vt}{\sqrt{1 - \dfrac{v^2}{c^2}}}, \quad y' = y, \quad z' = z, \quad t' = \frac{t - \dfrac{v}{c^2}x}{\sqrt{1 - \dfrac{v^2}{c^2}}} \tag{9.26}$$

where c is the velocity of light in vacuum. The derivation of these formulae are detailed in electromagnetic theory or modern physics courses and are not the subject of discussions here. Our purpose here is to show that knowing the above transformations, we can deduce many interesting physical observations as a result thereof.

PREPARATORY EXERCISE

Pb. 9.17 Show that, upon a Lorentz transformation, we have the equality:

$$x'^2 + y'^2 + z'^2 - c^2t'^2 = x^2 + y^2 + z^2 - c^2t^2$$

This is referred to as the Lorentz invariance of the norm of the space–time four-vectors. What is the equivalent invariant in 3-D Euclidean geometry?

*The asterisk indicates more advanced material that may be skipped in a first reading.

If we rename our coordinates such that

$$x_1 = x, \quad x_2 = y, \quad x_3 = z, \quad x_4 = jct \tag{9.27}$$

the Lorentz transformation takes the following matricial form:

$$\mathbf{L}_\beta = \begin{bmatrix} \frac{1}{\sqrt{1-\beta^2}} & 0 & 0 & \frac{j\beta}{\sqrt{1-\beta^2}} \\ 0 & 1 & 0 & 0 \\ 0 & 0 & 1 & 0 \\ -\frac{j\beta}{\sqrt{1-\beta^2}} & 0 & 0 & \frac{1}{\sqrt{1-\beta^2}} \end{bmatrix} \tag{9.28}$$

where $\beta = \frac{v}{c}$, the relations that were given earlier relating the primed and unprimed coordinates can be summarized by

$$\begin{bmatrix} x_1' \\ x_2' \\ x_3' \\ x_4' \end{bmatrix} = \begin{bmatrix} \frac{1}{\sqrt{1-\beta^2}} & 0 & 0 & \frac{j\beta}{\sqrt{1-\beta^2}} \\ 0 & 1 & 0 & 0 \\ 0 & 0 & 1 & 0 \\ -\frac{j\beta}{\sqrt{1-\beta^2}} & 0 & 0 & \frac{1}{\sqrt{1-\beta^2}} \end{bmatrix} * \begin{bmatrix} x_1 \\ x_2 \\ x_3 \\ x_4 \end{bmatrix} \tag{9.29}$$

In-Class Exercises

Pb. 9.18 Write the above transformation for the case that the two coordinate systems are moving from each other at half the speed of light, and find (x', y', z', t') if

$$x = 2, \quad y = 3, \quad z = 4, \quad ct = 3$$

Pb. 9.19 Find the determinant of $\mathbf{L}_\beta$.

Pb. 9.20 Find the multiplicative inverse of $\mathbf{L}_\beta$, and compare it to the transpose.

Pb. 9.21 Find the approximate expression of $\mathbf{L}_\beta$ for $\beta \ll 1$. Give a physical interpretation to your result using Newtonian mechanics.

9.4.2 Addition Theorem for Velocities

The physical problem of interest here is: assuming that a point mass is moving in the primed system in the x'–y' plane with uniform speed u' and its trajectory is making an angle θ' with the x'-axis, what is the speed of this particle, as viewed in the unprimed system, and what is the angle that its trajectory makes with the x-axis, as observed in the unprimed system?

In the unprimed and primed systems, the parametric equations for the point particle motion are respectively given by

$$x = ut\cos(\theta), \qquad y = ut\sin(\theta) \tag{9.30}$$

$$x' = u't'\cos(\theta'), \qquad y' = u't'\sin(\theta') \tag{9.31}$$

where u and u' are the speeds of the particle in the unprimed and primed systems, respectively. Note that if the primed system moves with velocity v with respect to the unprimed system, then the unprimed system moves with a velocity $-v$ with respect to the primed system, and using the Lorentz transformation, we can write the following equalities:

$$ut\cos(\theta) = \frac{(u'\cos(\theta') + v)}{\sqrt{1-\beta^2}}t' \tag{9.32}$$

$$ut\sin(\theta) = u't'\sin(\theta') \tag{9.33}$$

$$t = \frac{[1 + (u'v / c^2)\cos(\theta')]}{\sqrt{1-\beta^2}}t' \tag{9.34}$$

Dividing Eq. (9.32) and Eq. (9.33) by Eq. (9.34), we obtain

$$u\cos(\theta) = \frac{(u'\cos(\theta') + v)}{[1 + (u'v / c^2)\cos(\theta')]} \tag{9.35}$$

$$u\sin(\theta) = \frac{u'\sin(\theta')\sqrt{1-\beta^2}}{[1 + (u'v / c^2)\cos(\theta')]} \tag{9.36}$$

From this we can deduce the magnitude and direction of the velocity of the particle, as measured in the unprimed system:

$$u^2 = \frac{u'^2 + v^2 + 2u'v\cos(\theta') - (u'^2v^2/ c^2)\sin^2(\theta')}{[1 + (u'v / c^2)\cos(\theta')]^2} \tag{9.37}$$

$$\tan(\theta) = \frac{u'\sin(\theta')\sqrt{1-\beta^2}}{u'\cos(\theta') + v} \tag{9.38}$$

PREPARATORY EXERCISES

Pb. 9.22 Find the velocity of a photon (the quantum of light) in the unprimed system if its velocity in the primed system is $u' = c$.

(Note the constancy of the velocity of light, if measured from either the primed or the unprimed system. As mentioned previously, this constituted one of only two postulates in Einstein's formulation of the theory of special relativity, which determined uniquely the form of the dynamical boost transformation.)

Pb. 9.23 Show that if u' is parallel to the x'-axis, then the velocity addition formula takes the following simple form:

$$u = \frac{u' + v}{1 + \dfrac{u'v}{c^2}}$$

Pb. 9.24 Find the approximate form of the above expression for u when $\beta \ll 1$, and show that it reduces to the expression of velocity addition in Newtonian mechanics.

IN-CLASS EXERCISES

Pb. 9.25 Find the angle θ, if $\theta' = \dfrac{\pi}{2}$ and $u' = v = \dfrac{c}{2}$.

Pb. 9.26 Plot the angle θ as a function of θ', when $v/c = 0.99$ and $u'/c = 1$.

Pb. 9.27 Let the variable ϕ be defined such that $\tanh(\phi) = \beta$. Write the Lorentz transformation matrix as function of ϕ. Can you give the Lorentz transformation a geometric interpretation in non-Euclidean geometry?

Pb. 9.28 Using the result of Pb. 9.27, write the resultant transformation from a boost with parameter ϕ_1, followed by another boost with parameter ϕ_2. Does this rule for composition of Lorentz transformations remind you of a similar transformation that you have studied previously in this chapter?

9.5 Iterative Constructs*

9.5.1 The Koch Curve

In the previous sections of this chapter, we discussed the matrix representation for geometrical and velocity boosts transformations. In Chapter 6, taking advantage of the isomorphism that exists between complex numbers and points in 2-D planes, we were able to reduce the representation of all 2-D

* The asterisk indicates more advanced material that may be skipped in a first reading.

geometry transformations to algebraic transformations on complex numbers. In this section, we will illustrate a mixed method, using both transformations of complex numbers and matrix iterative constructions, to generate fractal graphs in 2-D geometry. We will show in detail, how to construct the Nth-order iteration of the Koch curve, whose geometrical constructs we had discussed earlier in Section 3.3.

To assist in understanding the algorithm, we reproduce in Figure 9.3, the first- and second-order iterations of the Koch curves with the vertices in each order numbered from 0 to 4^N. We have chosen the initial point and the final point at each order to be at the points (0, 0) and (1, 0), respectively.

1. We represent the vertices numbered 1 to 4^n at the nth iteration by a row vector of complex numbers, each representing the coordinates of the corresponding vertex.
2. The coordinates of the vertices for all iterations up to the Nth order will then be represented by a $(N \otimes 4^N)$ matrix, with the elements of the first 4^n columns of the nth row ($n \leqslant N$) equal to the corresponding elements of the row vector of (1), and the other elements of the row padded with zeros.
3. The positively numbered vertices of the first Koch iteration are represented by the row:

$$Z(1, :) = [1/3 \quad (1/3)\ \exp(j\pi/3) \quad 2/3 \quad 1 \quad 0 \quad \cdots \quad 0]$$

4. The elements of the nth row are obtained from the elements of the $(n-1)$th row as follows:
   a. The first group of 4^{n-1} elements are the same as the nonzero elements of the $(n-1)$th row scaled down by a factor of 3.

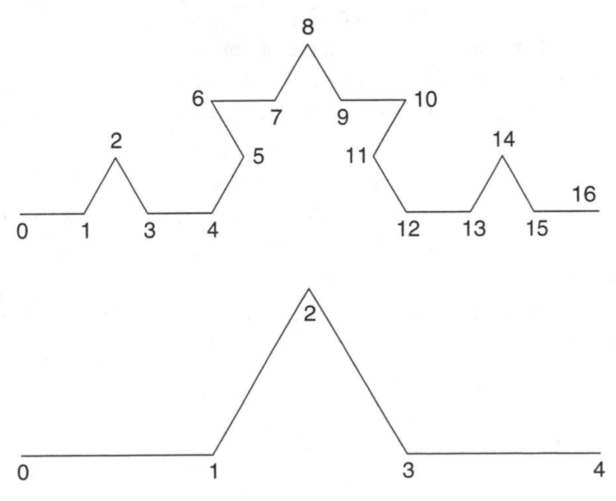

FIGURE 9.3
The first two iterations of the Koch curve with their vertices numbered.

b. The second group of 4^{n-1} elements are the same complex numbers as the nonzero elements of the $(n-1)$th row scaled down by a factor of 3, rotated around the origin by the angle $\pi/3$, and then translated by the vector (1/3, 0).

c. The third group of 4^{n-1} elements are the same complex numbers as the nonzero elements of the $(n-1)$th row scaled down by a factor of 3, rotated around the origin by the angle $(-\pi/3)$, and then translated by the vector $\left(\frac{1}{3}(1+\cos(\pi/3)), \frac{1}{3}\sin(\pi/3)\right)$.

d. The fourth group of 4^{n-1} elements are the same complex numbers as the nonzero elements of the $(n-1)$th row scaled down by a factor of 3, and then translated by the vector (2/3, 0).

Recalling that the geometric transformations on a point in 2-D space representing respectively a rotation by the angle (θ) around the origin and a translation by the vector (T_x, T_y) are represented through the following relations between old and new coordinates:

$$z' = z\exp(j\theta)$$

$$z' = z + T_x + jT_y$$

The *m-file* generating the Nth iteration of the Koch curve is

```
N=?;
Z=zeros(N,4^N);
Z(1,1:4)=[1/3 (1/3)*(1+exp(i*pi/3)) 2/3 1];
for n=2:N
    for m=1:4^n
        if 1<=m & m<=4^(n-1)
            Z(n,m)=(1/3)*Z(n-1,m);
        elseif (4^(n-1)+1)<=m & m<=2*(4^(n-1))
            Z(n,m)=(1/3)*(Z(n-1,m-4^(n-1)))*...
                exp(j*pi/3)+1/3;
        elseif (2*(4^(n-1))+1)<=m & m<=(3*(4^(n-1)))
            Z(n,m)=(1/3)*(Z(n-1,m-2*4^(n-1)))*...
                exp(-j*pi/3)+(1/3)*(1+exp(j*pi/3));
        else
            Z(n,m)=(1/3)*(Z(n-1,m-3*4^(n-1)))+2/3;
        end
    end
end
K=[0 Z(N,:)];
plot(K,'-')
axis([0 1 -0.5 0.5])
axis square
```

Homework Problems

Pb. 9.29 Rewrite the above program for the case that the initial and final points of the Koch curve are located at (0, 0) and (cos(θ), sin(θ)).

Pb. 9.30 Write a program to plot a snowflake.

9.5.2 The Serpenski Curve

Barnsley generated the Serpinski curve by repeated affine transformations on a point in a 2-D plane. The transformations are all of the form

$$\mathbf{x}(n) = \mathbf{A}\mathbf{x}(n-1) + \mathbf{B}$$

where **x** is a two-dimensional column vector representing the coordinates of a point in a 2-D plane, **A** a $(2 \otimes 2)$ matrix, and **B** a two-dimensional column vector.

The **A** matrix is in all instances given by

$$\mathbf{A} = \begin{bmatrix} 0.5 & 0 \\ 0 & 0.5 \end{bmatrix}$$

while the column vector **B** respectively may have with probability (0.33, 0.33, 0.34) one of the following three values:

$$\mathbf{B1} = \begin{bmatrix} 1 \\ 1 \end{bmatrix} \quad \mathbf{B2} = \begin{bmatrix} 1 \\ 50 \end{bmatrix} \quad \mathbf{B3} = \begin{bmatrix} 50 \\ 50 \end{bmatrix}$$

The initial point for the iteration is [0; 0].

The program to generate this curve is

```
N=100000;
r=rand(1,N);
x=zeros(2,N);
x(:,1)=[0;0];
A=[0.5 0; 0 0.5];
B1=[1;1];
B2=[1;50];
B3=[50;50];
for n=2:N
    if r(n)<=0.33;
```

```
            x(:,n)=A*x(:,n-1)+B1;
        elseif r(n)>0.33 & r(n)<=0.66
            x(:,n)=A*x(:,n-1)+B2;
        else
            x(:,n)=A*x(:,n-1)+B3;
        end
    end
plot(x(1,:),x(2,:),'.')
axis([0 100 0 100])
axis square
axis off
```

Homework Problems

Pb. 9.31 Rewrite the above Barnsley construct for the Serpinski curve using complex numbers to represent points in a 2-D plane.

Pb. 9.32 The Fractal Fern can also be generated by an affine transformation but where the **A** matrix and the column vector **B** can have one of the following four forms:

$$\mathbf{A1} = [0.85\ \ 0.04;\ \ -0.04\ \ 0.85];\quad \mathbf{A2} = [0.20\ \ -0.26;\ \ 0.23\ \ 0.22];$$

$$\mathbf{A3} = [-0.15\ \ 0.28;\ \ 0.26\ \ 0.24];\quad \mathbf{A4} = [0\ \ 0;\ \ 0\ \ 0.16];$$

$$\mathbf{B1}=[0;1.6];\quad \mathbf{B2}=[0;1.6];\quad \mathbf{B3}=[0;0.44];\quad \mathbf{B4}=[0;0].$$

The probability for each of the different transformations is given respectively by 85, 7, 7, and 1%. Take the initial point at [0.5; 0.5].
Plot this curve.

Pb. 9.33 The Fractal Tree can also be generated by an affine transformation but where the **A** matrix and the column vector **B** can have one of the following four forms:

$$\mathbf{A1} = [0.42\ \ -0.42;\ \ 0.42\ \ 0.42];\quad \mathbf{A2} = [0.42\ \ 0.42;\ \ -0.42\ \ 0.42];$$

$$\mathbf{A3} = [0.1\ \ 0;\ \ 0\ \ 0.1];\quad \mathbf{A4} = [0\ \ 0;\ \ 0\ \ 0.5];$$

$$\mathbf{B1}=[0;0.2];\quad \mathbf{B2}=[0;0.2];\quad \mathbf{B3}=[0;0.2];\quad \mathbf{B4}=[0;\ 0].$$

The probability for each of the different transformations is given respectively by 40, 40, 15, and 5%. Take the initial point at [0; 0].
Plot this curve.

9.6 MATLAB Commands Review

`colormap`	Control the color mix of an image.
`fliplr`	Flip a matrix left to right.
`flipud`	Flip a matrix in the up-to-down direction.
`imagesc`	Create a pixel intensity map from data stored in a matrix.
`load`	Import data files from outside MATLAB.
`rot90`	Rotate a matrix by 90°.
`toeplitz`	Specialized matrix constructor that describes, *inter alia,* the operation of a blur in an image.

10

*A Taste of Probability Theory**

10.1 Introduction

In addition to its everyday use in all aspects of our public, personal, and leisure lives, probability plays an important role in electrical engineering practice in at least three important aspects. It is the mathematical tool to deal with three broad areas.

1. The problems associated with the inherent uncertainty in the input of certain systems. The random arrival time of certain inputs to a system cannot be predetermined; for example, the log-on and the log-off times of terminals and workstations connected to a computer network, or the data packets' arrival time to a computer network node.
2. The problems associated with the distortion of a signal due to noise. The effects of noise have to be dealt with satisfactorily at each stage of a communication system from the generation, to the transmission, to the detection phases. The source of this noise may be due to either fluctuations inherent in the physics of the problem (e.g., quantum effects and thermal effects) or due to random distortions due to externally generated uncontrollable parameters (e.g., weather and geography).
3. The problems associated with inherent human and computing machine limitations while solving very complex systems. Individual treatment of the dynamics of very large number of molecules in a material, in which more than 10^{22} molecules may exist in a quart-size container, is not possible at this time, and we have to rely on statistical averages when describing the behavior of such systems. This is the field of statistical physics and thermodynamics.

Furthermore, probability theory provides the necessary mathematical tools for error analysis in all experimental sciences. It permits estimation of the

---

* The asterisk indicates more advanced material that may be skipped in a first reading.

error bars and the confidence level for any experimentally obtained result, through a methodical analysis and reduction of the raw data.

In future courses in probability, random variables, stochastic processes (which is random variables theory with time as a parameter), information theory, and statistical physics, you will study techniques and solutions to the different types of problems from the above list. In this very brief introduction to the subject, we introduce only the very fundamental ideas and results — where more advanced courses seem to almost always start.

10.2 Basics

Probability theory is best developed mathematically based on a set of axioms from which a well-defined deductive theory can be constructed. This is referred to as the axiomatic approach. We concentrate, in this section, on developing the basics of probability theory, using a physical description of the underlying concepts of probability and related simple examples, to lead us intuitively to what is usually the starting point of the set theoretic axiomatic approach.

Assume that we conduct n independent trials under identical conditions, in each of which, depending on chance, a particular event A of particular interest either occurs or does not occur. Let $n(A)$ be the number of experiments in which A occurs. Then, the ratio $n(A)/n$, called the relative frequency of the event A to occur in a series of experiments, clusters for $n \to \infty$ about some constant. This constant is called the probability of the event A, and is denoted by

$$P(A) = \lim_{n\to\infty} \frac{n(A)}{n} \tag{10.1}$$

From this definition, we know specifically what is meant by the statement that the probability for obtaining a head in the flip of a fair coin is $1/2$.

Let us consider the rolling of a single die as our prototype experiment.

1. The possible outcomes of this experiment are elements belonging to the set:

$$S = \{1, 2, 3, 4, 5, 6\} \tag{10.2}$$

If the die is fair, the probability for each of the elementary elements of this set to occur in the roll of a die is equal to

$$P(1) = P(2) = P(3) = P(4) = P(5) = P(6) = \frac{1}{6} \tag{10.3}$$

2. The observer may be interested not only in the elementary elements' occurrence, but in finding the probability of a certain event which may consist of a set of elementary outcomes; for example,
   a. An event may consist of "obtaining an even number of spots on the upward face of a randomly rolled die." This event then consists of all successful trials having as experimental outcomes any member of the set

$$E = \{2, 4, 6\} \tag{10.4}$$

   b. Another event may consist of "obtaining three or more spots" (hence, we will use this form of abbreviated statement, and not keep repeating: on the upward face of a randomly rolled die). Then, this event consists of all successful trials having experimental outcomes any member of the set

$$B = \{3, 4, 5, 6\} \tag{10.5}$$

   Note that, in general, events may have overlapping elementary elements.

For a fair die, using the definition of the probability as the limit of a relative frequency, it is possible to conclude, based on experimental trials, that

$$P(E) = P(2) + P(4) + P(6) = \frac{1}{2} \tag{10.6}$$

while

$$P(B) = P(3) + P(4) + P(5) + P(6) = \frac{2}{3} \tag{10.7}$$

and

$$P(S) = 1 \tag{10.8}$$

The last equation [Eq. (10.8)] is the mathematical expression for the statement that the probability of the event that includes all possible elementary outcomes is 1 (i.e., certainty).

It should be noted that if we define the events O and C to mean the events of "obtaining an odd number" and "obtaining a number smaller than 3," respectively, we can obtain these events' probabilities by enumerating the elements of the subsets of S that represent these events; namely,

$$P(O) = P(1) + P(3) + P(5) = \frac{1}{2} \tag{10.9}$$

$$P(C) = P(1) + P(2) = \frac{1}{3} \tag{10.10}$$

However, we also could have obtained these same results by noting that the events E and O (B and C) are disjoint and that their union spanned the set S. Therefore, the probabilities for events O and C could have been deduced, as well, through the relations

$$P(O) = 1 - P(E) \tag{10.11}$$

$$P(C) = 1 - P(B) \tag{10.12}$$

From the above and similar observations, it would be a satisfactory representation of the physical world if the above results were codified and elevated to the status of axioms for a formal theory of probability. However, the question arises how many of these basic results (the axioms) one really needs to assume, such that it will be possible to derive all other results of the theory from this seed. This is the starting point for the formal approach to the probability theory.

The following axioms were proven to be a satisfactory starting point. Assign to each event A, consisting of elementary occurrences from the set S, a number $P(A)$, which is designated as the probability of the event A, and such that

1. $$0 \leqslant P(A) \tag{10.13}$$

2. $$P(S) = 1 \tag{10.14}$$

3. If $A \cap B = \varnothing$, where $\varnothing$ is the empty set then $P(A \cup B) = P(A) + P(B)$ (10.15)

In the following examples, we illustrate some common techniques for finding the probabilities for certain events. Look around, and you will find plenty more.

Example 10.1

Find the probability for getting three sixes in a roll of three dice.

Solution: First, compute the number of elements in the total sample space. We can describe each roll of the dice by a 3-tuplet (a, b, c), where a, b, and c can take the values 1, 2, 3, 4, 5, 6. There are $6^3 = 216$ possible 3-tuplets. The event that we are seeking is realized only in the single elementary occurrence when the 3-tuplet (6, 6, 6) is obtained; therefore, the probability for this event, for fair dice, is

$$P(A) = \frac{1}{216} \quad \square$$

Example 10.2

Find the probability of getting only two sixes in a roll of three dice.

Solution: The event in this case consists of all elementary occurrences having the following forms: $(a, 6, 6)$, $(6, b, 6)$, $(6, 6, c)$, where $a = 1, \ldots, 5$; $b = 1, \ldots, 5$; and $c = 1, \ldots, 5$. Therefore, the event A consists of elements corresponding to 15 elementary occurrences, and its probability is

$$P(A) = \frac{15}{216} \quad \square$$

Example 10.3

Find the probability that, if three individuals are asked to guess a number from 1 to 10, their guesses will be different numbers.

Solution: There are 1000 distinct equiprobable 3-tuplets (a, b, c), where each component of the 3-tuplet can have any value from 1 to 10. The event A occurs when all components have unequal values. Therefore, while a can have any of 10 possible values, b can have only 9, and c can have only 8. Therefore, $n(A) = 8 \times 9 \times 10$, and the probability for the event A is

$$P(A) = \frac{8 \times 9 \times 10}{1000} = 0.72 \quad \square$$

Example 10.4

An inspector checks a batch of 100 microprocessors, 5 of which are defective. He examines 10 items selected at random. If none of the 10 items is defective, he accepts the batch. What is the probability that he will accept the batch?

Solution: The number of ways of selecting 10 items from a batch of 100 items is

$$N = \frac{100!}{10!(100-10)!} = \frac{100!}{10!90!} = C_{10}^{100}$$

where C_k^n is the binomial coefficient and represents the number of combinations of n objects taken k at a time without regard to order. It is equal to $n!/k!(n-k)!$. All these combinations are equally probable.

If the event A is that where the batch is accepted by the inspector, then A occurs when all 10 items selected belong to the set of acceptable quality units. The number of elements in A is

$$N(A) = \frac{95!}{10!85!} = C_{10}^{95}$$

and the probability for the event A is

$$P(A) = \frac{C_{10}^{95}}{C_{10}^{100}} = \frac{86 \times 87 \times 88 \times 89 \times 90}{96 \times 97 \times 98 \times 99 \times 100} = 0.5837$$

□

In-Class Exercises

Pb. 10.1 A cube whose faces are colored is split into 125 smaller cubes of equal size.

a. Find the probability that a cube drawn at random from the batch of randomly mixed smaller cubes will have three colored faces.

b. Find the probability that a cube drawn from this batch will have two colored faces.

Pb. 10.2 An urn has three blue balls and six red balls. One ball was randomly drawn from the urn and then a second ball, which was blue. What is the probability that the first ball drawn was blue?

Pb. 10.3 Find the probability that the last two digits of the cube of a random integer are 1. Solve the problem analytically, and then compare your result to a numerical experiment that you will conduct and where you compute the cubes of all numbers from 1 to 1000.

Pb. 10.4 From a lot of n resistors, p are defective. Find the probability that k resistors out of a sample of m selected at random are found defective.

Pb. 10.5 Three cards are drawn from a deck of cards.

a. Find the probability that these cards are the ace, the king, and the queen of hearts.

b. Would the answer change if the statement of the problem was "an ace, a king, and a queen"?

Pb. 10.6 Show that

$$P(\bar{A}) = 1 - P(A)$$

where $\bar{A}$, the complement of A, are all events in S having no element in common with A.

NOTE In solving certain category of probability problems, it is often convenient to solve for $P(A)$ by computing the probability of its complement and then applying the above relation.

Pb. 10.7 Show that if $A_1, A_2, \ldots, A_n$ are mutually exclusive events, then

$$P(A_1 \cup A_2 \cup \ldots \cup A_n) = P(A_1) + P(A_2) + \cdots + P(A_n)$$

[*Hint*: Use mathematical induction and Eq. (10.15).]

10.3 Addition Laws for Probabilities

We start by reminding the reader of the key results of elementary set theory.

- The Commutative law states that

$$A \cap B = B \cap A \tag{10.16}$$

$$A \cup B = B \cup A \tag{10.17}$$

- The Distributive laws are written as

$$A \cap (B \cup C) = (A \cap B) \cup (A \cap C) \tag{10.18}$$

$$A \cup (B \cap C) = (A \cup B) \cap (A \cup C) \tag{10.19}$$

- The Associative laws are written as

$$(A \cup B) \cup C = A \cup (B \cup C) = A \cup B \cup C \tag{10.20}$$

$$(A \cap B) \cap C = A \cap (B \cap C) = A \cap B \cap C \tag{10.21}$$

- De Morgan's laws are

$$\overline{(A \cup B)} = \overline{A} \cap \overline{B} \tag{10.22}$$

$$\overline{(A \cap B)} = \overline{A} \cup \overline{B} \tag{10.23}$$

- The Duality principle states that if in an identity, we replace unions by intersections, intersections by unions, S by $\varnothing$, and $\varnothing$ by S, then the identity is preserved.

THEOREM 1 If we define the difference of two events $A_1 - A_2$ to mean the events in which A_1 occurs but not A_2, the following equalities are valid:

$$P(A_1 - A_2) = P(A_1) - P(A_1 \cap A_2) \tag{10.24}$$

$$P(A_2 - A_1) = P(A_2) - P(A_1 \cap A_2) \tag{10.25}$$

$$P(A_1 \cup A_2) = P(A_1) + P(A_2) - P(A_1 \cap A_2) \tag{10.26}$$

PROOF From the basic set theory algebra results, we can deduce the following equalities:

$$A_1 = (A_1 - A_2) \cup (A_1 \cap A_2) \tag{10.27}$$

$$A_2 = (A_2 - A_1) \cup (A_1 \cap A_2) \tag{10.28}$$

$$A_1 \cup A_2 = (A_1 - A_2) \cup (A_2 - A_1) \cup (A_1 \cap A_2) \tag{10.29}$$

Further note that the events $(A_1 - A_2)$, $(A_2 - A_1)$, and $(A_1 \cap A_2)$ are mutually exclusive. Using the results from Pb. 10.7, Eq. (10.27) and Eq. (10.28), and the preceding comment, we can write

$$P(A_1) = P(A_1 - A_2) + P(A_1 \cap A_2) \tag{10.30}$$

$$P(A_2) = P(A_2 - A_1) + P(A_1 \cap A_2) \tag{10.31}$$

which establish Eq. (10.24) and Eq. (10.25). Next, consider Eq. (10.29); because of the mutual exclusivity of each event represented by each of the parenthesis on its LHS, we can use the results of Pb. 10.7, to write

$$P(A_1 \cup A_2) = P(A_1 - A_2) + P(A_2 - A_1) + P(A_1 \cap A_2) \tag{10.32}$$

using Eq. (10.30) and Eq. (10.31), this can be reduced to Eq. (10.26).

THEOREM 2 Given any n events $A_1, A_2, \ldots, A_n$ and defining $P_1, P_2, P_3, \ldots, P_n$ to mean

$$P_1 = \sum_{i=1}^{n} P(A_i) \tag{10.33}$$

$$P_2 = \sum_{1 \le i < j \le n} P(A_i \cap A_j) \tag{10.34}$$

$$P_3 = \sum_{1 \le i < j < k \le n} P(A_i \cap A_j \cap A_k) \tag{10.35}$$

etc., then

$$P\left(\bigcup_{k=1}^{n} A_k\right) = P_1 - P_2 + P_3 - P_4 + \cdots + (-1)^{n-1} P_n \tag{10.36}$$

This theorem can be proven by mathematical induction (we do not give the details of this proof here).

EXAMPLE 10.5

Using the events E, O, B, C as defined in Section 10.2, use Eq. (10.36) to show that: $P(E \cup O \cup B \cup C) = 1$.

Solution: Using Eq. (10.36), we can write

$$\begin{aligned} P(E \cup O \cup B \cup C) &= P(E) + P(O) + P(B) + P(C) \\ &- [P(E \cap O) + P(E \cap B) + P(E \cap C) + P(O \cap B) + P(O \cap C) + P(B \cap C)] \\ &+ [P(E \cap O \cap B) + P(E \cap O \cap C) + P(E \cap B \cap C) + P(O \cap B \cap C)] \\ &- P(E \cap O \cap B \cap C) \\ &= \left[\frac{1}{2} + \frac{1}{2} + \frac{2}{3} + \frac{1}{3}\right] - \left[0 + \frac{2}{6} + \frac{1}{6} + \frac{2}{6} + \frac{1}{6} + 0\right] + [0+0+0+0] - [0] = 1 \end{aligned}$$

□

EXAMPLE 10.6

Show that for any n events $A_1, A_2, \ldots, A_n$, the following inequality holds:

$$P\left(\bigcup_{k=1}^{n} A_k\right) \leq \sum_{k=1}^{n} P(A_k)$$

Solution: We prove this result by mathematical induction:

- For $n = 2$, the result holds because by Eq. (10.26) we have

$$P(A_1 \cup A_2) = P(A_1) + P(A_2) - P(A_1 \cap A_2)$$

  and since any probability is a nonnegative number, this leads to the inequality

$$P(A_1 \cup A_2) \leq P(A_1) + P(A_2)$$

- Assume that the theorem is true for $(n-1)$ events, then we can write

$$P\left(\bigcup_{k=2}^{n} A_k\right) \leq \sum_{k=2}^{n} P(A_k)$$

- Using associativity, Eq. (10.26), the result for $(n-1)$ events, and the nonnegativity of the probability, we can write

$$\begin{aligned} P\left(\bigcup_{k=1}^{n} A_k\right) &= P\left(A_1 \cup \left(\bigcup_{k=1}^{n} A_k\right)\right) = P(A_1) + P\left(\bigcup_{k=2}^{n} A_k\right) - P\left(A_1 \cap \left(\bigcup_{k=2}^{n} A_k\right)\right) \\ &\leq P(A_1) + \sum_{k=2}^{n} P(A_k) - P\left(A_1 \cap \left(\bigcup_{k=2}^{n} A_k\right)\right) \leq \sum_{k=1}^{n} P(A_k) \end{aligned}$$

which is the desired result. □

In-Class Exercises

Pb. 10.8 Show that if the events $A_1, A_2, \ldots, A_n$ are such that

$$A_1 \subset A_2 \subset \ldots \subset A_n$$

then

$$P\left(\bigcup_{k=1}^{n} A_k\right) = P(A_n)$$

Pb. 10.9 Show that if the events $A_1, A_2, \ldots, A_n$ are such that

$$A_1 \supset A_2 \supset \ldots \supset A_n$$

then

$$P\left(\bigcap_{k=1}^{n} A_k\right) = P(A_n)$$

Pb. 10.10 Find the probability that a positive integer randomly selected will be divisible by

a. 2 and 3.

b. 2 or 3.

Pb. 10.11 Show that the expression for Eq. (10.36) simplifies to

$$P(A_1 \cup A_2 \cup \ldots \cup A_n) = C_1^n P(A_1) - C_2^n P(A_1 \cap A_2) + C_3^n P(A_1 \cap A_2 \cap A_3) - \cdots + (-1)^{n-1} P(A_1 \cap A_2 \cap \ldots \cap A_n)$$

when the probability for the intersection of any number of events is independent of the indices.

Pb. 10.12 A filing stack has n drawers, and a secretary randomly files m letters in these drawers.

a. Assuming that $m > n$, find the probability that there will be at least one letter in each drawer.

b. Plot this probability for $n = 12$, and $15 \leqslant m \leqslant 50$.

(*Hint:* Take the event A_j to mean that no letter is filed in the jth drawer and use the result of Pb. 10.11.)

10.4 Conditional Probability

The conditional probability of an event A assuming C and denoted by $P(A|C)$ is, by definition, the ratio

$$P(A|C) = \frac{P(A \cap C)}{P(C)} \tag{10.37}$$

EXAMPLE 10.7

Considering the events E, O, B, C as defined in Section 10.2 and the above definition for conditional probability, find the probability that the number of spots showing on the die is even, assuming that it is equal to or greater than 3.

Solution: In the above notation, we are asked to find the quantity $P(E|B)$. Using Eq. (10.37), this is equal to

$$P(E|B) = \frac{P(E \cap B)}{P(B)} = \frac{P(\{4,6\})}{P(\{3,4,5,6\})} = \frac{\left(\frac{2}{6}\right)}{\left(\frac{4}{6}\right)} = \frac{1}{2}$$

In this case, $P(E|B) = P(E)$. When this happens, we say that the two events E and B are independent. □

Example 10.8

Find the probability that the number of spots showing on the die is even, assuming that it is larger than 3.

Solution: Call D the event of having the number of spots larger than 3. Using Eq. (10.37), $P(E|D)$ is equal to

$$P(E|D) = \frac{P(E \cap D)}{P(D)} = \frac{P(\{4,6\})}{P(\{4,5,6\})} = \frac{\left(\frac{2}{6}\right)}{\left(\frac{3}{6}\right)} = \frac{2}{3}$$

In this case, $P(E|D) \neq P(E)$; and thus the two events E and D are not independent. □

Example 10.9

Find the probability of picking a blue ball first, then a red ball from an urn that contains five red balls and four blue balls.

Solution: From the definition of conditional probability [Eq. (10.37)], we can write

$$P(\text{Blue ball first and Red ball second}) = P(\text{Red ball second}|\text{Blue ball first}) \times P(\text{Blue ball first})$$

The probability of picking a blue ball first is

$$P(\text{Blue ball first}) = \frac{\text{Original number of Blue balls}}{\text{Total number of balls}} = \frac{4}{9}$$

The conditional probability is given by

$$P(\text{Red ball second}|\text{Blue ball first}) = \frac{\text{Number of Red balls}}{\text{Number of balls remaining after first pick}} = \frac{5}{8}$$

giving

$$P(\text{Blue ball first and Red ball second}) = \frac{4}{9} \times \frac{5}{8} = \frac{5}{18}$$ □

10.4.1 Total Probability Theorem

If $[A_1, A_2, \ldots, A_n]$ is a partition of the total elementary occurrences set S, that is,

$$\bigcup_{i=1}^{n} A_i = S \quad \text{and} \quad A_i \cap A_j = \varnothing \quad \text{for} \quad i \neq j$$

and B is an arbitrary event, then

$$P(B) = P(B|A_1)P(A_1) + P(B|A_2)P(A_2) + \cdots + P(B|A_n)P(A_n) \qquad (10.38)$$

PROOF From the algebra of sets, and the definition of a partition, we can write the following equalities:

$$\begin{aligned} B = B \cap S &= B \cap (A_1 \cup A_2 \cup \ldots \cup A_n) \\ &= (B \cap A_1) \cup (B \cap A_2) \cup \ldots \cup (B \cap A_n) \end{aligned} \qquad (10.39)$$

Since the events $(B \cap A_i)$ and $(B \cap A_j)$ and are mutually exclusive for $i \neq j$, then using the results of Pb. 10.7, we can deduce that

$$P(B) = P(B \cap A_1) + P(B \cap A_2) + \cdots + P(B \cap A_n) \qquad (10.40)$$

Now, using the conditional probability definition [Eq. (10.38)], Eq. (10.40) can be written as

$$P(B) = P(B|A_1)P(A_1) + P(B|A_2)P(A_2) + \cdots + P(B|A_n)P(A_n) \qquad (10.41)$$

This result is known as the total probability theorem.

10.4.2 Bayes Theorem

The statement of Bayes theorem is as follows:

$$P(A_i|B) = \frac{P(B|A_i)P(A_i)}{P(B|A_1)P(A_1) + P(B|A_2)P(A_2) + \cdots + P(B|A_n)P(A_n)} \qquad (10.42)$$

PROOF From the definition of the conditional probability [Eq. (10.37)], we can write

$$P(B \cap A_i) = P(A_i|B)P(B) \qquad (10.43)$$

Again, using Eq. (10.37) and Eq. (10.43), we have

$$P(A_i|B) = \frac{P(B|A_i)P(A_i)}{P(B)} \tag{10.44}$$

Now, substituting Eq. (10.41) in the denominator of Eq. (10.44), we obtain Eq. (10.42).

EXAMPLE 10.10

A digital communication channel transmits the signal as a collection of ones (1s) and zeros (0s). Assume (statistically) that 40% of the 1s and 33% of the 0s are changed upon transmission. Suppose that, in a message, the ratio between the transmitted 1 and the transmitted 0 was 5/3. What is the probability that the received signal is the same as the transmitted signal if

a. The received signal was a 1?
b. The received signal was a 0?

Solution: Let O be the event that 1 was received, and Z be the event that 0 was received. If H_1 is the hypothesis that 1 was received and H_0 is the hypothesis that 0 was received, then from the statement of the problem, we know that

$$\frac{P(H_1)}{P(H_0)} = \frac{5}{3} \quad \text{and} \quad P(H_1) + P(H_0) = 1$$

giving

$$P(H_1) = \frac{5}{8} \quad \text{and} \quad P(H_0) = \frac{3}{8}$$

Furthermore, from the text of the problem, we know that

$$P(O|H_1) = \frac{3}{5} \quad \text{and} \quad P(Z|H_1) = \frac{2}{5}$$

$$P(O|H_0) = \frac{1}{3} \quad \text{and} \quad P(Z|H_0) = \frac{2}{3}$$

From the total probability result [Eq. (10.41)], we obtain

$$\begin{aligned} P(O) &= P(O|H_1)P(H_1) + P(O|H_0)P(H_0) \\ &= \frac{3}{5} \times \frac{5}{8} \times \frac{1}{3} \times \frac{3}{8} = \frac{1}{2} \end{aligned}$$

and

$$P(Z) = P(Z|H_1)P(H_1) + P(Z|H_0)P(H_0)$$
$$= \frac{2}{5} \times \frac{5}{8} + \frac{2}{3} \times \frac{3}{8} = \frac{1}{2}$$

The probability that the received signal is 1 if the transmitted signal was 1 from Bayes theorem

$$P(H_1|O) = \frac{P(H_1)P(O|H_1)}{P(O)} = \frac{\frac{5}{3}\frac{3}{5}}{\frac{1}{2}} = \frac{3}{4}$$

Similarly, we can obtain the probability that the received signal is 0 if the transmitted signal is 0

$$P(H_0|Z) = \frac{P(H_0)P(Z|H_0)}{P(Z)} = \frac{\frac{3}{8}\frac{2}{3}}{\frac{1}{2}} = \frac{1}{2} \quad \square$$

IN-CLASS EXERCISES

Pb. 10.13 Show that when two events A and B are independent, the addition law for probability becomes

$$P(A \cup B) = P(A) + P(B) - P(A)P(B)$$

Pb. 10.14 Consider four boxes, each containing 1000 resistors. Box 1 contains 100 defective items; Box 2 contains 400 defective items; Box 3 contains 50 defective items; and Box 4 contains 80 defective items.

a. What is the probability that a resistor chosen at random from any of the boxes is defective?

b. What is the probability that if the resistor is found defective, it came from Box 2?

(*Hint*: The randomness in the selection of the box means that $P(B_1) = P(B_2) = P(B_3) = P(B_4) = 0.25$.)

10.5 Repeated Trials

Bernoulli trials refer to identical, successive, and independent trials, in which an elementary event A can occur with probability

$$p = P(A) \tag{10.45}$$

or fail to occur with probability

$$q = 1 - p \tag{10.46}$$

In the case of n consecutive Bernoulli trials, each elementary event can be described by a sequence of 0s and 1s, such as in the following:

$$\omega = \underbrace{1\,0\,0\,0\,1 \ldots 0\,1}_{n\ digits\,-\,k\ ones} \tag{10.47}$$

where n is the number of trials, k the number of successes, and $(n - k)$ the number of failures. Because the trials are independent, the probability for the above single occurrence is

$$P(\omega) = p^k q^{n-k} \tag{10.48}$$

The total probability for the event with k successes in n trials is going to be the probability of the single event multiplied by the number of configurations with a given number of digits and a given number of 1s. The number of such configurations is given by the binomial coefficient C_k^n. Therefore

$$P(k \text{ successes in } n \text{ trials}) = C_k^n p^k q^{n-k} \tag{10.49}$$

Example 10.11

Find the probability that the number 3 will appear twice in five independent rolls of a die.

Solution: In a single trial, the probability of success (i.e., 3 showing up) is

$$p = \frac{1}{6}$$

Therefore, the probability that it appears twice in five independent rolls will be

$$P(2 \text{ successes in 5 trials}) = C_2^5 p^2 q^5 = \frac{5!}{2!3!}\left(\frac{1}{6}\right)^2\left(\frac{5}{6}\right)^3 = 0.16075 \quad \square$$

EXAMPLE 10.12

Find the probability that in a roll of two dice, three occurrences of snake-eyes (one spot on each die) are obtained in 10 rolls of the two dice.

Solution: The space S of the roll of two dice consists of 36 elementary elements (6×6), only one of which results in a snake-eyes configuration; therefore

$$p = 1/36; \quad k = 3; \quad n = 10$$

and

$$P(3 \text{ successes in 10 trials}) = C_3^{10} p^3 q^7 = \frac{10!}{3!7!}\left(\frac{1}{36}\right)^3 \left(\frac{35}{36}\right)^7 = 0.00211 \quad \square$$

IN-CLASS EXERCISES

Pb. 10.15 Assuming that a batch of manufactured components has an 80% chance of passing an inspection, what is the chance that at least 16 batches in a lot of 20 would pass the inspection?

Pb. 10.16 In an experiment, we keep rolling a fair die until it comes up showing three spots. What are the probabilities that this will take

a. exactly four rolls?

b. at least four rolls?

c. at most four rolls?

Pb. 10.17 Let X be the number of successes in a Bernoulli trials experiment with n trials and the probability of success p in each trial. If the mean number of successes m, also called average value $\bar{X}$ and expectation value $E(X)$, is defined as

$$m \equiv \bar{X} \equiv E(X) \equiv \sum XP(X)$$

and the variance is defined as

$$\text{Var}(X) \equiv E((X - \bar{X})^2)$$

show that

$$\bar{X} = np \quad \text{and} \quad \text{Var}(X) = np(1-p)$$

10.6 Generalization of Bernoulli Trials

In the above Bernoulli trials, we considered the case of whether or not a single event A was successful (i.e., two choices). This was the simplest partition of the set S.

In cases where we partition the set S in r subsets $S = \{A_1, A_2, \ldots, A_r\}$, and the probabilities for these single events are, respectively $\{p_1, p_2, \ldots, p_r\}$, where $p_1 + p_2 + \cdots + p_r = 1$, it can be easily proven that the probability in n independent trials for the event A_1 to occur k_1 times, the event A_2 to occur k_2 times, etc., is given by

$$P(k_1, k_2, \ldots, k_r; n) = \frac{n!}{k_1! k_2! \ldots k_r!} p_1^{k_1} p_2^{k_2} \ldots p_r^{k_r} \tag{10.50}$$

where $k_1 + k_2 + \cdots + k_r = n$

Example 10.13

Consider the sum of the spots in a roll of two dice. We partition the set of outcomes {2, 3, …, 11, 12} into the three events $A_1 = \{2, 3, 4, 5\}$, $A_2 = \{6, 7\}$, $A_3 = \{8, 9, 10, 11, 12\}$. Find $P(1, 7, 2; 10)$.

Solution: The probabilities for each of the events are, respectively,

$$p_1 = \frac{10}{36}, \quad p_2 = \frac{11}{36}, \quad p_3 = \frac{15}{36}$$

and

$$P(1,7,2;10) = \frac{10!}{1!7!2!} \left(\frac{10}{36}\right)^1 \left(\frac{11}{36}\right)^7 \left(\frac{15}{36}\right)^2 = 0.00431 \quad \square$$

10.7 The Poisson and the Normal Distributions

In this section, we obtain approximate expressions for the binomial distribution in different limits. We start by considering the expression for the probability of k successes in n Bernoulli trials with two choices for outputs; that is, Eq. (10.49).

10.7.1 The Poisson Distribution

Consider the limit when $p << 1$, but $np \equiv a \approx O(1)$. Then

$$P(k=0) = \frac{n!}{0!n!} p^0 (1-p)^n = \left(1 - \frac{a}{n}\right)^n \tag{10.51}$$

But in the limit $n \to \infty$,

$$\left(1 - \frac{a}{n}\right)^n = e^{-a} \tag{10.52}$$

giving

$$P(k=0) = e^{-a} \tag{10.53}$$

Now consider $P(k=1)$; it is equal to

$$\lim_{n\to\infty} P(k=1) = \frac{n!}{1!(n-1)!} p^1 (1-p)^{n-1} \approx a\left(1 - \frac{a}{n}\right)^n \approx ae^{-a} \tag{10.54}$$

For $P(k=2)$, we obtain

$$\lim_{n\to\infty} P(k=2) = \frac{n!}{2!(n-2)!} p^2 (1-p)^{n-2} \approx \frac{a^2}{2!}\left(1 - \frac{a}{n}\right)^n \approx \frac{a^2}{2!} e^{-a} \tag{10.55}$$

Similarly,

$$\lim_{n\to\infty} P(k) \approx \frac{a^k}{k!} e^{-a} \tag{10.56}$$

We compare in Figure 10.1 the exact with the approximate expression for the probability distribution, in the region of validity of the Poisson approximation.

EXAMPLE 10.14

A massive parallel computer system contains 1000 processors. Each processor fails independently of all others and the probability of its failure is 0.002 over a year. Find the probability that the system has no failures during 1 year of operation.

Solution: This is a problem of Bernoulli trials with $n = 1000$ and $p = 0.002$

$$P(k=0) = C_0^{1000} p^0 (1-p)^{1000} = (0.998)^{1000} = 0.13506$$

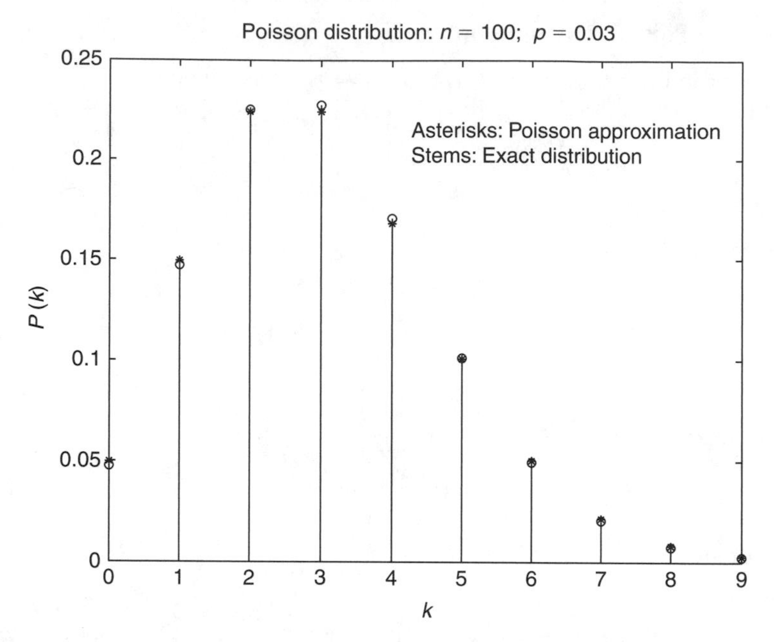

FIGURE 10.1
The Poisson distribution.

or, using the Poisson approximate formula, with $a = np = 2$

$$P(k = 0) \approx e^{-a} = e^{-2} \approx 0.13533 \quad \square$$

EXAMPLE 10.15

Due to the random vibrations affecting its supporting platform, a recording head introduces glitches on the recording medium at the rate of $n = 100$ glitches per min. What is the probability that $k = 3$ glitches are introduced in the recording over any interval of time $\Delta t = 1\,\text{s}$?

Solution: If we choose an interval of time equal to 1 min, the probability for an elementary event to occur in the subinterval Δt in this 1 min interval is

$$p = \frac{1}{60}$$

The problem reduces to finding the probability of $k = 3$ in $n = 100$ trials.

The Poisson formula gives this probability as

$$P(3) = \frac{1}{3!}\left(\frac{100}{60}\right)^3 \exp\left(-\frac{100}{60}\right) = 0.14573$$

where $a = 100/60$. (For comparison, the exact value for this probability, obtained using the binomial distribution expression, is 0.1466.) $\square$

Homework Problem

Pb. 10.18 Let $A_1, A_2, \ldots, A_{m+1}$ be a partition of the set S, and let $p_1, p_2, \ldots, p_{m+1}$ be the probabilities associated with each of these events. Assuming that n Bernoulli trials are repeated, show, using Eq. (10.50), that the probability that the event A_1 occurs k_1 times, the event A_2 occurs k_2 times, etc., is given in the limit $n \to \infty$ by

$$\lim_{n\to\infty} P(k_1, k_2, \ldots, k_{m+1}; n) = \frac{(a_1)^{k_1} e^{-a_1}}{k_1!} \frac{(a_2)^{k_2} e^{-a_2}}{k_2!} \cdots \frac{(a_m)^{k_m} e^{-a_m}}{k_m!}$$

where $a_i = np_i$.

10.7.2 The Normal Distribution

Prior to considering the derivation of the normal distribution, let us recall Sterling's formula, which is the approximation of $n!$ when $n \to \infty$

$$\lim_{n\to\infty} n! \approx \sqrt{2\pi n}\, n^n e^{-n} \tag{10.57}$$

We seek the approximate form of the binomial distribution in the limit of very large n and $npq \gg 1$. Using Eq. (10.57), the expression for the probability given in Eq. (10.49), reduces to

$$P(k \text{ successes in } n \text{ trials}) = \frac{1}{\sqrt{2\pi}} \sqrt{\frac{n}{k(n-k)}} \left(\frac{np}{k}\right)^k \left(\frac{nq}{(n-k)}\right)^{n-k} \tag{10.58}$$

Now examine this expression in the neighborhood of the mean (see Pb. 10.17). We define the distance from this mean, normalized to the square root of the variance, as

$$x = \frac{k - np}{\sqrt{npq}} \tag{10.59}$$

Using the leading two terms of the power expansion of ($\ln(1 + \varepsilon) = \varepsilon - \varepsilon^2/2 + \ldots$), the natural logarithm of the two parentheses on the RHS of Eq. (10.58) can be approximated by

$$\ln\left(\frac{k}{np}\right)^{-k} \approx -(np + \sqrt{npq}\, x)\left(\sqrt{\frac{q}{np}}x - \frac{1}{2}\frac{q}{np}x^2\right) \tag{10.60}$$

$$\ln\left(\frac{n-k}{nq}\right)^{-(n-k)} \approx -(nq - \sqrt{npq}\, x)\left(-\sqrt{\frac{p}{nq}}x - \frac{1}{2}\frac{p}{nq}x^2\right) \tag{10.61}$$

Adding Eq. (10.61) and Eq. (10.62), we deduce that

$$\lim_{n\to\infty}\left(\frac{np}{k}\right)^{k}\left(\frac{nq}{(n-k)}\right)^{n-k} = e^{-x^2} \tag{10.62}$$

Furthermore, we can approximate the square root term on the RHS of Eq. (10.58) by its value at the mean; that is

$$\sqrt{\frac{n}{n(n-k)}} \approx \frac{1}{\sqrt{npq}} \tag{10.63}$$

Combining Eq. (10.62) and Eq. (10.63), we can approximate Eq. (10.58), in this limit, by the Gaussian distribution

$$P(k \text{ successes in } n \text{ trials}) = \frac{1}{\sqrt{2\pi npq}}\exp\left[-\frac{(k-np)^2}{2npq}\right] \tag{10.64}$$

This result is known as the De Moivre–Laplace theorem. We compare in Figure 10.2 the binomial distribution and its Gaussian approximation in the region of the validity of the approximation.

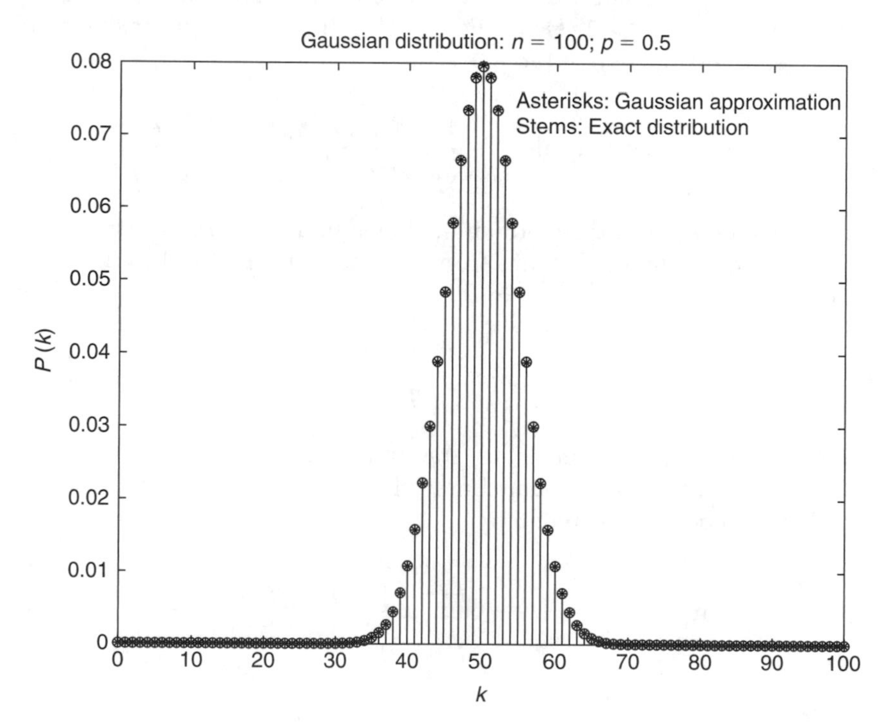

FIGURE 10.2
The normal (Gaussian) distribution.

EXAMPLE 10.16

A fair die is rolled 400 times. Find the probability that an even number of spots show up 200, 210, 220, and 230 times.

Solution: In this case, $n = 400$; $p = 0.5$; $np = 200$; and $\sqrt{npq} = 10$.

Using Eq. (10.65), we get $\begin{cases} P(200 \text{ even}) = 0.03989; & P(210 \text{ even}) = 0.02419 \\ P(220 \text{ even}) = 0.00540; & P(230 \text{ even}) = 4.43 \times 10^{-4} \end{cases}$

□

Homework Problems

Pb. 10.19 Using the results of Pb. 4.34, relate in the region of validity of the Gaussian approximation the quantity

$$\sum_{k=k_1}^{k_2} P(k \text{ successes in } n \text{ trials})$$

to the Gaussian integral, specifying each of the parameters appearing in your expression. (*Hint:* First show that in this limit, the summation can be approximated by an integration.)

Pb. 10.20 Let $A_1, A_2, \ldots, A_r$ be a partition of the set S, and let $p_1, p_2, \ldots, p_r$ be the probabilities associated with each of these events. Assuming n Bernoulli trials are repeated, show that, in the limit $n \to \infty$ and where k_i are in the vicinity of $np_i >> 1$, the following approximation is valid:

$$P(k_1, k_2, \ldots, k_r; n) = \frac{\exp\left\{-\frac{1}{2}\left[\frac{(k_1 - np_1)^2}{np_1} + \cdots + \frac{(k_r - np_r)^2}{np_r}\right]\right\}}{\sqrt{(2\pi n)^{r-1} p_1 \ldots p_r}}$$

Appendix A: Review of Elementary Functions

In this appendix, we review the basic features and characteristics of the simple elementary functions.

A.1 Affine Functions

By an affine function, we mean an expression of the form

$$y(x) = ax + b \tag{A.1}$$

In the special case where $b = 0$, we say that y is a linear function of x.

We can interpret the parameters in the above function as representing the slope–intercept form of a straight line (Figure A.1). Here, a is the slope, which is a measure of the steepness of a line; and b is the y-intercept (i.e., the line intersects the y-axis at the point $(0, b)$).

The following cases illustrate the different possibilities:

1. $a = 0$: this specifies a horizontal line at a height b above the x-axis and which has zero slope.
2. $a > 0$: the height of a point on the line (i.e., the y-value) increases as the value of x increases.
3. $a < 0$: the height of the line decreases as the value of x increases.
4. $b > 0$: the line y-intercept is positive.
5. $b < 0$: the line y-intercept is negative.
6. $x = k$: this function represents a vertical line passing through the point $(k, 0)$.

It should be noted that

- If two lines have the same slope, they are parallel.
- Two nonvertical lines are perpendicular if and only if their slopes are negative reciprocals of each other. (It is easy to deduce this

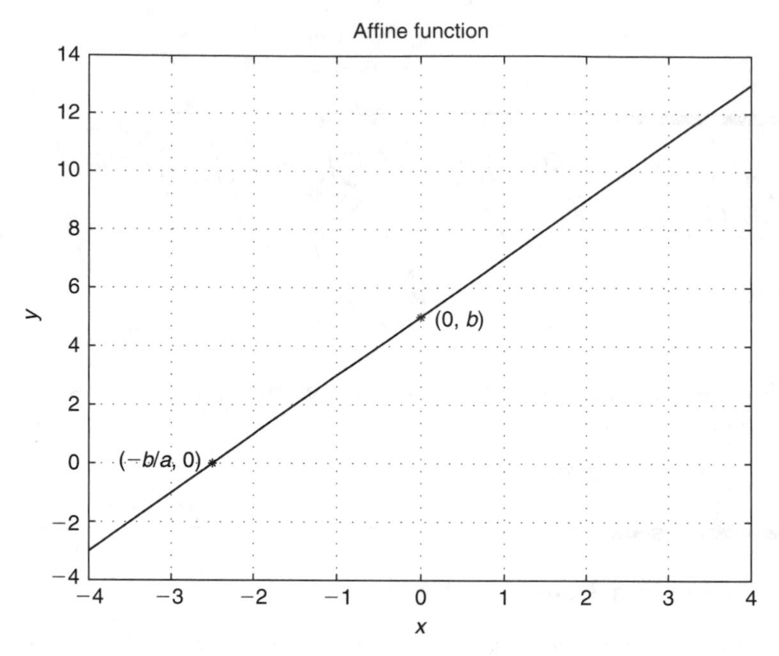

FIGURE A.1
Graph of the line $y = ax + b$ ($a = 2$, $b = 5$).

property if you remember the relationship that you learned in trigonometry relating the sine and cosine of two angles that differ by $\pi/2$.) See Section A.4 for more details.

A.2 Quadratic Functions

Parabola

A quadratic parabolic function is an expression of the form

$$y(x) = ax^2 + bx + c \quad \text{where } a \neq 0 \tag{A.2}$$

Any x for which $ax^2 + bx + c = 0$ is called a root or a zero of the quadratic function. The graphs of quadratic functions are called parabolas.

If we plot these parabolas, we note the following characteristics:

1. For $a > 0$, the parabola opens up (convex curve) as shown in Figure A.2.
2. For $a < 0$, the parabola opens down (concave curve) as shown in Figure A.2.

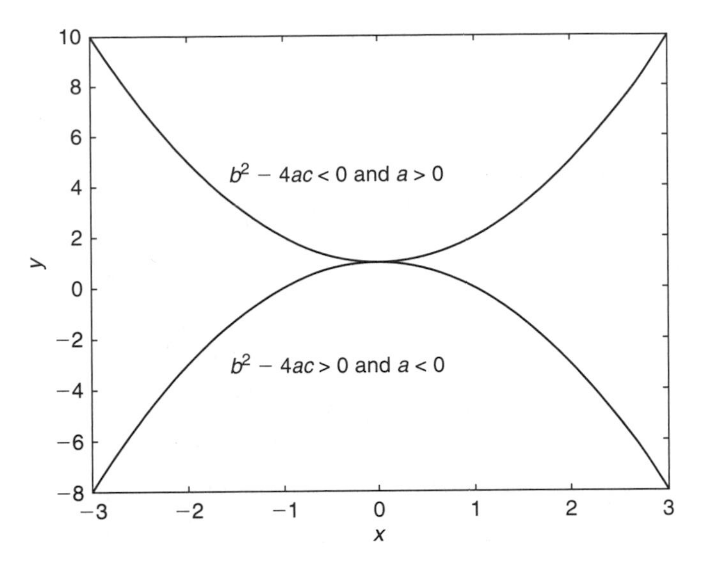

FIGURE A.2
Graph of a quadratic parabolic (second-order polynomial) function with 0 or 2 roots.

3. The parabola does not always intersect the x-axis; but where it does, this point's abscissa is a real root of the quadratic equation.

A parabola can cross the x-axis in either 0 or 2 points, or the x-axis can be tangent to it at one point. If the vertex of the parabola is above the x-axis and the parabola opens up, there is no intersection, and hence, no real roots. If, on the other hand, the parabola opens down, the curve will intersect at two values of x equidistant from the vertex position. If the vertex is below the x-axis, we reverse the convexity conditions for the existence of two real roots. We recall that the roots of a quadratic equation are given by

$$x_{\pm} = \frac{-b \pm \sqrt{b^2 - 4ac}}{2a} \tag{A.3}$$

When $b^2 - 4ac < 0$, the parabola does not intersect the x-axis. There are no real roots; the roots are said to be complex conjugates. When $b^2 - 4ac = 0$, the x-axis is tangent to the parabola and we have one double root.

Geometrical Description of a Parabola

The parabola can also be described through the following geometric construction: a parabola is the locus of all points P in a plane that are equidistant from a fixed line (called the directrix) and a fixed point (called the focus) not situated on the line (Figure A.3).

$$d_1 = d_2 \tag{A.4}$$

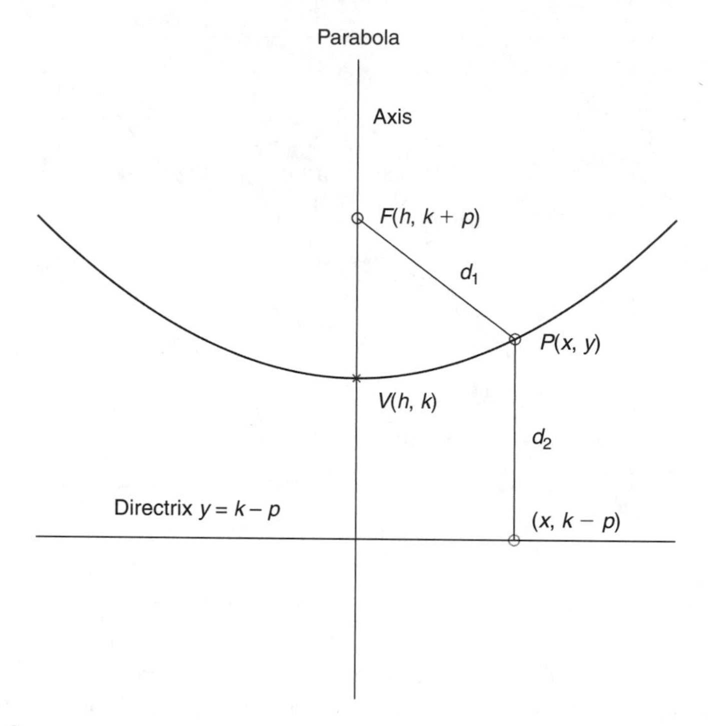

FIGURE A.3
Graph of a parabola defined through geometric parameters (parameter values: $h = 2$, $k = 2$, $p = 1$).

The algebraic expression for the parabola, using the above geometric parameters, can be obtained by specifically writing and equating the expressions for the distances of a point on the parabola from the focus and from the directrix

$$\sqrt{(x-h)^2+(y-(k+p))^2} = |y-(k-p)| \tag{A.5}$$

Squaring both sides of this equation, this equality reduces to

$$(x-h)^2 = 4p(y-k) \tag{A.6}$$

or in standard form, it can be written as

$$y = \frac{x^2}{4p} - \frac{h}{2p}x + \left(\frac{h^2+4pk}{4p}\right) \tag{A.7}$$

Ellipse

The standard form of the equation describing an ellipse is given by

$$\frac{(x-h)^2}{a^2} + \frac{(y-k)^2}{b^2} = 1 \tag{A.8}$$

The ellipse's center is located at (h, k), and assuming $a > b$, the major axis length is equal to $2a$, the minor axis length is equal to $2b$, the foci are located at $(h - c, k)$ and $(h + c, k)$, and those of the vertices at $(h - a, k)$ and $(h + a, k)$, where

$$c^2 = a^2 - b^2 \tag{A.9}$$

Geometric Definition of an Ellipse

An ellipse is the locus of all points P such that the sum of the distance between P and two distinct points (called the foci) is constant and greater than the distance between the two foci.

$$d_1 + d_2 = 2a \tag{A.10}$$

The center of the ellipse is the midpoint between foci, and the two points of intersection of the line through the foci and the ellipse are called the vertices (Figure A.4).

The eccentricity of an ellipse is the ratio of the distance between the center and a focus to the distance between the center and a vertex; that is

$$\varepsilon = c/a \tag{A.11}$$

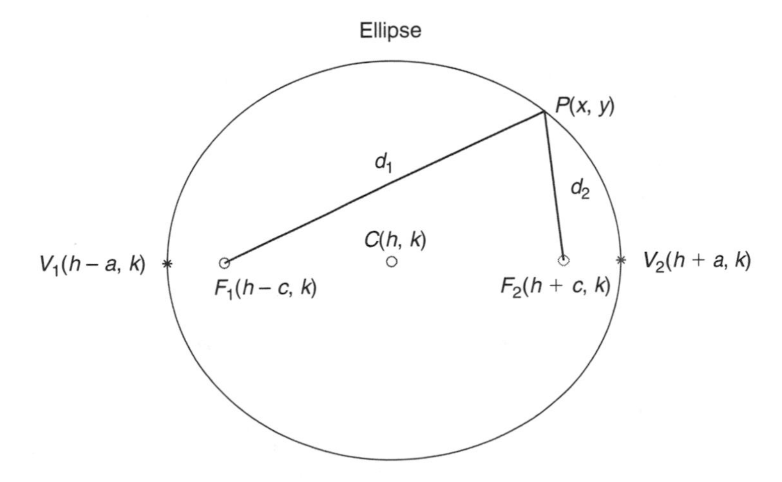

FIGURE A.4
Graph of an ellipse defined through geometric parameters (parameter values: $h = 2$, $k = 2$, $a = 3$, $b = 2$).

Hyperbola

The standard form of the equation describing a hyperbola is given by

$$\frac{(x-h)^2}{a^2} - \frac{(y-k)^2}{b^2} = 1 \tag{A.12}$$

The center of the hyperbola is located at (h, k), and assuming $a > b$, the major axis length is equal to $2a$, the minor axis length is equal to $2b$, the foci are located at $(h - c, k)$ and $(h + c, k)$, and those of the vertices at $(h - a, k)$ and $(h + a, k)$. In this case, $c > a > 0$ and $c > b > 0$ and

$$c^2 = a^2 + b^2 \tag{A.13}$$

Geometric Definition of a Hyperbola

A hyperbola is the locus of all points P in a plane such that the absolute value of the difference of the distances between P and the two foci is constant and is less than the distance between the two foci (Figure A.5); that is

$$|d_1 - d_2| = 2a \tag{A.14}$$

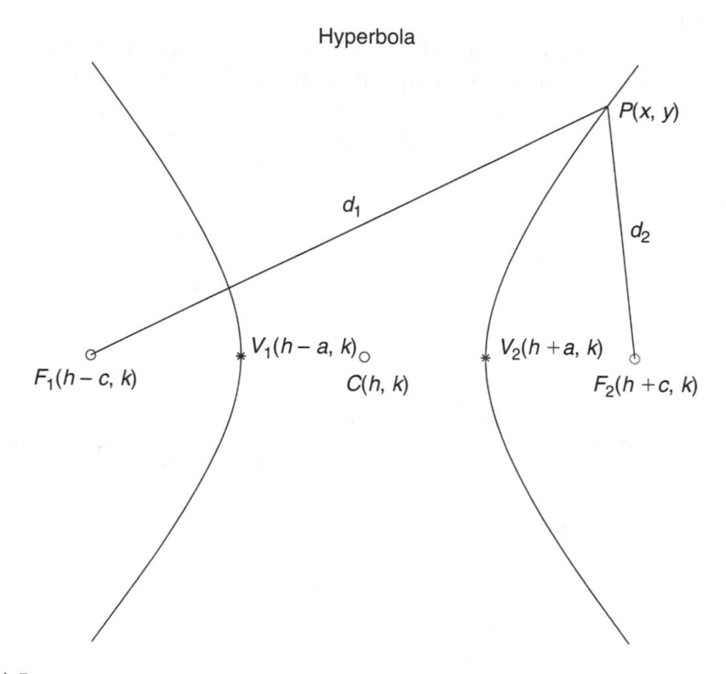

FIGURE A.5
Graph of a hyperbola defined through geometric parameters (parameter values: $h = 2$, $k = 2$, $a = 1$, $b = 3$).

The center of the hyperbola is the midpoint between foci, and the two points of intersection of the line through the foci and the hyperbola are called the vertices.

A.3 Polynomial Functions

A polynomial function is an expression of the form

$$p(x) = a_n x^n + a_{n-1} x^{n-1} + \cdots + a_1 x + a_0 \tag{A.15}$$

where $a_n \neq 0$ for an nth-degree polynomial.

The fundamental theorem of algebra states that, for the above polynomial, there are exactly n complex roots; furthermore, if all the polynomial coefficients are real, then the complex roots always come in pairs consisting of a complex number and its complex conjugate.

A.4 Trigonometric Functions

The trigonometric circle is defined as the circle with center at the origin of the coordinates axes and having radius 1 (Figure A.6).

The trigonometric functions are defined as functions of the components of a point P on the trigonometric circle. Specifically, if we define the angle θ as the angle between the x-axis and the line OP, then

- $\cos(\theta)$ is the x-component of point P.
- $\sin(\theta)$ is the y-component of point P.

Using the Pythagorean theorem in the right-angle triangle OQP, one deduces that

$$\sin^2(\theta) + \cos^2(\theta) = 1 \tag{A.16}$$

Using the above definitions for the sine and cosine functions and elementary geometry, it is easy to note the following properties for the trigonometric functions:

$$\sin(-\theta) = -\sin(\theta) \quad \text{and} \quad \cos(-\theta) = \cos(\theta) \tag{A.17}$$

$$\sin(\theta + \pi) = -\sin(\theta) \quad \text{and} \quad \cos(\theta + \pi) = -\cos(\theta) \tag{A.18}$$

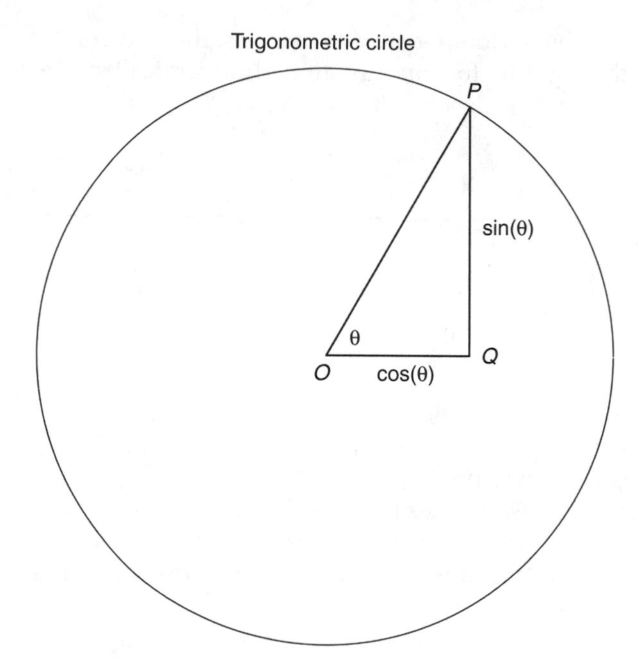

FIGURE A.6
The trigonometric circle.

$$\sin(\theta + \pi/2) = \cos(\theta) \quad \text{and} \quad \cos(\theta + \pi/2) = -\sin(\theta) \tag{A.19}$$

$$\sin(\pi/2 - \theta) = \cos(\theta) \quad \text{and} \quad \cos(\pi/2 - \theta) = \sin(\theta) \tag{A.20}$$

The tangent and cotangent functions are defined as

$$\tan(\theta) = \frac{\sin(\theta)}{\cos(\theta)} \quad \text{and} \quad \cot(\theta) = \frac{1}{\tan(\theta)} \tag{A.21}$$

Other important trigonometric relations relate the angles and sides of a triangle. These are the so-called law of cosines and law of sines in a triangle

$$c^2 = a^2 + b^2 - 2ab\cos(\gamma) \tag{A.22}$$

$$\frac{\sin(\alpha)}{a} = \frac{\sin(\beta)}{b} = \frac{\sin(\gamma)}{c} \tag{A.23}$$

where the sides of the triangle are a, b, and c, and the angles opposite of each of these sides are denoted by α, β, and γ, respectively.

A.5 Inverse Trigonometric Functions

The inverse of a function $y = f(x)$ is a function, denoted by $x = f^{-1}(y)$, having the property that $y = f(f^{-1}(y))$. It is important to note that a function $f(x)$ that is single-valued (i.e., to each element x in its domain, there corresponds one, and only one, element y in its range) may have an inverse that is multivalued (i.e., many x values may correspond to the same y). Typical examples of multivalued inverse functions are the inverse trigonometric functions. In such instances, a single-valued inverse function can be defined if the range of the inverse function is defined on a more limited region of space. For example, the $\cos^{-1}$ function (called arc cosine) is single-valued if $0 \leqslant x \leqslant \pi$.

Note that the above notation for the inverse of a function should not be confused with the negative-one power of the function $f(x)$, which should be written as

$$(f(x))^{-1} \quad \text{or} \quad 1/f(x)$$

Also note that because the inverse function reverses the role of the x- and y-coordinates, the graphs of $y = f(x)$ and $y = f^{-1}(x)$ are symmetric with respect to the line $y = x$ (i.e., the first bisector of the coordinate axes).

A.6 The Natural Logarithmic Function

The natural logarithmic function is defined by the following integral:

$$\ln(x) = \int_1^x \frac{1}{t}\,dt \tag{A.24}$$

The following properties of the logarithm can be directly deduced from the above definition:

$$\ln(ab) = \ln(a) + \ln(b) \tag{A.25}$$

$$\ln(a^r) = r\ln(a) \tag{A.26}$$

$$\ln\left(\frac{1}{a}\right) = -\ln(a) \tag{A.27}$$

$$\ln\left(\frac{a}{b}\right) = \ln(a) - \ln(b) \tag{A.28}$$

To illustrate the technique for deriving any of the above relations, let us consider the first of them:

$$\ln(ab) = \int_1^{ab} \frac{1}{t}\,dt = \int_1^{a} \frac{1}{t}\,dt + \int_a^{ab} \frac{1}{t}\,dt \tag{A.29}$$

The first term on the RHS is $\ln(a)$, while the second term through the substitution $u = t/a$ reduces to the definition of $\ln(b)$.

Note that

$$\ln(1) = 0 \tag{A.30}$$

$$\ln(e) = 1 \tag{A.31}$$

where $e = 2.71828$.

A.7 The Exponential Function

The exponential function is defined as the inverse function of the natural logarithmic function; that is

$$\exp(\ln(x)) = x \quad \text{for all } x > 0 \tag{A.32}$$

$$\ln(\exp(y)) = y \quad \text{for all } y \tag{A.33}$$

The following properties of the exponential function hold for all real numbers:

$$\exp(a)\exp(b) = \exp(a + b) \tag{A.34}$$

$$(\exp(a))^b = \exp(ab) \tag{A.35}$$

$$\exp(-a) = \frac{1}{\exp(a)} \tag{A.36}$$

$$\frac{\exp(a)}{\exp(b)} = \exp(a - b) \tag{A.37}$$

It should be pointed out that any of the above properties can be directly obtained from the definition of the exponential function and the properties

of the logarithmic function. For example, the first of these relations can be derived as follows:

$$\ln(\exp(a)\exp(b)) = \ln(\exp(a)) + \ln(\exp(b)) = a + b \tag{A.38}$$

Taking the exponential of both sides of this equation, we obtain:

$$\exp(\ln(\exp(a)\exp(b))) = \exp(a)\exp(b) = \exp(a + b) \tag{A.39}$$

which is the desired result.

Useful Features of the Exponential Function

If the exponential function is written in the form

$$y(x) = \exp(-bx) \tag{A.40}$$

the following features are apparent:

1. If $b > 0$, then the function is convergent at $(+\infty)$ and goes to zero there.
2. If $b < 0$, then the function blows up at $(+\infty)$.
3. If $b = 0$, then the function is equal everywhere to a constant $y = 1$.
4. The exponential functions are monotonically increasing for $b < 0$, and monotonically decreasing for $b > 0$.
5. If $b_1 > b_2 > 0$, then everywhere on the positive x-axis, $y_1(x) < y_2(x)$.
6. The exponential function has no roots.
7. For $b > 0$, the product of the exponential function with any polynomial goes to zero at $(+\infty)$.

We plot in Figure A.7 and Figure A.8 examples of the exponential function for different values of the parameters. The first six properties above are clearly exhibited in these figures.

A.8 The Hyperbolic Functions

The hyperbolic cosine function is defined by

$$\cosh(x) = \frac{\exp(x) + \exp(-x)}{2} \tag{A.41}$$

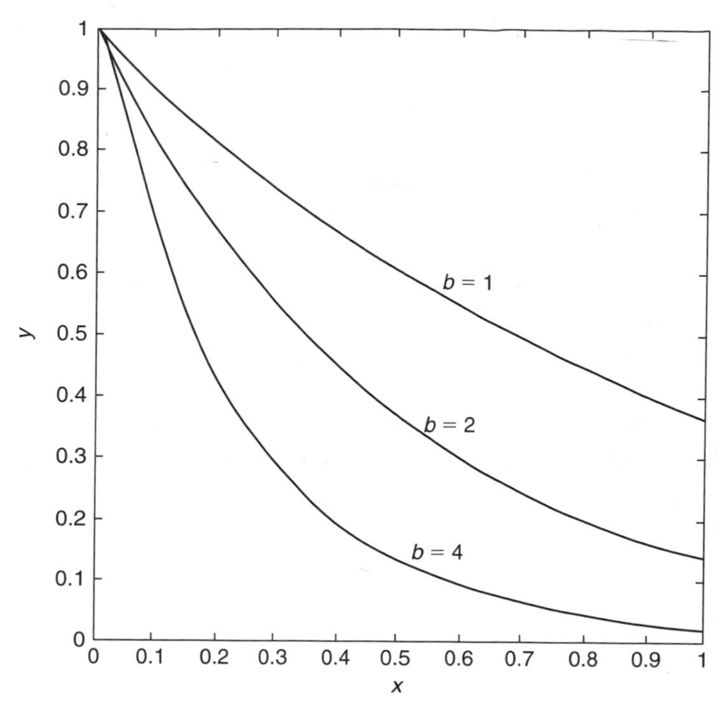

FIGURE A.7
The graph of the function $y = \exp(-bx)$ for different positive values of b.

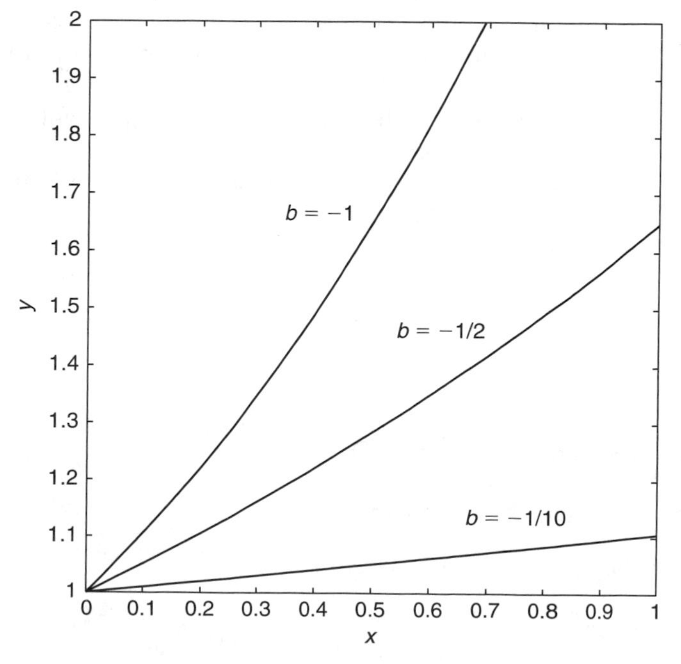

FIGURE A.8
The graph of the function $y = \exp(-bx)$ for different negative values of b.

and the hyperbolic sine function is defined by

$$\sinh(x) = \frac{\exp(x) - \exp(-x)}{2} \tag{A.42}$$

Using the above definitions, it is straightforward to derive the following relations:

$$\cosh^2(x) - \sinh^2(x) = 1 \tag{A.43}$$

$$1 - \tanh^2(x) = \text{sech}^2(x) \tag{A.44}$$

A.9 The Inverse Hyperbolic Functions

$$y = \sinh^{-1}(x) \quad \text{if } x = \sinh(y) \tag{A.45}$$

Using the definition of the hyperbolic functions, we can write the inverse hyperbolic functions in terms of logarithmic functions. For example, considering the inverse hyperbolic sine function from above, we obtain

$$e^y - 2x - e^{-y} = 0 \tag{A.46}$$

multiplying by e^y everywhere, we obtain a second-degree equation in e^y:

$$e^{2y} - 2xe^y - 1 = 0 \tag{A.47}$$

Solving this quadratic equation, and choosing the plus term in front of the discriminant, since e^y is positive everywhere, we obtain

$$e^y = x + \sqrt{x^2 + 1} \tag{A.48}$$

giving, for the inverse hyperbolic sine function, the expression

$$y = \sinh^{-1}(x) = \ln(x + \sqrt{x^2 + 1}) \tag{A.49}$$

In a similar manner, one can show the following other identities:

$$\cosh^{-1}(x) = \ln(x + \sqrt{x^2 - 1}) \tag{A.50}$$

$$\tanh^{-1}(x) = \frac{1}{2}\ln\left(\frac{1+x}{1-x}\right) \tag{A.51}$$

$$\text{sech}^{-1}(x) = \frac{1}{2}\ln\left(\frac{1+\sqrt{1-x^2}}{x}\right) \tag{A.52}$$

Appendix B: Determinants

In this appendix, we shall summarize the basic properties of determinants. The proofs of these properties are not given here and the reader is referred to any standard book of linear algebra for a more comprehensive treatment of this subject. Our purpose here is to give a quick reference of the basic properties of determinants to allow for their practical use.

Definition: The determinant of a square matrix **M**, of size $(N \otimes N)$, is a number equal to

$$\det(\mathbf{M}) = \sum_{P} \varepsilon_{abc\cdots n} M_{1a}\, M_{2b}\, M_{3c} \cdots M_{Nn}$$

where $\varepsilon_{abc\cdots n}$ is the antisymmetric symbol over N objects and its value is given by

$$\varepsilon_{abc\cdots n} = \begin{cases} 1 & \text{if } (a, b, c, \ldots, n) \text{ is an even permutation of } (1, 2, 3, \ldots, N) \\ -1 & \text{if } (a, b, c, \ldots, n) \text{ is an odd permutation of } (1, 2, 3, \ldots, N) \\ 0 & \text{otherwise} \end{cases}$$

1. An even (odd) permutation consists of an even (odd) number of transpositions. (A transposition is a permutation where only two elements exchange positions.)
2. The total number of permutations of N objects is $N!$, and the determinant has this number of terms in the sum.
3. Each term in the sum consists of the product of one element from each row and each column. Products containing more than a single term from a given column or row do not contribute to the determinant because the corresponding antisymmetric symbol is zero.

Properties: The fundamental properties of determinants are summarized by the following theorems:

THEOREM 1 A determinant does not change value if a common factor of each element of a column (row) is factored out as an overall multiplicative constant.

THEOREM 2 If the elements of any column (row) are all 0s, then the determinant of that matrix is 0.

THEOREM 3 The determinant of the identity matrix is equal to 1.

THEOREM 4 Let the determinant D be of the form

$$D = \begin{vmatrix} M_{11} & M_{12} & \cdots & \cdots & \cdots & \cdots & M_{1N} \\ M_{21} & M_{22} & & & & & \vdots \\ \vdots & & & \ddots & & & \vdots \\ \vdots & & & & & & \vdots \\ M_{i1}+P_{i1} & M_{i2}+P_{i2} & \cdots & \cdots & & & M_{iN}+P_{iN} \\ \vdots & \vdots & & & & & \vdots \\ M_{N1} & M_{N2} & \cdots & \cdots & & \cdots & M_{NN} \end{vmatrix}$$

then

$$D = \begin{vmatrix} M_{11} & M_{12} & \cdots & \cdots & \cdots & \cdots & M_{1N} \\ M_{21} & M_{22} & & & & & \vdots \\ \vdots & & \ddots & & & & \vdots \\ \vdots & & & & & & \vdots \\ M_{i1} & M_{i2} & \cdots & \cdots & & & M_{iN} \\ \vdots & \vdots & & & & & \vdots \\ M_{N1} & M_{N2} & \cdots & \cdots & & \cdots & M_{NN} \end{vmatrix} + \begin{vmatrix} M_{11} & M_{12} & \cdots & \cdots & \cdots & \cdots & M_{1N} \\ M_{21} & M_{22} & & & & & \vdots \\ \vdots & & & \ddots & & & \vdots \\ \vdots & & & & & & \vdots \\ P_{i1} & P_{i2} & \cdots & \cdots & & & P_{iN} \\ \vdots & \vdots & & & & & \vdots \\ M_{N1} & M_{N2} & \cdots & \cdots & & \cdots & M_{NN} \end{vmatrix}$$

THEOREM 5 Interchanging any two columns (rows) changes the sign of the determinant.

THEOREM 6 If any two columns (rows) are multiple of each other, the determinant is 0.

THEOREM 7 The determinant of the product of two matrices is the product of their determinants.

THEOREM 8 A scalar multiple of a column (row) may be added to another column (row) without changing the value of a determinant.

If we define the cofactor of an element of the matrix **M**, $\text{cof}(M_{ij})$ as the product of $(-1)^{i+j}$ by the determinant of the $(N-1 \otimes N-1)$ matrix obtained by eliminating both the ith row and jth column of the matrix **M**, then the following theorems are valid:

THEOREM 9 The value of a determinant can be obtained by summing over the product of each element of the row multiplied by its cofactor.

$$\det(\mathbf{M}) = \sum_j M_{ij}\text{cof}(M_{ij})$$

(Note that the summation is over j only.)

Definition: The classical adjoint (not to be confused with the Hermitian adjoint) of an $(N \otimes N)$ matrix is the $(N \otimes N)$ matrix which elements are defined as

$$(\text{adj } \mathbf{M})_{ij} = \text{cof}(M_{ji})$$

THEOREM 10 The matrix $\mathbf{M}$ has an inverse denoted by $\mathbf{M}^{-1}$ if and only if $\det(\mathbf{M}) \neq 0$, and the value of this inverse is given by

$$\mathbf{M}^{-1} = \left(\frac{\text{adj } \mathbf{M}}{\det(\mathbf{M})}\right)$$

THEOREM 11 The solution of the system of linear equations:

$$\mathbf{MX} = \mathbf{B}$$

where both $\mathbf{X}$ and $\mathbf{B}$ are $(N \otimes 1)$ columns, is given by Cramer's rule

$$X_i = \frac{1}{\det(\mathbf{M})}\sum_j B_j \text{ cof}(M_{ji})$$

EXAMPLE B.1

Compute the cofactors, determinant, and inverse of the following matrix:

$$\mathbf{M} = \begin{pmatrix} 1 & 3 & 5 \\ 7 & 11 & 13 \\ 17 & 19 & 23 \end{pmatrix}$$

Solution:

1. The cofactors are

$$\text{cof}(M_{11}) = (-1)^{1+1}\det\begin{pmatrix} 11 & 13 \\ 19 & 23 \end{pmatrix} = 6$$

$$\text{cof}(M_{12}) = (-1)^{1+2}\det\begin{pmatrix} 7 & 13 \\ 17 & 23 \end{pmatrix} = 60$$

$$\text{cof}(M_{13}) = (-1)^{1+3}\det\begin{pmatrix} 7 & 11 \\ 17 & 19 \end{pmatrix} = -54$$

$$\text{cof}(M_{21}) = (-1)^{2+1}\det\begin{pmatrix} 3 & 5 \\ 19 & 23 \end{pmatrix} = 26$$

$$\text{cof}(M_{22}) = (-1)^{2+2}\det\begin{pmatrix} 1 & 5 \\ 17 & 23 \end{pmatrix} = -62$$

$$\text{cof}(M_{23}) = (-1)^{2+3}\det\begin{pmatrix} 1 & 3 \\ 17 & 19 \end{pmatrix} = 32$$

$$\text{cof}(M_{31}) = (-1)^{3+1}\det\begin{pmatrix} 3 & 5 \\ 11 & 13 \end{pmatrix} = -16$$

$$\text{cof}(M_{32}) = (-1)^{3+2}\det\begin{pmatrix} 1 & 5 \\ 7 & 13 \end{pmatrix} = 22$$

$$\text{cof}(M_{33}) = (-1)^{3+3}\det\begin{pmatrix} 1 & 3 \\ 7 & 11 \end{pmatrix} = -10$$

2. The determinant of **M** is

$$\begin{aligned}\det(\mathbf{M}) &= M_{11}\ \text{cof}(M_{11}) + M_{12}\ \text{cof}(M_{12}) + M_{13}\ \text{cof}(M_{13}) \\ &= (1\times 6) + (3\times 60) - (5\times 54) = -84\end{aligned}$$

3. The inverse of the matrix **M** is

$$\mathbf{M}^{-1} = \begin{pmatrix} \dfrac{\text{cof}(M_{11})}{\det(\mathbf{M})} & \dfrac{\text{cof}(M_{21})}{\det(\mathbf{M})} & \dfrac{\text{cof}(M_{31})}{\det(\mathbf{M})} \\ \dfrac{\text{cof}(M_{12})}{\det(\mathbf{M})} & \dfrac{\text{cof}(M_{22})}{\det(\mathbf{M})} & \dfrac{\text{cof}(M_{32})}{\det(\mathbf{M})} \\ \dfrac{\text{cof}(M_{13})}{\det(\mathbf{M})} & \dfrac{\text{cof}(M_{23})}{\det(\mathbf{M})} & \dfrac{\text{cof}(M_{33})}{\det(\mathbf{M})} \end{pmatrix} = \begin{pmatrix} \dfrac{-6}{84} & \dfrac{-26}{84} & \dfrac{16}{84} \\ \dfrac{-60}{84} & \dfrac{62}{84} & \dfrac{-22}{84} \\ \dfrac{54}{84} & \dfrac{-32}{84} & \dfrac{10}{84} \end{pmatrix} \quad \square$$

Appendix C: Symbolic Calculations with MATLAB

In this text, we have used MATLAB as a tool for numerical computation; however the Symbolic Math Toolbox based on the Maple® V software package gives MATLAB the ability to also perform symbolic calculations and obtain in certain cases answers in closed form.

To facilitate the use of the Symbolic Math Toolbox, the MATLAB student edition has incorporated within its kernel all the functions of the Symbolic Math Toolbox; however the full accessibility to the Maple Kernel requires the Toolbox.

A word of caution: the syntax used in the Symbolic Math Toolbox is not identical to that used in the Maple commercial products.

The reader proposing to use Symbolic MATLAB is advised to keep in mind, the following:

1. A closed-form solution to a problem is not always possible. Actually, the solutions that can be written in closed form constitute a very small subset of the much larger set of solutions that can be obtained numerically. For example, we cannot always find in a closed form the integral of a function, while it is easy to find numerically the definite integral of any function.
2. The availability of the Symbolic Math Toolbox should in no way preclude the student from learning the analytical techniques and methodologies available to solve certain classes of problems. A deeper understanding of any new material and the ability of the student to develop practical skills for analyzing new problems can only be achieved through mastering these gymnastics.
3. The Symbolic Toolbox can be used to great advantage, at least in the early stages of studying a new topic, to verify the answers obtained by hand, or to give hints to how to proceed when stuck.
4. One should not hesitate to use the Symbolic Toolbox once a certain level of familiarity and mastery of a particular subject has been achieved and the problem at hand has an analytical solution.

We shall review in this appendix different functions from the Symbolic Math Toolbox which are suitable for the following tasks: symbolic manipulation of algebraic expressions, solving algebraic equations, symbolic 1-D calculus including finding limits, derivatives, integrals, Taylor's expansion of a function, sums of series and solutions to ordinary differential equations, symbolic linear algebra including finding determinants, matrix products, matrix inverse, eigenvalues, and eigenfunctions. We shall finally introduce at a very elementary level the Laplace and the z-transforms, key techniques used in more advanced studies in engineering to solve, *inter alia*, constant coefficients linear differential and difference equations, respectively.

The material of this appendix is arranged, so as to permit reader's easy reference to specific commands. Fuller details on the syntax of each command can be obtained by entering **`help 'name of command'`**.

The commands have been grouped here according to their functionality.

C.1 Symbolic Manipulation

In this section, we shall cover those commands that are suitable for simplifying algebraic expressions and making them prettier (i.e., more transparent to further analysis and manipulation).

It should be noted that the same symbols used for algebraic and logical operations in numerical MATLAB are still valid here.

C.1.1 Creating Symbolic Expressions

1. MATLAB is instructed that you are proceeding with Symbolic Math when you enter the command

   ```
   >> syms x y z real
   ```

   This command specifies that you desire MATLAB to consider x, y, and z as real symbolic variables.

2. In case you would also like MATLAB to use the symbol π or other irrational quantities such as $\sqrt{2}$ in subsequent calculations rather than their floating-point approximations, MATLAB has to be instructed this at the beginning of the calculation.

Example C.1

Entering

```
pi=sym('pi')
```

returns

```
pi =
     pi
```

□

3. If you wish to instruct MATLAB that the function f is a function of the variable t, you use the syntax

```
f=sym('f(t)')
```

4. If you wish to have in an expression E, the variable x replaced by y, where y can be either a symbolic variable or a number, you use the command

```
subs(E,x,y)
```

EXAMPLE C.2

Entering

```
syms x y
E=(x^3+y^2+x*y+y+4);
subs(E,x,y)
```

returns

```
ans =
     y^3+2*y^2+y+4
```

While, entering

```
syms t
f=sym('f(t)');
h=subs(f,t,t+2)-subs(f,t,t-1)-f
```

returns

```
h =
    f(t+2)-f(t-1)-f(t)
```

□

5. MATLAB command **`pretty`** displays an expression on the screen to resemble typeset mathematics.

EXAMPLE C.3

Entering

```
syms x y
E=((x^2)+(y^3)/(3+sin(x)))/(cos(y)^3+x^2*y^2);
pretty(E)
```

returns

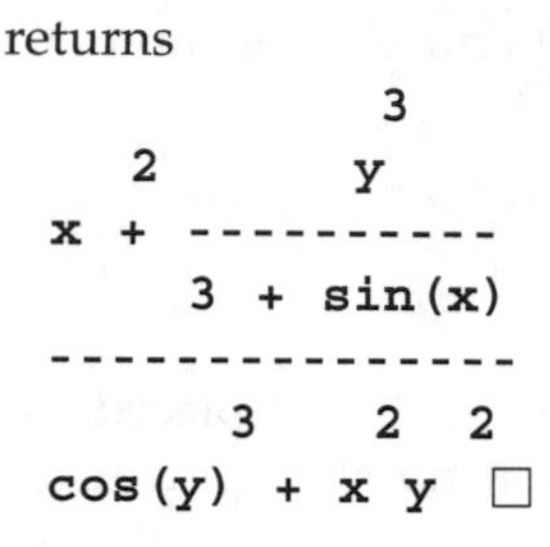

Usually, this command is used only after all algebraic manipulations have been completed; however it can also be used to advantage to double check visually the accuracy of the expressions entered. The logic being that most people's eyes are better trained to detect typos when the expression is in the typeset format.

6. If at any time, you wish to obtain the numerical value of an expression E, the command to use is

```
double(E)
```

EXAMPLE C.4

Entering

```
syms x
E=x^2*sin(x)^3+cos(x)^2+1;
F=subs(E,x,1/2);
double(F)
```

returns

```
ans =
     1.7977
```

□

The use of the word **double** here stands for floating-point double precision. You should be careful not to use this command except having previously used the command **subs** to give values to the different symbolic variables.

7. When not sure which are the symbolic variables in an expression, use the command

```
findsym(E)
```

This gives a list of the symbolic variables in expression E in alphabetical order.

C.1.2 Algebraic Manipulation

In addition to those commands that generated and modified symbolic expressions, MATLAB Symbolic Math Toolbox allows the algebraic manipulation of expressions. The principal functions available are:

8. `collect(E,x)`. This command collects together terms in the expression E having the same power of x.

Example C.5

Entering

```
syms x
E=(exp(x)+2*x)*(x+3);
F=collect(E,x)
```

returns

```
F =
  2*x^2+(exp(x)+6)*x+3*exp(x)
```
□

9. `expand(E)`. This command expands positive integer powers and products in the expression E. This command also uses the addition and/or product properties of elementary functions.

Example C.6

Entering

```
syms x y a
E=((a+x)^2)*((x+3)^2)*cos(x+y);
F=expand(E)
```

returns

```
F =
        a^2*x^2*cos(x)*cos(y)-
        a^2*x^2*sin(x)*sin(y)+6*a^2*x*cos(x)*cos(y)-
        6*a^2*x*sin(x)*sin(y)+9*a^2*cos(x)*cos(y)-
        9*a^2*sin(x)*sin(y)+2*a*x^3*cos(x)*cos(y)-
        2*a*x^3*sin(x)*sin(y)+12*a*x^2*cos(x)*cos(y)-
        12*a*x^2*sin(x)*sin(y)+18*a*x*cos(x)*cos(y)-
        18*a*x*sin(x)*sin(y)+x^4*cos(x)*cos(y)-
        x^4*sin(x)*sin(y)+6*x^3*cos(x)*cos(y)-
        6*x^3*sin(x)*sin(y)+9*x^2*cos(x)*cos(y)-
        9*x^2*sin(x)*sin(y)
```
□

10. `factor(E)`. This command factors the expression E.

Example C.7

Entering

```
syms x
E=x^2-9+(x+3)^2+(x^2+7*x+12);
F=factor(E)
```

returns

```
F =
        (x+3)*(3*x+4)
```
□

11. **simplify(E)**. This command performs a series of algebraic transformations on the expression E, and returns the simplest form that it finds.

Example C.8

Entering

```
syms x y
E=exp(x)*exp(y)*(sin(x)*cos(y)+sin(y)*cos(x));
simplify(E)
```

returns

```
ans =
        (sin(x)*cos(y)+sin(y)*cos(x))*exp(x+y)
```

As you note that the program recognized the exponential product rule but did not seem to recognize that the first parenthesis is simply **sin(x+y)**. Another command **simple** will on the other hand give the form with the least number of characters, after a number of built-in intermediate manipulations not displayed here

```
simple(E)
ans =
        sin(x+y)*exp(x+y)
```
□

12. **poly2sym(p,x)**. This command takes a polynomial, written as a polynomial coefficient vector, and writes it instead as a symbolic polynomial in the variable x.

Example C.9

Find the polynomial in x associated with the polynomial coefficient vector

$$[4 \quad 3 \quad a \quad 2 \quad 7]$$

Solution:
Entering

```
syms a x
p=[4 3 a 2 7];
E=poly2sym(p,x)
```

returns

```
E =
      4*x^4+3*x^3+x^2*a+2*x+7
```
□

13. **sym2poly(E)**. This command converts a polynomial symbolic expression E into a polynomial coefficient vector.

EXAMPLE C.10

Find the polynomial coefficient vector associated with the polynomial

$$5x^3 + 2x^2 + x + \frac{1}{3}$$

Solution:
Entering

```
syms x
E=5*x^3+2*x^2+x+1/3;
p=sym2poly(E)
```

returns

```
p =
      5.0000    2.0000    1.0000    0.3333
```
□

14. **[num,den]=numden(E)**. This command performs all the operations on the different rational forms in the expression E (i.e., reduce to common denominator, etc.) and gives as output the resulting numerator and denominator of the final result.

EXAMPLE C.11

Write the following expression as a quotient of two polynomials:

$$\frac{3}{(x+3)} + \frac{5}{(x^2-9)} + (x^2+4)$$

Solution:
Entering

```
syms x
E=3/(x+3)+5/(x^2-9)+(x^2+4);
[num,den]=numden(E)
```

returns

```
num =
        -12*x^2-120-31*x+x^5-5*x^3+3*x^4
den =
        (x+3)*(x^2-9)
```

□

C.1.3 Plotting Symbolic Expressions

The Symbolic Math Toolbox graphics are implemented by the prefix **ez** (for easy) series. These serve the same functions as the corresponding graphics commands of the numerical MATLAB kernel, but do not require explicitly specific data points. The input arguments of these functions can be function handles, string expressions or symbolic math expressions.

15. **ezplot** is the principal command for 2-D graphics. It can be used in a variety of ways:
    - **ezplot(f, [xmin xmax])** plots $f = f(x)$ over the specified domain.

EXAMPLE C.12

Plot the function

$$f(x) = x^2 + 3x \quad \text{for } 0 \leq x \leq 3$$

Solution:
Enter

```
syms x;
f=x^2+3*x;
ezplot(f,[0 3])
```

□

- **ezplot(f,[xmin,xmax,ymin,ymax])** plots the implicitly defined function $f(x, y) = 0$ over the specified x–y domain.

EXAMPLE C.13

Consider the ellipse whose implicit equation is

$$\frac{x^2}{9} + \frac{y^2}{4} = 1$$

Plot this curve in the domain $-4 \leq x \leq 4$ and $-3 \leq y \leq 3$.

Solution:
Enter

```
syms x y
f=(x^2)/9+(y^2)/4-1;
ezplot(f,[-4,4,-3,3])
```
□

- **ezplot(x,y,[tmin, tmax])** is the command to be used when plotting the parametric equations $x(t)$ and $y(t)$ over the interval $t_{min} \leq t \leq t_{max}$.

Example C.14

Consider the ellipse described by the parametric equations

$$x(t) = 3\cos(t), \quad y(t) = 2\sin(t)$$

Plot this curve for $0 \leq t \leq 2\pi$.

Solution:
Enter

```
syms t
x=3*cos(t);
y=2*sin(t);
ezplot(x,y,[0,2*pi])
```
□

16. **ezpolar(f,[thetain thetafin])** is the polar coordinate plotter of the function f over the interval $\theta_{in} \leq \theta \leq \theta_{fin}$.

Example C.15

Create a polar plot of the function f over the indicated interval

$$f = 1/(1 - 0.5\cos(\theta)), \qquad 0 \leq \theta \leq 2\pi$$

Solution:
Enter

```
syms theta
f=1/(1-0.5*cos(theta));
ezpolar(f,[0 2*pi])
```
□

17. **ezplot3(x,y,z,[tmin,tmax])** is the 3-D parametric curve plotter. It plots the curve described by the parametric equations $x(t)$, $y(t)$, and $z(t)$ over the domain $t_{min} \leq t \leq t_{max}$.

Example C.16

Plot the helix described by the following parametric equations over the indicated domain:

$$x = 3\cos(t), \quad y = 3\sin(t), \quad z = 0.1\,t, \qquad 0 \leqslant t \leqslant 10\,\pi$$

Solution:
Enter

```
syms t
x=3*cos(t);
y=3*sin(t);
z=0.1*t;
ezplot3(x,y,z,[0,10*pi]) □
```

18. **`ezcontour(f,[xmin,xmax],[ymin,ymax])`** plots the contour lines of $f(x, y)$, where f is a symbolic expression that represents a mathematical function of the two variables x and y, over the domain $x_{min} \leq x \leq x_{max}$ and $y_{min} \leq y \leq y_{max}$.

Example C.17

Consider the function f given by

$$f = (1 + x^2)\exp(-x^2 - y^2) + \cos(4y)\exp(-x^2)$$

Obtain its contour lines over the interval $0 \leq x \leq 2$, $0 \leq y \leq 2\pi$.

Solution:
Enter

```
syms x y
f=(1+x^2)*exp(-x^2-y^2)+cos(4*y)*exp(-x^2);
ezcontour(f,[0,2],[0,2*pi]) □
```

19. **`ezmesh(f,[xmin,xmax],[ymin,ymax])`**. This command plots the mesh 3-D graph of $f(x, y)$, where f is a symbolic expression that represents a mathematical function of the two variables x and y, over the domain $x_{min} \leq x \leq x_{max}$ and $y_{min} \leq y \leq y_{max}$.

Example C.18

Plot the mesh of the function f given by

$$f = (1 + x^2)\exp(-x^2 - y^2) + \cos(4y)\exp(-x^2)$$

over the interval $0 \leq x \leq 2$, $0 \leq y \leq 2\pi$.

Solution: The following program plots the function f 3-D mesh over the given domain:

```
syms x y
f=(1+x^2)*exp(-x^2-y^2)+cos(4*y)*exp(-x^2);
ezmesh(f,[0,2],[0,2*pi]) □
```

20. `ezsurf(f,[xmin,xmax],[ymin,ymax])`. This command plots the 3-D graph of the surface $f(x, y)$, where f is a symbolic expression that represents a mathematical function of the two variables x and y, over the domain $x_{min} \le x \le x_{max}$ and $y_{min} \le y \le y_{max}$.

EXAMPLE C.19

Plot the surface associated with the function f given by

$$f = (1 + x^2)\exp(-x^2 - y^2) + \cos(4y)\exp(-x^2)$$

over the interval $0 \le x \le 2$, $0 \le y \le 2\pi$.

Solution:
Enter

```
syms x y
f=(1+x^2)*exp(-x^2-y^2)+cos(4*y)*exp(-x^2);
ezsurf(f,[0,2],[0,2*pi]) □
```

C.2 Symbolic Solution of Algebraic and Transcendental Equations

21. `solve(eq1,eq2,...,var1,var2,...)`. This command is suitable for solving one or a set of algebraic or transcendental equations. The number of equations should be equal to the number of variables.

The syntax allows that the functional forms of the equations be specifically included in the command, or that they be called by reference to previous definitions for the equations.

EXAMPLE C.20

Solve the equation

$$x^2 - 5x + 6 = 0$$

Solution: We include here the specific form of the equation in the **solve** command:
Entering

```
syms x
solve('x^2-5*x+6=0',x)
```

returns

```
ans =
        3
        2
```

□

EXAMPLE C.21

Solve the system of equations

$$3x + 4y = 25$$

$$y - x = 1$$

Solution: Here we shall include the equations by reference in the **solve** command:
Entering

```
syms x y
eq1='3*x+4*y=25';
eq2='-x+y=1';
[x,y]=solve(eq1,eq2,x,y)
```

returns

```
x =
      3
y =
      4
```

□

C.3 Symbolic Calculus

22. **limit**. This command finds the limit of a symbolic expression at a point. It can compute either the left or right one-sided limits. Its syntax is

```
limit(F,x,a,'right') or limit(F,x,a,'left')
```

where x is the variable and a the point at which we desire to find the limit.

EXAMPLE C.22

Find $\lim_{x \to 0} \frac{\sin(x)}{x}$

Solution:
Entering

```
syms x
leftlimit=limit(sin(x)/x,x,0,'left')
rightlimit=limit(sin(x)/x,x,0,'right')
```

returns

```
leftlimit =
          1
rightlimit =
           1
```
□

EXAMPLE C.23

Find $\lim_{x \to \infty} \left(1 + \frac{a}{x}\right)^x$.

Solution:
Entering

```
syms x a
limit((1 + a/x)^x,x,inf,'left')
```

returns

```
ans =
      exp(a)
```
□

23. **diff(f,x,n)**. This command finds the nth-order derivative with respect to the variable x of a symbolic expression.

EXAMPLE C.24

Find $\partial^2 f/\partial x^2$ and $\partial f/\partial t$ of the expression

$$f = t \sin(xt) + x^2 \tan(t)$$

Solution:
Entering

```
syms x t
f=t*sin(x*t)+(x^2)*tan(t);
fx2=diff(f,x,2)
ft1=diff(f,t,1)
```

returns

```
fx2 =
        -t^3*sin(x*t)+2*tan(t)
ft1 =
        sin(x*t)+t*cos(x*t)*x+x^2*(1+tan(t)^2)
```
□

24. **taylor(f,n,x,a)**. This command finds the Taylor series of a symbolic expression at a point. It returns the $(n-1)$-order Maclaurin polynomial approximation to f, where f is a symbolic expression representing a function, x specifies the independent variable in the expression, and where a can be a symbol or a numerical value.

Example C.25

Find the seventh-order power expansion at $x = 0$ of the function

$$f = \ln(1 + x)$$

Solution:
Entering

```
syms x
taylor(log(1+x),8,x,0)
```

returns

```
ans =
      x-1/2*x^2+1/3*x^3-1/4*x^4+1/5*x^5-1/
      6*x^6+1/7*x^7
```
□

25. **symsum(f,n,a,b)**. This is the symbolic summation command. It gives the definite summation of the symbolic expression f from $n = a$ to $n = b$.

Example C.26

Find $\sum_{n=1}^{\infty} \frac{1}{n^2}$.

Solution:
Entering

```
syms n
s=symsum(1/n^2,n,1,Inf)
```

returns

```
s =
     1/6*pi^2
```
□

26. **int**. This command integrates with respect to the variable x a symbolic expression. Its syntaxes are
    - **int(f,x)**. This command returns the indefinite integral of f with respect to the symbolic scalar variable x.
    - **int(f,x,a,b)**. This command returns the definite integral of f with respect to x from a to b.

EXAMPLE C.27

Find the indefinite integral

$$\int \frac{x}{(1+x^2)^2}\,dx$$

Solution:
Entering

```
syms x
g=int(x/(1+x^2)^2,x)
```

returns

```
g =
      -1/2/(1+x^2)
```
□

EXAMPLE C.28

Find the definite integral

$$\int_0^1 x^4 \exp(-ax^2)\,dx$$

Solution:
Entering

```
syms x a
h=int(x^4*exp(-a*x^2),x,0,1)
```

returns

```
h =
      1/8*(3*pi^(1/2)*erf(a^(1/2))*a^2*exp(a)-
      6*a^(5/2)-4*a^(7/2))/a^(9/2)/exp(a)
```
□

27. **dsolve('eq1','eq2',...,'cond1','cond2',...,'x')**. This command provides the symbolic solution for a set of ordinary differential

equations. Here **eq1, eq2,...** are the ordinary differential equations we want to solve, **x** is the independent variable and the boundary and/or initial condition(s) are specified by **cond1,cond2,...**.

In writing the differential equations, the letter **D** denotes differentiation with respect to the independent variable. A **D** followed by a digit denotes repeated differentiation.

The initial/boundary conditions are specified with equations like **y(a)=b** and/or **Dy(a)=d**, where y is a dependent variable and b and d are constants. If the number of initial conditions specified is less than the number of dependent variables, the resulting solutions will contain the arbitrary constants **C1, C2,...**.

EXAMPLE C.29

Solve the ODE

$$\frac{dy}{dx} = -ax$$

Solution:
Entering

```
syms x y a
dsolve('Dy=-a*x','x')
```

returns

```
ans =
        -1/2*a*x^2+C1
```
□

EXAMPLE C.30

Solve the ODE

$$\frac{d^2y}{dx^2} = -y, \quad y(0) = 1, \quad \left.\frac{dy}{dx}\right|_{x=0} = 0$$

Solution:
Entering

```
syms x y
dsolve('D2y=-y','y(0)=1','Dy(0)=0','x')
```

returns

```
ans =
        cos(x)
```
□

EXAMPLE C.31

Solve the system of ODEs given by

$$\frac{dx}{dt} = 3x + 2y$$
$$\frac{dy}{dt} = -4x + 3y$$

Solution:
Entering

```
syms x y t
[x,y]=dsolve('Dx=3*x+2*y','Dy=-4*x+3*y','t')
```

returns

```
x =
exp(3*t)*(C1*sin(2*2^(1/2)*t)+C2*cos(2*2^(1/2)*t))
y =
exp(3*t)*2^(1/2)*(cos(2*2^(1/2)*t)*C1-
sin(2*2^(1/2)*t)*C2)
```
□

If **dsolve** cannot find an analytic solution for an equation, it prints the warning:

```
Warning: explicit solution could not be found
```

and returns an empty symbolic object.

C.4 Symbolic Linear Algebra

All the matrix operations valid for numerical linear algebra are also valid for symbolic manipulation. The standard matrices **ones, eye, etc.** still have the same definitions. Below, we summarize the most frequently used commands in symbolic vector and matrix calculations.

28. **poly**. This command returns the characteristic polynomial of a matrix. Its syntax is
    - **p = poly(A)** If **A** is a numeric matrix, this command returns the coefficients of the characteristic polynomial of **A**.
    - **f = poly(sym(A),x)** This command returns the characteristic polynomial of **A** as a function of the variable x.

Example C.32

Entering

```
syms x
A=[2 3 4; 1 5 6; 2 5 1];
p=poly(A)
f=poly(sym(A),x)
```

returns

```
p =
     1.0000    -8.0000    -24.0000    37.0000
f =
     x^3-8*x^2-24*x+37
```

□

29. **det(A)**. This command calculates the determinant of a matrix. It returns a symbolic expression, if **A** is symbolic; a numeric value, if **A** is numeric.

Example C.33

Entering

```
syms a b c d
A=[a b; c d];
B=[2 3; 1 5];
f=det(A)
g=det(B)
```

returns

```
f =
     a*d-b*c
g =
     7
```

□

30. **[V,D] = eig(A)**. This command computes the symbolic matrix eigenvalues and eigenvectors. **V**'s columns are the eigenvectors and the diagonal matrix **D** elements are the corresponding eigenvalues of the matrix **A**.

Example C.34

Find the eigenvectors and eigenvalues of the matrix:

$$\mathbf{M} = \begin{pmatrix} a & 1 \\ a^2 & 2 \end{pmatrix}$$

Solution:
Entering

```
syms a
M=[a 1; a^2 2];
[V,D]=eig(M)
```

returns

```
V =
[(-1+1/2*a+1/2*(4-4*a+5*a^2)^(1/2))/a^2,
(-1+1/2*a-1/2*(4-4*a+5*a^2)^(1/2))/a^2]
[ 1,1]
D =
[ 1+1/2*a+1/2*(4-4*a+5*a^2)^(1/2), 0]
[ 0, 1+1/2*a-1/2*(4-4*a+5*a^2)^(1/2)]
```
□

NOTE: At present, the symbolic command **eigen** only works for a limited functional formats of the matrix. MATLAB Symbolic Math Toolbox can find the eigenvectors for matrix of rationals, rational functions, algebraic numbers, or algebraic functions.

31. **inv(M)**. This command gives the inverse of a symbolic matrix.

EXAMPLE C.35

Find the inverse of the matrix **M** given by

$$\mathbf{M} = \begin{pmatrix} a & 1 \\ a^2 & 2 \end{pmatrix}$$

Solution:
Entering

```
syms a
M=[a 1; a^2 2];
invM=inv(M)
```

returns

```
invM =
        [ -2/a/(-2+a),     1/a/(-2+a)]
        [a/(-2+a),          -1/(-2+a)]
```
□

32. **expm(M)**. This command gives the matrix exponential of a symbolic matrix.

Example C.36

Find the rotation matrices associated with the Pauli spin matrices

$$R_i(\theta) = \exp(-j\sigma_i\theta)$$

where the σ matrices are as defined in Section 8.9.4, and are given by

$$\sigma_1 = \begin{pmatrix} 0 & 1 \\ 1 & 0 \end{pmatrix} \quad \sigma_2 = \begin{pmatrix} 0 & -j \\ j & 0 \end{pmatrix} \quad \sigma_3 = \begin{pmatrix} 1 & 0 \\ 0 & -1 \end{pmatrix}$$

Solution:
Entering

```
syms theta
sigma1=[0 1; 1 0];
sigma2=[ 0 -j; j 0];
sigma3=[1 0; 0 -1];
R1=expm(-j*sigma1*theta)
R2=expm(-j*sigma2*theta)
R3=expm(-j*sigma3*theta)
```

returns

```
R1 =
      [      cos(theta),  -i*sin(theta)]
      [ -i*sin(theta),       cos(theta)]
R2 =
      [ cos(theta),   -sin(theta)]
      [ sin(theta),    cos(theta)]
R3 =
      [cos(theta)-i*sin(theta), 0]
      [0, cos(theta)+i*sin(theta)]
```
□

33. **`tril(M,k)`**. This command gives the lower triangle of a symbolic matrix. It returns a lower triangular matrix that retains the elements of **M** on and below the kth diagonal and sets the remaining elements to 0. The values $k = 0$, $k > 0$, and $k < 0$ correspond to the main, superdiagonals, and subdiagonals, respectively.

Example C.37

Entering

```
syms t
    for m=1:4
          for n=1:4
                M(m,n)=t^(m+n);
          end
    end
```

```
M
Ml0=tril(M,0)
Mlp1=tril(M,1)
Mln1=tril(M,-1)
```

returns

```
M =
     [ t^2, t^3, t^4, t^5]
     [ t^3, t^4, t^5, t^6]
     [ t^4, t^5, t^6, t^7]
     [ t^5, t^6, t^7, t^8]
Ml0 =
     [ t^2,   0,   0,   0]
     [ t^3, t^4,   0,   0]
     [ t^4, t^5, t^6,   0]
     [ t^5, t^6, t^7, t^8]
Mlp1 =
     [ t^2, t^3,   0,   0]
     [ t^3, t^4, t^5,   0]
     [ t^4, t^5, t^6, t^7]
     [ t^5, t^6, t^7, t^8]
Mln1 =
     [   0,   0,   0,   0]
     [ t^3,   0,   0,   0]
     [ t^4, t^5,   0,   0]
     [ t^5, t^6, t^7,   0]
```
□

34. **triu(M,k)**. This command gives the upper triangle of a symbolic matrix. It returns an upper triangular matrix that retains the elements of **M** on and above the k-th diagonal and sets the remaining elements to 0. The values $k = 0$, $k > 0$, and $k < 0$ correspond to the main, superdiagonals, and subdiagonals, respectively.

EXAMPLE C.38

Entering

```
syms t
     for m=1:4
          for n=1:4
               M(m,n)=t^(m+n);
          end
     end
M
Mu0=triu(M,0)
Mup1=triu(M,1)
Mun1=triu(M,-1)
```

returns

```
M =
      [ t^2, t^3, t^4, t^5]
      [ t^3, t^4, t^5, t^6]
      [ t^4, t^5, t^6, t^7]
      [ t^5, t^6, t^7, t^8]
Mu0 =
      [ t^2, t^3, t^4, t^5]
      [   0, t^4, t^5, t^6]
      [   0,   0, t^6, t^7]
      [   0,   0,   0, t^8]
Mup1 =
      [   0, t^3, t^4, t^5]
      [   0,   0, t^5, t^6]
      [   0,   0,   0, t^7]
      [   0,   0,   0, 0]
Mun1 =
      [ t^2, t^3, t^4, t^5]
      [ t^3, t^4, t^5, t^6]
      [   0, t^5, t^6, t^7]
      [   0,   0, t^7, t^8]  □
```

C.5 z-Transform and Laplace Transform

The z-transform and the Laplace transform techniques are the analytic tools par excellence for solving linear constant coefficients difference and differential equations, respectively. We shall quickly review their properties here, and discuss how they can be called within the Symbolic Math Toolbox environment.

C.5.1 z-Transform

The z-transform of a sequence $\{f(k)\}$, where $f(k) = 0$ for $k = -1$, $k = -2$, $k = -3, \ldots$ is defined by

$$Z[f(k)] \equiv F(z) = f(0) + \frac{f(1)}{z} + \frac{f(2)}{z^2} + \frac{f(3)}{z^3} + \cdots = \sum_{k=0}^{\infty} \frac{f(k)}{z^k}$$

This series is finite only when $|z| > R$, where R is called the radius of convergence. The value of R depends on the sequence.

It is easy to derive from the definition of the z-transform the following properties:

Properties of the z-transform

Property	Discrete sequence	z-Transform
1. Linearity	$\alpha f(k) + \beta g(k)$	$\alpha F(z) + \beta G(z)$
2. Right-shifting	$f(k - m)$	$z^{-m}F(z)$
3. Left-shifting	$f(k + m)$	$z^m F(z) - \sum_{i=0}^{m-1} f(i) z^{m-i}$
4. Convolution	$\sum_{i=0}^{k} f(k-i)g(i)$	$F(z)G(z)$
5. Summation	$\sum_{i=0}^{k} f(i)$	$\frac{z}{z-1}F(z)$
6. Multiplication by a^k	$a^k f(k)$	$F(z/a)$
7. Differentiating	$kf(k)$	$-z\frac{dF}{dz}$
8. Periodic sequence	$f(k) = f(k + N)$	$F(z) = \frac{z^N}{z^N - 1} F_1(z)$ where $F_1(z) = \sum_{k=0}^{N-1} f(k) z^{-k}$
9. Initial value theorem	$f(0) = \lim_{\lvert z \rvert \to \infty} F(z)$	
10. Final value theorem	$f(\infty) = \lim_{z \to 1} (z-1) F(z)$ if $(z - 1) F(z)$ is analytic for $\lvert z \rvert \geq 1$	

Using the above properties and the below expressions for the z-transforms of the simple sequences given, it is easy to obtain the z-transform of more complicated sequences.

Seed pairs

Discrete sequence	z-Transform	Region of convergence
1	$\frac{z}{z-1}$	$\lvert z \rvert > 1$
$\delta(k)$	1	$\lvert z \rvert > 0$
$\sin(k\omega T)$	$\frac{z\sin(\omega T)}{z^2 - 2z\cos(\omega T) + 1}$	$\lvert z \rvert > 1$
$\frac{a^k}{k!}$	$\exp\left(\frac{a}{z}\right)$	$\lvert z \rvert > 0$

EXAMPLE C.39

Using the above tables, find the z-transform for the sequence

$$f(k) = k$$

Solution: Combining Property 7 with knowledge of the z-transform for $f(k) = 1$. We deduce that

$$Z[k] = -z\frac{d}{dz}\left(\frac{z}{z-1}\right) = \frac{z}{(z-1)^2} \quad \square$$

EXAMPLE C.40

Using the above tables, find the z-transform for the sequence

$$f(k) = a^k$$

Solution: Combining Property 6 with knowledge of the z-transform for $f(k) = 1$. We deduce that

$$Z[a^k] = \frac{z/a}{(z/a)-1} = \frac{z}{z-a} \quad \square$$

The Symbolic Math Toolbox command for finding the z-transform of a sequence is:

35. **`ztrans(f,k,z)`**. The sequence is $\{f(k)\}$, and z is the variable in the z-transform.

EXAMPLE C.41

Use the Symbolic Math Toolbox to find the z-transform of the sequence

$$f(k) = k^4 a^k$$

Solution:
Entering

```
syms a k z
f=(k^4)*a^k;
F=ztrans(f,k,z)
```

returns

```
F =
        -z*a*(z^3+11*z^2*a+11*z*a^2+a^3)/(-z+a)^5
```

$\square$

The inverse z-transform can as well be obtained in the Symbolic Math Toolbox:

36. **`iztrans(F,z,k)`**. The function $F(z)$ is the function we desire to find its inverse z-transform, and k is the variable in the answer sequence.

EXAMPLE C.42

Find the inverse z-transform of the function

$$F(z) = \exp(a/z)$$

Solution:
Entering

```
syms z a k
F=exp(a/z);
f=iztrans(F,z,k)
```

returns

```
f =
      a^k/k! □
```

C.5.2 Solving Constant Coefficients Linear Difference Equations Using z-Transform

Using essentially the linearity and the right-shifting properties of the z-transform, any linear difference equation can be reduced to an algebraic relation between the z-transforms of the input and output signals. Obtaining the solution of the output then simply reduces to finding an inverse z-transform.

EXAMPLE C.43

Solve the difference equation

$$y(k) = \beta u(k) + \alpha y(k-1) \quad \text{for } k = 0, 1, 2, \ldots$$

for the two cases $u(k) = \delta(k)$ and $u(k) = 1$.

Solution: Applying the linearity property of the z-transform, we can write for the difference equation:

$$Z[y(k)] = \beta Z[u(k)] + \alpha Z[y(k-1)]$$

Using the right-shifting property of the z-transform, we can write the last term as $z^{-1} Z[y(k)]$. Denoting the z-transform for $y(k)$ and $u(k)$ by $Y(z)$ and $U(z)$ respectively, we deduce the following algebraic relation:

$$Y(z) = \left(\frac{\beta}{1-\alpha z^{-1}}\right)U(z) = \left(\frac{\beta z}{z-\alpha}\right)U(z)$$

Case 1:

$$U(z) = 1 \Rightarrow Y(z) = \left(\frac{\beta z}{z-\alpha}\right) \Rightarrow y(k) = \beta\alpha^k$$

Case 2:

$$U(z) = \frac{z}{(z-1)} \Rightarrow Y(z) = \left(\frac{\beta z}{z-\alpha}\right)\left(\frac{z}{z-1}\right) = \frac{\beta z^2}{(z-\alpha)(z-1)}$$

$$\Rightarrow y(k) = \frac{\beta\alpha}{\alpha-1}\alpha^k + \frac{\beta}{1-\alpha}$$

where in each case, we used the inverse z-transform command to find the functional forms of the y-sequences.

We could also have obtained this inverse z-transform by hand through decomposing the expression of $Y(z)$ in partial fractions. For case 1, it is already in that form, and for case 2, we have

$$\frac{\beta z^2}{(z-\alpha)(z-1)} = \frac{\beta\alpha}{\alpha-1}\left(\frac{z}{z-\alpha}\right) + \frac{\beta}{1-\alpha}\left(\frac{z}{z-1}\right)$$

The inverse z-transform expressions for each of these terms can be deduced from the results of Example C.40. □

C.5.3 Laplace Transform

The Laplace transform of the function $f(t)$ denoted by $F(s)$ is defined by the integral

$$F(s) = \int_0^\infty f(t)e^{-st}\,dt, \qquad \mathrm{Re}(s) > 0$$

The inverse Laplace transform is defined for analytic functions of order $O(s^{-k})$ at infinity with $k > 1$ by means of the inversion integral:

$$f(t) = \frac{1}{2\pi j}\int_{\gamma-j\infty}^{\gamma+j\infty} F(s)e^{st}\,ds$$

where γ is a real number that exceeds the real part of all singularities of $F(s)$. In future courses on complex functions, you will learn how to use residue theorems, and other tools of complex functions calculus to actually compute these inverses. At this point, we will use the Symbolic Math Tool to obtain these quantities when needed.

C.5.3.1 Basic Properties of Laplace Transforms

1. For α and β arbitrary constants:

$$L[\alpha f(t) + \beta g(t)] = \alpha F(s) + \beta G(s)$$

This is the linearity property.

2. The scale change property:

$$L[f(mt)] = \frac{1}{m} F(s/m)$$

3. The time delay property:

$$L[f(t - t_0)] = \exp(-st_0)\, F(s)$$

4. The shift property:

$$L\lfloor e^{-at} f(t) \rfloor = F(a + s)$$

5. Multiplication by t^n:

$$L[t^n f(t)] = (-1)^n \frac{d^n F(s)}{ds^n}$$

6. If $n > 0$ is an integer and $\lim_{t\to\infty} f(t)e^{-st} = 0$, then for $t > 0$,

$$L\lfloor f^{(n)}(t) \rfloor = s^n F(s) - s^{n-1} f(0) - s^{n-2} f^{(1)}(0) - \cdots - f^{(n-1)}(0)$$

This is the formula for the transform of an nth-order derivative.

7. If $\lim_{t\to\infty}\left(e^{-st} \int_0^t f(u)du \right) = 0,$

then $L\left[\int_0^t f(u)du \right] = \frac{1}{s} F(s)$

This is the formula for the transform of an integral.

8. The convolution theorem:

If $f * g(t) \equiv \int_0^t f(t-u)\, g(u)\, du,$

then $L[f * g(t)] = F(s)\, G(s)$

9. The initial value property:

$$\lim_{s \to \infty}(s\,F(s)) = \lim_{t=0}(f(t))$$

10. The final value property:

$$\lim_{s \to 0}(s\,F(s)) = \lim_{t=\infty}(f(t))$$

Analytic expressions for the Laplace transform for many useful functions can be obtained using the Symbolic Math Toolbox. Others not yet incorporated there can be found in tables of Laplace pairs.

The commands for obtaining the Laplace and inverse Laplace transforms in MATLAB are respectively:

37. **`laplace(f,t,s)`**. This command gives $F(s)$ the Laplace transform of the function $f(t)$.
38. **`ilaplace(F,s,t)`**. This command gives $f(t)$ the inverse Laplace transform of $F(s)$.

Example C.44

Use the Symbolic Math Toolbox to find

$$L\left\lfloor t\,e^{-at} \right\rfloor$$

Solution:
Entering

```
syms a t s
f=t*exp(-a*t);
F=laplace(f,t,s)
```

returns

```
F =
     1/(s+a)^2
```

□

Below, for reference purposes, is a short table of the Laplace transform pairs for some of the most commonly used functions in electrical engineering applications.

Laplace transform pairs

$f(t)$	$F(s)$
$\delta(t)$	1
$U(t)$ (the unit step function)	$\frac{1}{s}$
$\frac{t^n}{n!}U(t)$	$\frac{1}{s^{n+1}}$
$\exp(-at)\ U(t)$	$\frac{1}{s+a}$
$\sin(\omega t)\ U(t)$	$\frac{\omega}{s^2+\omega^2}$
$\cos(\omega t)\ U(t)$	$\frac{s}{s^2+\omega^2}$
$\exp(-at)\sin(\omega t)\ U(t)$	$\frac{\omega}{(s+a)^2+\omega^2}$
$\exp(-at)\cos(\omega t)\ U(t)$	$\frac{s+a}{(s+a)^2+\omega^2}$

EXAMPLE C.45

Find analytically the Laplace transform of the function

$$t^2 \exp(-at)U(t)$$

Solution: Using Property 5 and the Laplace transform for the exponential from the above table, we have

$$L[t^2 \exp(-at)U(t)] = (-1)^2 \frac{d^2}{ds^2}\left(\frac{1}{s+a}\right) = \frac{2}{(s+a)^3} \quad \square$$

Verify the answer using the **`laplace`** command in the Symbolic Math Toolbox.

EXAMPLE C.46

Use the Symbolic Math Toolbox to find

$$L^{-1}\left[\left(\frac{1}{(s+a)^4}\right)\right]$$

Solution:
Entering

```
syms s a t
F=1/(s+a)^4;
f=ilaplace(F,s,t)
```

returns

```
f =
        1/6*t^3*exp(-a*t)
```
□

C.5.4 Solving Constant Coefficients Linear ODE Using Laplace Transform

Using essentially the linearity and the differentiation properties of the Laplace transform, any linear differential equation can be reduced to an algebraic relation between the Laplace transforms of the source and the signal. Obtaining the solution for the output signal then simply reduces to finding an inverse Laplace transform.

EXAMPLE C.47

The differential equation relating the output voltage of a system with a 6-unit step source term is given by

$$\frac{d^2y}{dt^2} + 5\frac{dy}{dt} + 6y = 6U(t)$$

The initial conditions are $y(0) = 4$ and $\left.\frac{dy}{dt}\right|_{t=0} = -7$.

Solution: Using the linearity property and the differentiation property of the Laplace transform, we can write the Laplace transforms of the two sides of the ODE as

$$(s^2Y(s) - 4s + 7) + 5(sY(s) - 4) + 6Y(s) = \frac{6}{s}$$

$$\Rightarrow Y(s) = \frac{4s^2 + 13s + 6}{s(s^2 + 5s + 6)}$$

The roots of the denominator are respectively $s = 0, -2, -3$ and the partial fraction expansion for this quantity is

$$Y(s) = \left(\frac{1}{s}\right) + 2\left(\frac{1}{s+2}\right) + \left(\frac{1}{s+3}\right)$$

Using the above table for the Laplace transform pairs, we obtain

$$y(t) = U(t) + 2\exp(-2t)\,U(t) + \exp(-3t)U(t) \quad \square$$

Verify that you obtain the same answer as above using the **dsolve** command of the Symbolic Math Toolbox.

Appendix D: Some Useful Formulae

Sum of Integers and Their Powers

$$\sum_{k=1}^{n} k = \frac{n(n+1)}{2}$$

$$\sum_{k=1}^{n} k^2 = \frac{n(n+1)(2n+1)}{6}$$

$$\sum_{k=1}^{n} k^3 = \left[\frac{n(n+1)}{2}\right]^3$$

$$\sum_{k=1}^{n} k^4 = \frac{n(n+2)(2n+1)(3n^2+3n-1)}{30}$$

$$\sum_{k=1}^{n} (2k-1) = n^2$$

$$\sum_{k=1}^{n} (2k-1)^2 = \frac{n(4n^2-1)}{3}$$

$$\sum_{k=1}^{n} (2k-1)^3 = n^2(2n^2-1)$$

$$\sum_{k=1}^{n} k(k+1)^2 = \frac{n(n+1)(n+2)(3n+5)}{12}$$

Arithmetic Series

$$\sum_{k=0}^{n-1} (a + kr) = \frac{n}{2}[2a + (n-1)r]$$

Geometric Series

$$\sum_{k=1}^{n} aq^{k-1} = \frac{a(q^n - 1)}{q - 1} \qquad q \neq 1$$

Arithmo-Geometric Series

$$\sum_{k=0}^{n-1} (a + kr)q^k = \frac{a - [a + (n-1)r]q^n}{(1-q)} + \frac{rq(1 - q^{n-1})}{(1-q)^2} \qquad q \neq 1$$

Taylor's Series

$$f(x + a) = \sum_{k=0}^{\infty} f^{(k)}(x)\frac{a^k}{k!}$$

$$f(x + a, y + b) = f(x, y) + a\frac{\partial f}{\partial x} + b\frac{\partial f}{\partial y} + \frac{1}{2!}\left[a^2\frac{\partial^2 f}{\partial x^2} + b^2\frac{\partial^2 f}{\partial y^2} + 2ab\frac{\partial^2 f}{\partial x \partial y}\right] + \cdots$$

Trigonometric Functional Relations

$$\sin(x) \pm \sin(y) = 2\sin\left[\frac{1}{2}(x \pm y)\right]\cos\left[\frac{1}{2}(x \mp y)\right]$$

$$\cos(x) + \cos(y) = 2\cos\left[\frac{1}{2}(x+y)\right]\cos\left[\frac{1}{2}(x-y)\right]$$

$$\cos(x) - \cos(y) = 2\sin\left[\frac{1}{2}(x+y)\right]\sin\left[\frac{1}{2}(y-x)\right]$$

$$\sin\left(\frac{1}{2}x\right) = \pm\sqrt{\frac{1}{2}(1-\cos(x))}$$

$$\cos\left(\frac{1}{2}x\right) = \pm\sqrt{\frac{1}{2}(1+\cos(x))}$$

$$\sin(2x) = 2\sin(x)\cos(x)$$

$$\sin(3x) = 3\sin(x) - 4\sin^3(x)$$

$$\sin(4x) = \cos(x)[4\sin(x) - 8\sin^3(x)]$$

$$\cos(2x) = 2\cos^2(x) - 1$$

$$\cos(3x) = 4\cos^3(x) - 3\cos(x)$$

$$\cos(4x) = 8\cos^4(x) - 8\cos^2(x) + 1$$

Relation of Trigonometric and Hyperbolic Functions

$$\sin(x) = -j\sinh(jx)$$

$$\cos(x) = \cosh(jx)$$

$$\tan(x) = \frac{1}{j}\tanh(jx)$$

Expansion of Elementary Functions in Power Series

$$e^x = \sum_{k=0}^{\infty}\frac{x^k}{k!}$$

$$\sin(x) = \sum_{k=0}^{\infty} (-1)^k \frac{x^{2k+1}}{(2k+1)!}$$

$$\cos(x) = \sum_{k=0}^{\infty} (-1)^k \frac{x^{2k}}{(2k)!}$$

$$\sinh(x) = \sum_{k=0}^{\infty} \frac{x^{2k+1}}{(2k+1)!}$$

$$\cosh(x) = \sum_{k=0}^{\infty} \frac{x^{2k}}{(2k)!}$$

Appendix E: Text Formatting

MATLAB Character Sequences for Common Symbols

Characters	Symbols	Characters	Symbols	Characters	Symbols
		Greek Letters			
\alpha	α	\kappa	κ	\rho	ρ
\beta	β	\lambda	λ	\sigma	σ
\chi	χ	\mu	μ	\varsigma	ς
\delta	δ	\nu	ν	\tau	τ
\epsilon	ε	\o	ο	\upsilon	υ
\phi	ϕ	\pi	π	\omega	ω
\gamma	γ	\varpi	ϖ	\xi	ξ
\eta	η	\theta	θ	\psi	ψ
\iota	ι	\vartheta	ϑ	\zeta	ζ
		Mostly Capital Greek Letters			
\Gamma	Γ	\Xi	Ξ	\Phi	Φ
\Delta	Δ	\Pi	Π	\Psi	Ψ
\Theta	Θ	\Sigma	Σ	\Omega	Ω
\Lambda	Λ	\Upsilon	Υ	\infty	∞
		Operations\Ordering			
\equiv	≡	\pm	±	\cdot	·
\neq	≠	\oplus	⊕	\div	÷
\approx	≈	\times	×	\leq	≤
\propto	∝	\otimes	⊗	\geq	≥
		Set Theoretic			
\forall	∀	\oslash	∅	\supset	⊃
\exists	∃	\cup	∪	\subseteq	⊆
\in	∈	\cap	∩	\supsetq	⊇
\ni	∍	\subset	⊂	\cong	≅
\neg	¬			\sim	~
		Others			
\nabla	∇	\partial	∂	\int	∫
\circ	∘	\bullet	•	\ldots	…
\wedge	∧	\langle	⟨	\rangle	⟩
\mid	|	\perp	⊥	\surd	√
\leftrightarrow	↔	\leftarrow	←	\rightarrow	→
\uparrow	↑	\downarrow	↓	\prime	′

The above table includes the MATLAB character sequences for the most common symbols. The complete listing can be accessed from the online Help. Other commands of practical importance include:

- All TeX strings are prefixed with the backslash \
- All characters grouping is achieved with the curly brackets { }
- Superscripts are specified by a carat ^
- Subscripts are specified by an underscore _
- Text font is specified by \fontname such as \fontname{courier}
- Text size is specified by \fontsize such as \fontsize{16}
- Font style are specified by using \bf, \it, \s, \rm, to refer to bold, italic, slanted, and normal roman, respectively.

The MATLAB command **`latex(E)`**, from the Symbolic Math Toolbox, allows us to convert any symbolic expression into its LaTeX format.

Selected References

Many excellent books discuss in further detail and depth the different topics covered in this book. Therefore, it is not possible to have an exhaustive listing of all the literature available. The following list includes those books relevant to the material covered, to which the author refers often and from which he learned a great deal.

1. Anton, H., *Elementary Linear Algebra*, 7th ed., Wiley, New York, 1994.
2. Barnsley, M., *Fractals Everywhere*, Academic Press, New York, 1993.
3. Bodie, Z., Kane, A., and Markus, A. J., *Investments*, Hill/Irwin, 2002.
4. Cadzow, J.A., *Discrete-Time Systems*, Prentice-Hall, Englewood Cliffs, NJ, 1973.
5. Cullen, C.G., *Matrices and Linear Transformation*, Dover, New York, 1990.
6. Davis, P. and Rabinowitz, P., *Methods of Numerical Integration*, 2nd ed., Academic Press, New York, 1984.
7. Einstein, A. et al., *Principles of Relativity*, Dover, New York, 1924.
8. Feynman, R.P., Leighton, R.B., and Sands, M., *The Feynman Lectures on Physics*, Addison-Wesley, Reading, MA, 1976.
9. Goldberg, S., *Introduction to Difference Equations*, Dover, New York, 1986.
10. Hager, W.W., *Applied Numerical Linear Algebra*, Prentice-Hall, Englewood Cliffs, NJ, 1988.
11. Hamming, R.W., *Digital Filters*, 3rd ed., Prentice-Hall, Englewood Cliffs, NJ, 1989.
12. Hamming, R.W., *Numerical Methods for Scientists and Engineers*, 2nd ed., Dover, New York, 1986.
13. Hildebrand, F.B., *Introduction to Numerical Analysis*, 2nd ed., Dover, New York, 1987.
14. Hoffman, K. and Kunze, R., *Linear Algebra*, 2nd ed., Prentice-Hall, Englewood Cliffs, NJ, 1971.
15. Lanczos, C., *Applied Analysis*, Prentice-Hall, Englewood Cliffs, NJ, 1956.
16. Maclane, S. and Birkhoff, G., *Algebra*, Macmillan, New York, 1967.
17. Mandelbrot, B.B., *The Fractal Geometry of Nature*, Freeman, New York, 1983.
18. *MATLAB — The Language of Technical Computing*: *Using MATLAB*, The MathWorks, Inc., Natick, MA, 2005.
    - *Optimization Toolbox: User's Guide Version 3*, The MathWorks, Inc., Natick, MA, 2004.
    - *Symbolic Math Toolbox: User's Guide Version 3*, The MathWorks, Inc., Natick, MA, 2004.

19. Morse, P.M. and Feshback, H., *Methods of Theoretical Physics*, McGraw-Hill, New York, 1953.
20. Peebles, P.J.E., *Quantum Mechanics*, Princeton University Press, Princeton, NJ, 1992.
21. Peitgen, H.O., Jurgens, H., and Saupe, D., *Chaos and Fractals: New Frontiers of Science*, Springer-Verlag, New York, 1992.
22. Proakis, J.G. and Manolakis, D.G., *Digital Signal Processing*, 3rd ed., Prentice-Hall, Upper Saddle River, NJ, 1996.
23. Ramo, S., Whinnery, J.R., and Van Duzer, T., *Fields and Waves in Communication Electronics*, Wiley, New York, 1994.
24. Saaty, T.L. and Bram, J., *Nonlinear Mathematics*, Dover, New York, 1981.
25. Schwerdtfeger, H., *Geometry of Complex Numbers*, Dover, New York, 1979.
26. Taub, H. and Schilling, D., *Principles of Communication Systems*, 2nd ed., McGraw-Hill, New York, 1986.
27. Van Loan, C.F., *Introduction to Scientific Computing*, Prentice-Hall, Upper Saddle River, NJ, 1997.
28. Yariv, A., *Quantum Electronics*, 3rd ed., Wiley, New York, 1989.

Index

RECEIVED
MAR - 9 2007